MICROBIAL BIOTECHNOLOGY

Technological Challenges and
Developmental Trends

MICROBIAL BIOTECHNOLOGY

Technological Challenges and Developmental Trends

Edited by

Bhima Bhukya, PhD
Anjana Devi Tangutur, PhD

AAP | APPLE ACADEMIC PRESS

Apple Academic Press Inc.	Apple Academic Press Inc.
3333 Mistwell Crescent	9 Spinnaker Way
Oakville, ON L6L 0A2	Waretown, NJ 08758
Canada	USA

© 2017 by Apple Academic Press, Inc.

First issued in paperback 2021

Exclusive worldwide distribution by CRC Press, a member of Taylor & Francis Group

No claim to original U.S. Government works

ISBN-13: 978-1-77463-605-3 (pbk)
ISBN-13: 978-1-77188-332-0 (hbk)

Library and Archives Canada Cataloguing in Publication

Microbial biotechnology : technological challenges and developmental trends / edited by Bhima Bhukya, PhD, Anjana Devi Tangutur, PhD.

Includes bibliographical references and index.
Issued in print and electronic formats.
ISBN 978-1-77188-332-0 (hardcover).--ISBN 978-1-77188-333-7 (pdf)
1. Microbial biotechnology. 2. Anti-infective agents. 3. Microorganisms.
4. Microbial enzymes. I. Bhukya, Bhima, author, editor II. Tangutur, Anjana Devi, author, editor

| TP248.27.M53M53 2016 | 660.62 | C2016-903453-4 | C2016-903454-2 |

Library of Congress Cataloging-in-Publication Data

Names: Bhukya, Bhima, editor. | Tangutur, Anjana Devi, editor.
Title: Microbial biotechnology : technological challenges and developmental   trends / editors, Bhima Bhukya, Anjana Devi Tangutur.
Other titles: Microbial biotechnology (Bhukya)
Description: Toronto ; New Jersey : Apple Academic Press, 2015. | Includes   bibliographical references and index.
Identifiers: LCCN 2016021768 (print) | LCCN 2016022096 (ebook) | ISBN   9781771883320 (hardcover : alk. paper) | ISBN 9781771883337 (eBook) | ISBN   9781771883337 ()
Subjects: | MESH: Microbiological Phenomena | Biotechnology Classification: LCC QR41.2 (print) | LCC QR41.2 (ebook) | NLM QW 4 | DDC 579--dc23
LC record available at https://lccn.loc.gov/2016021768

Apple Academic Press also publishes its books in a variety of electronic formats. Some content that appears in print may not be available in electronic format. For information about Apple Academic Press products, visit our website at **www.appleacademicpress.com** and the CRC Press website at **www.crcpress.com**

ABOUT THE EDITORS

Dr. B. Bhima

Bhima Bhukya, PhD, is Head and Assistant Professor at the Department of Microbiology, Osmania University, Hyderabad, India. Dr. Bhima, during his early scientific career, made fundamental contributions in the area of animal probiotics. Subsequently, he contributed towards the development of efficient stress-tolerant probiotic yeast and economically viable medium for large-scale production of yeast. His present research interests lie in development of recombinant probiotics using tools of proteomics and genomics, animal nutrition, biofuels, and bioprocess engineering. He has fifteen years of post graduate teaching and research experience. He was awarded the prestigious Raman postdoctoral fellowship/award to carry out post doctoral studies in the USA.

At present 12 PhD scholars and one post doctoral fellow are working under his supervision. Dr. Bhima has published twenty research articles in peer-reviewed journals and four book chapters and has presented over thirty five papers at various national and international conferences.

He gained experience in the area of animal nutrition-gas production technology at ICRISAT, Hyderabad, in the area of proteomics at the University of Hyderabad, Hyderabad, and in the area of plant biotechnology at the Center for Plant Molecular Biology, Osmania University, Hyderabad. He is a recipient of several awards, including the HUPO Young Scientist Award conferred by the World Human Proteome Organization, the SAB-Young Scientist Award by the Society for Applied Biotechnology, the Research Excellence award by the Indus Foundation and the state best teacher award 2015 (Young Teacher Award) by the government of Telangana, India. He has visited countries abroad for scientific interaction and has chaired scientific sessions during several prestigious national and international conferences.

Dr. Bhima is a life member of AMI, ISCA, BRSI and SAB. He is the Principal Investigator for several ongoing projects funded by UGC, DBT, DST etc. He has made significant contributions to university by working as a warden for PhD scholars hostel and has also been instrumental in several developmental activities at the department of Microbiology and faculty club of Osmania University. As a convener, Dr. Bhima successfully organized a three-day national conference on "Recent Trends in Microbial Biotechnology" and a two-day national seminar on "Advances in Microbial Technology" in 2015 at Osmania University.

Dr. Anjana Devi Tangutur

Anjana Devi Tangutur, PhD, is a Scientist at Council of Scientific and Industrial Research-Indian Institute of Chemical Technology (CSIR-IICT), Hyderabad, India. Dr. Anjana has over ten years of research experience. She has to date, a brilliant academic career, and pursued her doctoral degree at the Centre for Cellular & Molecular Biology-CSIR, Hyderabad. Later she joined the Chemical Biology Division of IICT as a Scientist. During her doctoral tenure, she has had significant exposure to various molecular biology and biochemistry approaches and to the mass spectrometry-based proteomics techniques. Dr. Tangutur's research activities at IICT involve screening of small molecules as anticancer agents (NFk-β, HDAC inhibitors, antitubulin agents, etc.) using various *in vitro* and *in vivo* tests. Her research interests lie mainly in applying molecular biology and proteomics based approaches to understand the underlying mechanisms behind disease pathology (in different cancers) or a particular mechanism in a biological system. Her research efforts have been published in reputed international and national journals and conference proceedings. She was selected for the AOHUPO/KSMS young scientist award (HUPO 6th Annual World Congress, Seoul, Korea). She has several ongoing projects funded by DBT, which include the BioCARe award for women scientists (2013–2016) and also the Rapid grant for young investigator (2013-2016). She served as a subject expert, external examiner, and also a member of the several interview selection committees in colleges, universities, and at CSIR-IICT.

She has co-guided several MPharm/MSc students for partial fulfillment of the post-graduation degree, and presently several qualified research scholars are pursuing PhD under her guidance. She is also a member of several editorial boards and a reviewer for several journals.

CONTENTS

List of Contributors..*xi*

List of Abbreviations .... *xv*

List of Symbols ....*xxi*

Preface ....*xxiii*

Introduction..*xxv*

**Part I: Antimicrobial Agents: Role and Applications in
Medicine and Healthcare** 1

1. **Antimicrobial Peptides from Plants and Their Application** 3
    B. Bhima and Mohammed Al Saiqali

2. **Development of Antitubercular Agents through Hybridization
    Strategies: Future Challenges and Perspectives** 27
    Srinivas Kantevari, Anvesh Jallapally, Anjana Devi Tangutur, and Dinesh Kumar

3. **Novel Bioactive Compounds from an Endophytic *Chaetomium
    arcuatum* Strain SAF-2 Isolated from *Semecarpus anacardium*** 63
    C. Ganesh Kumar, Narender Reddy Godumagadda, Poornima Mongolla,
    Sujitha Pombala, Preethi Badrinarayan, Cheruku Ravindra Reddy, Biswanath Das,
    K.V.S. Ramakrishna, Balasubramanian Sridhar, and Ahmed Kamal

4. **Understanding the Role of Biomarkers in the Pathogenesis of Sepsis**.....89
    S. Burgula, M. Swathi Raju, A. Nichita, A. Pawan Kumar, and Karthik Rajkumar

5. **Role of Copper in Modifying Cisplatin Induced Cytotoxicity in Yeast** 109
    B. Sreedhar and B. Vijaya Lakshmi

**Part II: Role of Microorganisms in Agriculture and
Plant Biotechnology** 131

6. **Role of Microbiome: Insights into the Kin Recognition Process
    in *Oryza sativa* by Proteomic and Metabolomic Studies** 133
    Anjana Devi Tangutur, Kommalapati Vamsi Krishna, Amrita Dutta Chowdhury, and
    Neelamraju Sarla

7. Associative Effect of Arbuscular Mycorrhizal Fungi and
    Rhizobium on Plant Growth and Biological Control of
    Charcoal Rot in Green Gram [*Vigna radiata* L. (Wilczek)] 155
    A. Hindumathi, B. N. Reddy, A. Sabitha Rani, and A. Narsimha Reddy

8. Morphological Changes in *Vigna radiata* Root Under
    Cadmium Induced Stress in the Presence of Plant
    Growth Promoting *Enterobacter* sp. C1D 171
    Rakesh Kumar Sharma, Kavita Barot, and G. Archana

9. Potential Use of *Trichoderma* Species as Promising Plant
    Growth Stimulator in Tomato (*Lycopersicum esculantum L.*).................185
    B. N. Reddy, K. Sarita, and A. Hindumathi

**Part III: Microbial Enzymes and Their Potential
Industrial Applications**...199

10. Strain Improvement of *Aspergillus niger* for the Enhanced
    Production of Cellulase in Solid State Fermentation.............................201
    G. Praveen Kumar Reddy, A. Sridevi, Kanderi Dileep Kumar,
    G. Ramanjaneyulu, A. Ramya, B.S. Shanthi Kumari, and B. Rajasekhar Reddy

11. Screening, Quantification, and Purification of Cellulases
    from Soil Actinomycetes 219
    Payal Das, Renu Solanki, and Monisha Khanna

12. Kinetic Studies of Alkali Pretreated Sweet Sorghum
    Bagasse Using Cellulase...249
    L. Saida and K. Venkata Sri Krishna

13. Secretion of Ligninolytic Enzymes by the White Rot Fungus
    Stereum ostrea Immobilized on Polyurethane Cubes Under
    the Influence of Chlorpyrifos 265
    B. S. Shanthi Kumari, K. Dileep Kumar, K. Y. Usha, A. Ramya, and B. Rajasekhar Reddy

14. Structure and Gas Diffusion Path Analysis of Hydrogenase
    Enzymes by Homology Modelling 277
    Nivedita Sahu, Anirudh Nelabhotla, and Pradhan Nityananda

**Part IV: Microorganisms in the Environment: Role and
Industrial Applications**...295

15. Yeast *Saccharomyces cerevisiae* as an Efficient Biological Agent
    for Decolorization of Reactive Dyes used in Textile Industry 297
    B. Bhima, N. Hanumalal, and B. Chandrasekhar

16. **Metabolite Profiling and Biological Activities of Extrolites from *Aspergillus turcosus* Strain KZR131 Isolated from Kaziranga National Park, Assam, India** 309

C. Ganesh Kumar, Poornima Mongolla, Jagadeesh Babu Nanubolu, Pathipati Usha Rani, and Kumar Katragunta

17. **Morphological, Cultural and Molecular Diversity of the Salt-Tolerant Alkaliphilic Actinomycetes from Saline Habitats** 337

S. D. Gohel and S. P. Singh

18. **Microbial Population Dynamics of Eastern Ghats of Andhra Pradesh for Xylanase Production** 355

G. Ramanjaneyulu, A. Ramya, and B. Rajasekhar Reddy

19. **Phylogenetic and Phenogram Based Diversity of Haloalkaliphilic Bacteria from the Saline Desert** 373

Hitarth B. Bhatt and Satya P. Singh

20. **Fungal Biotransformation of Drugs: Potential Applications in Pharma Industry** 387

G. Shyam Prasad and B. Sashidhar Rao

Index 409

LIST OF CONTRIBUTORS

G. Archana
Department of Microbiology and Biotechnology Center, The Maharaja Sayajirao University of Baroda, Vadodara, Gujarat – 390002, India, E-mail: archanagayatri@yahoo.com

Preethi Badrinarayan
Center for Molecular Modelling, CSIR-Indian Institute of Chemical Technology, Uppal Road, Hyderabad 500007, Telangana, India

Kavita Barot
Department of Microbiology and Biotechnology Center, The Maharaja Sayajirao University of Baroda, Vadodara, Gujarat – 390002, India

Hitarth B. Bhatt
UGC-CAS Department of Biosciences, Saurashtra University, Rajkot, Gujarat – 360005, India

B. Bhima
Department of Microbiology, University College of Science, Osmania University, Hyderabad, Telangana – 500007, India, E-mail: bhima.ou@gmail.com

S. Burgula
Department of Microbiology, Osmania University, Hyderabad, Telangana – 500007, India, E-mail: s_burgula@osmania.ac.in

B. Chandrasekhar
Department of Microbiology, University College of Science, Osmania University, Hyderabad, Telangana – 500007, India

Amrita Dutta Chowdhury
Biotechnology Laboratory, Directorate of Rice Research, Rajendranagar, Hyderabad, Telangana – 500030, India

Biswanath Das
Natural Products Chemistry Division, CSIR-Indian Institute of Chemical Technology, Uppal Road, Hyderabad 500007, Telangana, India

Payal Das
Acharya Narendra Dev College, University of Delhi, Govindpuri, Kalkaji, New Delhi – 110019, India.

Narender Reddy Godumagadda
Medicinal Chemistry and Pharmacology Division, CSIR-Indian Institute of Chemical Technology, Uppal Road, Hyderabad 500007, Telangana, India

S. D. Gohel
UGC-CAS, Department of Biosciences, Saurashtra University, Rajkot, Gujarat – 360005, India

N. Hanumalal
Department of Microbiology, University College of Science, Osmania University, Hyderabad, Telangana – 500007, India

A. Hindumathi
Mycology and Plant Pathology Laboratory, Department of Botany, Osmania University, Hyderabad, Telangana, – 500007, India

Anvesh Jallapally
CPC Division (Organic Chemistry Division-II), CSIR-Indian Institute of Chemical Technology, Hyderabad, Telangana – 500007, India

Ahmed Kamal
Medicinal Chemistry and Pharmacology Division, CSIR-Indian Institute of Chemical Technology, Uppal Road, Hyderabad 500007, Telangana, India

Srinivas Kantevari
CPC Division (Organic Chemistry Division-II) and Academy of Scientific and Innovative Research, CSIR-Indian Institute of Chemical Technology, Hyderabad, Telangana – 500007, India, E-mail: kantevari@yahoo.com

Kumar Katragunta
Natural Products Chemistry Division, CSIR-Indian Institute of Chemical Technology, Uppal Road, Hyderabad 500007, Telangana, India

Monisha Khanna
Acharya Narendra Dev College, University of Delhi, Govindpuri, Kalkaji, New Delhi – 110 019, India, E-mail: monishaandc@gmail.com

K. Venkata Sri Krishna
Center for Biotechnology, Institute of Science and Technology, Jawaharlal Nehru Technological University Hyderabad, Hyderabad, Telangana – 500085, India

Kommalapati Vamsi Krishna
Center for Chemical Biology, CSIR-Indian Institute of Chemical Technology, Hyderabad, Telangana – 500607, India

A. Pawan Kumar
Department of Microbiology, Osmania University, Hyderabad, Telangana – 500007, India

C. Ganesh Kumar
Medicinal Chemistry and Pharmacology Division, CSIR-Indian Institute of Chemical Technology, Uppal Road, Hyderabad, Telangana – 500007, India, E-mail: cgkumar@iict.res.in

Dinesh Kumar
Centre for Chemical Biology, CSIR-Indian Institute of Chemical Technology, Hyderabad, Telangana – 500007, India

Kanderi Dileep Kumar
Department of Microbiology, Sri Krishnadevaraya University, Anantapur, Andhra Pradesh – 515591, India

B. Vijaya Lakshmi
Assistant Professor, Institute of Genetics and Hospital for Genetic Diseases, Osmania University, Hyderabad, Telangana – 500007, India, E-mail: bodigavijayasri@gmail.com

Poornima Mongolla
Medicinal Chemistry and Pharmacology Division, CSIR-Indian Institute of Chemical Technology, Uppal Road, Hyderabad 500007, Telangana, India.
Acharya Nagarjuna University, Guntur, Andhra Pradesh – 522510, India

Jagadeesh Babu Nanubolu
Laboratory of X-ray Crystallography, CSIR-Indian Institute of Chemical Technology, Uppal Road, Hyderabad 500007, Telangana, India

Anirudh Nelabhotla
CSIR-Indian Institute of Chemical Technology, Hyderabad, Telangana – 500007, India

A. Nichita
Department of Microbiology, Osmania University, Hyderabad, Telangana – 500007, India

Pradhan Nityananda
AcSIR- CSIR-Indian Institute of Chemical Technology, Hyderabad, Telangana – 500007, India.

Sujitha Pombala
Medicinal Chemistry and Pharmacology Division, CSIR-Indian Institute of Chemical Technology, Uppal Road, Hyderabad 500007, Telangana, India

G. Shyam Prasad
Department of Biochemistry, University College of Science, Osmania University, Hyderabad, Telangana – 500007, India

Karthik Rajkumar
Department of Microbiology, Osmania University, Hyderabad, Telangana – 500007, India

M. Swathi Raju
Department of Microbiology, Osmania University, Hyderabad, Telangana – 500007, India

K. V. S. Ramakrishna
Nuclear Magnetic Resonance Center, CSIR-Indian Institute of Chemical Technology, Uppal Road, Hyderabad 500007, Telangana, India

G. Ramanjaneyulu
Department of Microbiology, Sri Krishnadevaraya University, Anantapur – 515591, Andhra Pradesh, India, E-mail: ramanj.003@gmail.com

A. Ramya
Department of Microbiology, Sri Krishnadevaraya University, Anantapur, Andhra Pradesh – 515591, India

A. Sabitha Rani
Mycology and Plant Pathology Laboratory, University College for Women, Koti, Hyderabad, Telangana, – 500095, India

Pathipati Usha Rani
Biology and Biotechnology Division, CSIR-Indian Institute of Chemical Technology, Uppal Road, Hyderabad 500007, Telangana, India

B. Sashidhar Rao
Department of Biochemistry, University College of Science, Osmania University, Hyderabad, Telangana – 500007, India, E-mail: sashi_rao@yahoo.com

A. Narsimha Reddy
Mycology and Plant Pathology Laboratory, Department of Botany, Osmania University, Hyderabad, Telangana, – 500007, India

B. N. Reddy
Mycology and Plant Pathology Laboratory, Deptartment of Botany, Osmania University, Hyderabad, Telangana – 500007, India, E-mail: reddybn1@yahoo.com

B. Rajasekhar Reddy
Department of Microbiology, Sri Krishnadevaraya University, Anantapur, Andhra Pradesh – 515591, India, E-mail: rajasekharb64@gmail.com

Cheruku Ravindra Reddy
Natural Products Chemistry Division, CSIR-Indian Institute of Chemical Technology, Uppal Road, Hyderabad 500007, Telangana, India

G. Praveen Kumar Reddy
Department of Microbiology, Sri Krishnadevaraya University, Anantapur, Andhra Pradesh – 515591, India

Nivedita Sahu
Department of Medicinal Chemistry and Pharmacology, Indian Institute of Chemical Technology, Hyderabad, Telangana – 500007, India, E-mail: nivedita@iict.res.in

L. Saida
Center for Biotechnology, Institute of Science and Technology, Jawaharlal Nehru Technological University Hyderabad, Hyderabad, Telangana – 500085, India, E-mail: lavudisaida@jntuh.ac.in

Mohammed Al Saiqali
Department of Microbiology, University College of Science, Osmania University, Hyderabad, Telangana – 500007, India

K. Sarita
Mycology and Plant Pathology Laboratory, Department of Botany, Osmania University College for Women, Koti, Hyderabad, Telangana – 500095, India

Neelamraju Sarla
Biotechnology Laboratory, Directorate of Rice Research, Rajendranagar, Hyderabad, Telangana – 500030, India, E-mail: sarla_neelamraju@yahoo.com

Rakesh Kumar Sharma
Department of Microbiology and Biotechnology Center, The Maharaja Sayajirao University of Baroda, Vadodara, Gujarat – 390002, India

Satya P. Singh
UGC-CAS, Department of Biosciences, Saurashtra University, Rajkot, Gujarat – 360005, India, E-mail: satyapsingh@yahoo.com

Renu Solanki
Acharya Narendra Dev College, University of Delhi, Govindpuri, Kalkaji, New Delhi – 110019, India.

B. Sreedhar
UGC-FRP Assistant Professor, Department of Biochemistry, Kakatiya University, Warangal, Telangana – 506009, India

A. Sridevi
Department of Applied Microbiology, Padmavati Mahila University, Tirupathi, Andhra Pradesh – 517502, India

Balasubramanian Sridhar
Laboratory of X-ray Crystallography, CSIR-Indian Institute of Chemical Technology, Uppal Road, Hyderabad 500007, Telangana, India

Anjana Devi Tangutur
Center for Chemical Biology, CSIR-Indian Institute of Chemical Technology, Hyderabad, Telangana – 500607, India, E-mail: anjdevi@gmail.com

K. Y. Usha
Department of Microbiology, Sri Krishnadevaraya University, Anantapur, Andhra Pradesh – 515591, India

LIST OF ABBREVIATIONS

AAS	atomic absorption spectroscopy
AChE	acetylcholine esterase
ACN	acetonitrile
ACP	anticancer peptides
AD	Alzheimer's disease
ADM	adrenomedullin
AG	arginine glycerol
AGP	acid-glycoprotein
AIA	actinomycetes isolation agar
AKI	acute kidney injury
AM	arbuscular mycorrhizal
AMF	arbuscular mycorrhizal fungi
AMP	antimicrobial peptides
ANOVA	analysis of variance
API	analytical profile index
APO	acute pulmonary oedema
APP	acute phase proteins
APR	acute phase response
ATCC	American type culture collection
ATCI	acetylthiocholine iodide
ATP	adenosine triphosphate
AUROC	area under receiver operating curve
BCG	Bacillus Calmette–Guérin
BCS	bathocuproiene disulfate
BH	Baylis–Hillmann
BLAST	basic local alignment search tool
BNF	biological nitrogen fixing
BSA	bovine serum albumin
BSR	basic scientific research
CA	*clostridium acetobutylicum*
CAS	career advancement scheme
CCD	charge coupled device
CCK	cyclic cystine knot
CDDP	cis-diamminedichloroplatinum

CFU	colony forming units
CL	cardiolipin
CMB	complex medium broth
CMC	carboxymethyl cellulose
COD	chemical oxygen demand
CpI	*clostridium pasteurianum*
CPP	cell-penetrating peptides
CRP	c-reactive protein
CSD	Cambridge Structural Database
CSIR	Council of Scientific and Industrial Research
CuRE	copper-responsive element
CV	coefficient of variation
CZA	Czapek dox agar
DAS	days after sowing
DCF	dichlorofluorescein
DEAE	diethylaminoethyl
DGGE	denaturing gradient gel electrophoresis
DIC	disseminated intravascular coagulation
DME	drug metabolism enzymes
DMPK	dystrophia myotonica protein kinase
DMR	digital mobile radio
DMSO	dimethyl sulfoxide
DNS	dinitrosalicylic
DTNB	beta dystrobrevin
DW	dry weight
EC	electric conductivity
ELISA	enzyme-linked immunosorbent assay
EMS	ethyl methane sulfonate
EtOH	ethyl alcohol
FDA	food and drug administration
FPU	filter paper unit
FW	fresh weight
GAST	gibberellic acid-stimulated arabidopsis
GC	gas chromatography
GI	gastrointestinal
GSH	glutathione
GSSG	glutathione disulfide
HPLC	high performance liquid chromatography
HTS	high throughput screening
ICU	intensive care unit

IEF	isoelectric focusing
IPG	immobilized pH gradient
ISP	internet service provider
ITS	internal transcribed spacer
IU	international units
IUPAC	International Union of Pure and Applied Chemistry
JHB	johannesburg
LAC	laccase
LBP	lipopolysaccharide-binding protein
LC	liquid chromatography
LCM	low copper medium
LiP	lignin peroxidase
LPS	lipopolysaccharides
LSD	least significant difference
LTP	lipid transfer proteins
MABA	*mycobacterium tuberculosis* β-ketoacyl-ACP reductase
MALDI	matrix assisted laser desorption/ionization
MBL	mannose-binding lectin
MBM	modified bristol medium
MCC	multicomponent cyclocondensation
MDR	multiple drug resistant
MEGA	molecular evolutionary genetic analysis
MELAS	myopathy, encephalopathy, lactic acidosis and stroke-like
MIC	minimum inhibitory concentration
MM	Michaelis-Menten
MNC	mononuclear cells
MnP	manganese peroxidase
MRP	multidrug resistance-associated
MRSA	methicillin resistant *staphylococcus aureus*
MS	Murashige-Skoog
MSM	mineral salts agar medium
MT	metallothionein
MTCC	microbial type culture collection
NAPI	nutritional and acute phase indicator
NCBI	National Center for Biotechnology Information
NGAL	neutrophil gelatinase-associated lipocalin
NMR	nuclear magnetic resonance
NPK	nutrient uptake
OD	optical density
OM	organic matter

OP	organophosphorus pesticides
ORTEP	Oak Ridge Thermal Ellipsoid Plot
OSDD	open source model for drug discovery
PAL	prealbumin
PAS	peripheral anionic site
PBS	phosphate buffered saline
PCR	polymerase chain reaction
PCT	procalcitonin
PDA	potato dextrose agar
PDB	potato dextrose broth
PE	phosphatidylethanolamine
PetF	photosynthetic ferredoxin
PFL	pyruvate formate lyase
PFOR	pyruvate ferredoxin (flavodoxin) oxido-reductase
PG	phosphatidylglycerol
PGPR	plant growth promoting rhizobacteria
PNPG	ρ-nitrophenyl-β-D- glucopyranoside
PQ	plastoquinone
PS	phosphatidylserine
PtpA	protein tyrosine phosphatase A
PtpB	protein tyrosine phosphatase B
PTS	postural tachycardia syndrome
PVDF	polyvinylidene fluoride
RBA	rose bengal agar
RC	regenerated cellulose
REMA	resazurin microtitre plate assay
RFSMS	research fellowship in science for meritorious students
RH	relative humidity
RMSD	root-mean-square deviation
RNA	ribonucleic acid
ROS	reactive oxygen species
SA	starch agar
SAA	serum amyloid A
SAR	structure-activity relationship
SCA	starch casein agar
SCAM	starch-casein agar medium
SD	synthetic dextrose medium
SDS	sodium dodecyl sulfate
SH	sulfhydryl group

SI	selectivity index
SIRS	systemic inflammatory response syndrome
SM	sphingomyelin
SmF	state fermentation
SO	*scenedesmus obliquus*
SOD	superoxide dismutase
SOFA	sequential organ failure assessment
SP	substance P
SPDB	Swiss-Pdb
SPR	structure-property relationship
SSB	sweet sorghum bagasse
SSE	sum of squared errors
SSF	solid state fermentation
TAE	tris-acetate-EDTA
TB	tuberculosis
TBA	thiobarbituric acid
TBARS	thiobarbituric acid reactive substances
TCA	tricyclic antidepressants
TFA	trifluoroacetic acid
TGGE	temperature gradient gel electrophoresis
TLC	thin-layer chromatography
TLR	toll like receptors
TMS	tetramethylsilane
TNF	tumor necrosis factor
TPZ	trifluoperazine
TSA	*trichoderma* specific agar
UGC	university grants commission
UNESCO	United Nations Educational, Scientific, and Cultural Organization
UPGMA	unweighted pair group mean averages
UV	ultraviolet
VERO	verda reno
VFA	volatile fatty acids
VIP	vasoactive intestinal peptide
VRE	vancomycin-resistant enterococci
VSV	vesicular stomatitis virus
WBC	white blood cells
WHO	World Health Organization
XDR	extensively drug-resistant
XP	extra precision

YEMA	yeast extract mannitol agar
YEME	yeast extract-malt extract
YEPD	yeast extract peptone and dextrose
YM	yeast mold
YPD	yeast extract/peptone/dextrose

LIST OF SYMBOLS

Ac	actinium
Ag	silver
Br	bromine
C	carbon
Ca	calcium
Cd	cadmium
Cl	chlorine
Cu	copper
F	fluorine
Fe	iron
H	hydrogen
Hg	mercury
K	potassium
Mg	magnesium
Mn	manganese
N	nitrogen
Na	sodium
Ni	nickel
O	oxygen
P	phosphorus
Pt	platinum
S	sulfur
Sn	tin
U	uranium
Zn	zinc

PREFACE

The microbial world which surrounds us affects each and every aspect of our lives. The plethora of microbes surrounding us is unexplored. Microbes present in our immediate environment, soil, water, or plant roots have direct or indirect influence leading to either a harmful or beneficial effect. This book is, therefore, a first comprehensive book designed for researchers to understand the role and recent implications of the diverse variety of microbes in the agriculture, environment, healthcare, or pharmaceutical industries. The book is also intended to provide an overall view of current diverse roles and new applications of microbial biotechnology in the above fields.

The book has four major parts. The first part (Part I) includes five chapters describing the role and application of microbes, antimicrobial agents, or bioactive compounds derived from them in medicine and healthcare. The first chapter provides a review of the antimicrobial peptides from various plant species and their applications. The second chapter describes the development of antitubercular agents by various synthesis strategies (using molecular hybridization technique) to overcome the drug resistance mechanisms and side effects associated with the present drugs, related future challenges, and perspectives. The third chapter explains the various novel bioactive compounds obtained from an endophytic strain isolated from *Semecarpus anacardium*. The fourth chapter describes the role of biomarkers in the pathogenesis of sepsis. The fifth chapter is on the role of copper in modifying cisplatin induced cytotoxicity in yeast.

The second part (Part II) contains four chapters describing the role of microorganisms in agriculture and plant biotechnology. The sixth chapter describes the role of microbiome in the context of kin recognition mechanism in *Oryza sativa*, which a the staple food crop; the authors have attempted to explore it further by using proteomic and metabolomic approaches. The seventh chapter provides details on the associative effects of arbuscular mycorrhizal fungi and *Rhizobium* on the plant growth and biological control of charcoal rot in green gram. Chapter 8 contains details on morphological changes in *Vigna radiata* root under cadmium-induced stress in the presence of plant growth promoting *Enterobacter* sp. C1D. The ninth chapter describes potential use of *Trichoderma* species as a promising plant growth stimulator in tomato (*Lycopersicum esculantum* L.).

The third part (Part III) comprises of five chapters describing microbial enzymes and their potential industrial applications. The tenth chapter covers the strain improvement of *Aspergillus niger* for the enhanced production of cellulase in solid state fermentation followed by screening, quantification, and purification of cellulases from soil actinomycetes in Chapter 11. Again the twelfth chapter is on kinetic studies of alkali-pretreated sweet sorghum bagasse using cellulase. The thirteenth chapter describes ligninolytic enzymes secreted by the white rot fungus *Stereum ostrea* immobilized on polyurethane cubes under the influence of chlorpyrifos. Chapter 14 is on the structure and gas diffusion path analysis of hydrogenase enzymes by homology modelling.

The fourth part (Part IV) is comprised of six chapters describing microorganisms in the environment with an emphasis on their role and industrial applications. It describes microbes isolated from unusual diverse habitats and their roles in different industries. The fifteenth chapter is on the yeast *Saccharomyces cerevisiae* as an efficient biological agent for the decolorization of reactive dyes used in the textile industry. The sixteenth chapter is on metabolite profiling and biological activities of extrolites from *Aspergillus turcosus* strain KZR131 isolated from the Kaziranga National Park, Assam, India, whereas Chapter 7 is on the morphological, cultural, and molecular diversity of the salt-tolerant alkaliphilic actinomycetes from saline habitats. Again, the eighteenth chapter is on microbial population dynamics of the Eastern Ghats of Andhra Pradesh for xylanase production followed by the nineteenth chapter which describes the phylogenetic and phenogram based diversity of haloalkaliphilic bacteria from the saline desert. Chapter 20 provides a view of the potential applications of fungal biotransformation of drugs in the pharma industry.

We hope this book proves to be a new and valuable resource to researchers and graduate students, agricultural scientists, food technologists, and microbiologists interested in biotechnological application of microbes in various fields. We also hope that this book will expand the existing knowledge and pave the way towards the utilization of the unexplored micro flora for the benefit of mankind.

INTRODUCTION

The book **Microbial Biotechnology: Technological Challenges and Developmental Trends** is the summation and compilation of research work by participants and delegates at the conference "Recent Trends in Microbial Biotechnology," convened by Dr. B. Bhima, Head, Department of Microbiology, Osmania University, Hyderabad, Telangana, India, from 26 to 28 February, 2015. It explores the recent implications, applications, and current trends of microbial biotechnology in pharma, food and other industries, healthcare, and disease. The book consists of 20 chapters and contains subject matter that addresses various issues in Microbial Biotechnology. The book covers research work carried out at different places and in different states in India.

The book brings together information on a broad range of topics in recent trends in microbial biotechnology in a single source. Each chapter is written in a concise manner and ideally suited for researchers and academicians. It discusses the related topics of microbial biotechnology in sufficient depth. The book is organized in a proper fashion and effectively presented. We have also made each chapter as independent as possible and have included numerous cross references at the end of each chapter. Data is also put in the form of tables wherever appropriate in the text. Elaborate figures/illustrations are provided for easy understanding of the subject.

The prime motto of the book is "Basic and Applied Aspects of Microbial Biotechnology," to address the challenges in creating a more secure, sustainable, and affordable system for food, feed, energy, and health through consolidating the underpinning biotechnology research platforms. This prepares a ground for seeding new ideas and nurturing knowledge through critical analysis and unabridged discussions on biotechnological developments. The research work carried out and published by authors from different spheres will serve to generate knowledge-based research in biotechnology, one of the major technologies of the twenty-first century. Its wide-ranging, multidisciplinary activities include recombinant DNA techniques, cloning, and the application of microbiology to the production of goods from bread to antibiotics to assist in protecting the natural environment and developing environmentally sustainable industries and institutions. The majority of current biotechnological applications are of microbial origin, and it is widely appreciated that the microbial world contains by far the greatest fraction of biodiversity in the biosphere. Because of their biotech impact, numerous efforts are being undertaken worldwide, with an ultimate goal to deliver new usable substances of microbial origin to the marketplace. It will be a good opportunity to inform stakeholders and decision makers about the environmental impacts and societal implications of emerging biotechnologies. Nevertheless, the book also aims to promote teaching methodology and research in a way to transform and aid economy.

This is the first book which highlights basic and applied aspects in microbial biotechnology. It provides a comprehensive view of the most current information on the implications and applications of microbes in various fields. It presents unbiased original research results on microbes by incorporating case studies wherever appropriate. Extensive references are provided at the end of each chapter to enhance further study.

On one side, the book covers a wide range of topics in current trends and challenges in microbial biotechnology ranging from *in vitro* screening of different medicinal plants for antimicrobial activity and studies on antimicrobial peptides from plants to strain improvement of *Aspergillus niger* for the enhanced production of cellulase in solid state fermentation; kinetic studies of alkali pretreated sweet sorghum bagasse using cellulase; screening, quantification, and purification of cellulases from soil actinomycetes; structure and gas diffusion path analysis of hydrogenase enzymes by homology modelling to phylogenetic and phenogram based diversity of haloalkaliphilic bacteria from the saline desert; and secretion of ligninolytic enzymes by the white rot fungus *Stereum ostrea* immobilized on polyurethane cubes under the influence of chlorpyrifos.

The book also includes chapters on unusual and precious microbes from unique and rare sources as described, screening and evaluation of antibacterial substances from the Indo-Gangetic plain harboring microbes; microbial population dynamics of the Eastern Ghats of Andhra Pradesh for xylanase production; morphological, cultural, and molecular diversity of the salt-tolerant alkaliphilic actinomycetes from the saline habitats; metabolite profiling and biological activities of the extrolites from *Aspergillus turcosus* strain KZR131 isolated from the Kaziranga National Park, Assam, India; and novel bioactive compounds from an endophytic *Chaetomium arcuatum* strain SAF-2 isolated from *Semecarpus anacardium*.

In addition, the associations, beneficial roles, and effects of microbes associated with plants are tremendous. There are chapters that describe the current research carried out in this angle where the authors describe the role of vesicular arbuscular mycorrhizae (VAM), PGPR, etc.; associative effect of arbuscular mycorrhizal fungi and *Rhizobium* on plant growth and the biological control of charcoal rot in green gram [*Vigna radiata* L. (Wilczek)]; morphological changes in *Vigna radiata* root under cadmium induced stress in the presence of plant growth promoting *Enterobacter* sp. C1D; and the role of microbiome: Insights into the kin recognition process in *Oryza sativa* by proteomic and metabolomic studies.

The book also provides an overview of the relevant topics of pharma, food, and other industries which include: understanding sepsis and its biomarkers; fungal biotransformation of drugs: potential applications in the pharma industry; role of copper in modifying cisplatin-induced cytotoxicity in yeast; isolation and characterization of yeast for the decolorization of reactive dyes; and development of antitubercular agents through hybridization strategies: future challenges and perspectives.

On the whole, the book deals with an avalanche of information available through research by taking a multidisciplinary approach and focusing on a broad spectrum of issues on different microorganisms and their recent applications and implications in agriculture, soil and forest, industry, public health/medicine, organic chemistry, biomass conversion, optimal production processes for different microbes, screening methods, and application of -omic approaches such as genomics, proteomics, and metabolomics, or other biotechnology tools to have a deeper understanding of the microbial based new and emerging products, trends, processes, and technologies.

PART I

Antimicrobial Agents: Role and Applications in Medicine and Healthcare

CHAPTER 1

ANTIMICROBIAL PEPTIDES FROM PLANTS AND THEIR APPLICATION

B. BHIMA[1*] and MOHAMMED AL SAIQALI[1]

[1]*Department of Microbiology, University College of Science, Osmania University, Hyderabad, Telangana – 500007, India.*

Corresponding author: B. Bhima. E-mail: bhima.ou@gmail.com

CONTENTS

Abstract 4

1.1   Introduction..4

1.2   Classification and Antimicrobial Activity of Plant Peptides................ 6

1.3   Mechanism of Action.. 11

1.4   Application of Plant Antimicrobial Peptides 14

1.5   Conclusion 19

Keywords 20

References.. 20

ABSTRACT

Antimicrobial peptides (AMPs) form part of the innate immunity establishing first line of defense against pathogen(s). Plant AMPs are a component of plant barrier defense system. They have been isolated from different parts of a variety of plant species and are known to have activities toward phytopathogens and also against human pathogens. All plant organs express AMPs constitutively or in response to microbial challenges. Plant AMPs are structurally and functionally diverse. Plant AMPs are grouped into several families and share general features such as positive charge and the presence of disulfide bonds which stabilize the structure. Thionins, defensins, lipid transfer proteins (LTPs), snakins, knottins, hevein-like AMPs and cyclotides represent the different plant AMPs. Besides targeting known bacterial, fungal, and other pathogens, some of them can be directed against other organisms, like insects. They have anticancer and antiviral activities. The biological activity of plant AMPs primarily depends on the interactions with membrane lipids, using different mechanisms to kill the targeted pathogens, but other modes of action exist as in the case of defensins with a-amylase activity or a defensin-like peptide that interacts with a receptor kinase. Plant AMPs are considered as promising antibiotic compounds with important applications in biotechnology, pharmacy and agriculture. AMPs, therefore, possess a high potential for therapeutic use in healthcare and can be used as natural antibiotics as an alternative for their chemical counterparts, protection of plants or animals against diseases. Six classes of plant AMPs will be discussed in this chapter and special attention will be given to their mechanism of action and applications. Therefore, this review can be considered as a snapshot of the progress in this field of research.

1.1   INTRODUCTION

Eukaryotic organisms including plants secrete a wide variety of AMPs produced by ribosomal (defensins and small bacteriocins) or non ribosomal (cyclopeptides and pseudo peptides) synthesis. Plants are a major source of diverse molecules with pharmacological potential. More than 300 plant AMP sequences have been described.[1,2] Plants produce small cysteine-rich AMPs as a mechanism of natural defense. AMPs may be expressed constitutively or induced in response to a pathogen attack. Plant AMPs are abundantly expressed in almost all plant species, and some are cysteine-rich peptides. Plant AMPs are produced in all organs and are more abundant in the outer layer. This abundance is consistent with their role as a constitutive host defense against microbial invaders attacking from the outside. Plant AMPs are released immediately after the initiation of infection. AMPs are expressed by a single gene and the expression requires less biomass and energy consumption.[3,4] A majority of plant AMPs are generally cationic with a net charge at neutral pH, varing from +2 to +9 and amphipathic. This enables the peptides to interact and disrupt lipid membranes. The molecular weight is between 2 and 10 kDa. These are basic, and contain 4, 6, 8, or 12 cysteines which form disulfide bonds conferring structural and thermodynamic stability. Many antimicrobial peptide families have been isolated from plants. One such example is Pp-Thionin, that shows activity against *Xanthomonas campestris, Rhizobium meliloti, Micrococcus luteus*. Moreover, Pp-AMP1 and Pp-AMP2 have a potent activity against many plant pathogens, including *Erwinia carotovora, Agrobacterium rhizogenes, Agrobacterium radiobacter, Clavibacter michiganensis*, and *Curtobacterium flaccumfaciens*, at low concentrations. Circulins A-B and Cyclopsychotride A from the Cyclotides family show antimicrobial activity against human pathogens such as *Micrococcus luteus, Staphylococcus aureus, Pseudomonas aeruginosa, Klebsiella oxytoca*, and *Escherichia coli*, at micromolar concentrations.[5,6] Plant AMPs are classified based on their primary and secondary structure, identity of their amino acid sequences and the number and position of cysteines forming disulfide bonds into different families thionins, defensins, cyclotides, lipid transfer proteins, snakins, hevein, and knottin-like peptides.[7] The amphipathicity of the AMPs is enhanced upon the induction of specific secondary structures, such as α-helices, β-sheets, or extended polyproline-like helices. This amphipathicity plays an important role in its mechanism of action. The mechanism of action of plant AMPs is of two types: (1) Membrane disruptive (2) Intracellular targets.[8,9] According to the first mechanism, AMPs cause membrane collapse by interacting with the lipid molecules on the bacterial cell surface. In this method, the cationic peptides are attracted electrostatically to negatively charged molecules such as anionic phospholipids, lipopolysaccharides (LPS) in Gram-negative bacteria and teichoic acid in Gram-positive bacteria, which are located asymmetrically in the membrane structure. The positively charged peptides can also interact with membrane lipids through specific receptors at the surface of the cell. Consequently, peptide binding to the membrane can activate several pathways that will cause cell death. According to the second mechanism, plant antimicrobial peptides interfere with protein and DNA synthesis, as well as DNA replication and cause cell death. The primary biological activities of plant AMPs are antifungal, antibacterial, and insecticidal.[6,10]

Plant AMPs exhibit enzyme inhibitory activities and have a role in heavy metal tolerance, abiotic stress, and development. Some plant AMPs show cytotoxic activity against mammalian cells and anticancer activity against cancer cells from different origins.[2] Defensins, thionins, and cyclotides have cytotoxic and anticancer properties.

1.2 CLASSIFICATION AND ANTIMICROBIAL ACTIVITY OF PLANT PEPTIDES

Based on the primary structure and amino acid composition, plant peptides are classified into cyclotides, defensins, hevein-like, impatiens, knottins, lipid-transfer proteins, snakins, thionins, etc.[5,6,11]

1.2.1 THIONINS

Thionins have been proposed to allocate a family of homologous peptides that includes purothionins, which were first isolated from wheat seeds. Thionins constitute a family of basic peptides, with low molecular weight (~5 kDa), rich in basic and sulfur-containing residues (arginine, lysine, and cysteine). Its diverse members have high sequential and structural similarities besides presenting toxic effects against bacteria, fungi, yeast, and animal and plant cells.[5,12,13] The first report about the presence of a lethal antimicrobial substance with potent activity was in wheat flour. In 1942, Balls and collaborators crystallized this toxic substance from the endosperm of *Triticum aestivum L.* It is a low molecular weight proteinaceous material with a high sulfur content, named purothionin. Proteins similar to purothionins are being isolated from other cereal leaves and endosperms of barley (α and β-hordothionins), oat (α and β-avenohtionins), and rye (secalethionins).[13] Thionins were found in the endosperm of all the species belonging to the genus *Triticum* and *Aegilops*.[11] Three small basic proteins purified from *Viscum album*,[14] named viscotoxins and their primary structures were identified, having similarities to purothionins. Other Proteins with similar properties have been detected from other members of the family *Viscacea*, such as phoratoxins A and B from *Phoradendron tomentosum*, denclatoxin B from *Dendrophtora clavata* and ligatoxins A and B from *Phoradendron liga*.[15] Crambin holds a special position among the thionins due to its high hydrophobicity. It presents a net charge of zero when compared with the others. Presently, no toxic activity has been demonstrated for crambin.

Peptides were isolated from the seeds of *Pyrularia pubera*. It is a parasite plant in the same group of mistletoe. Thionins from *Pyrularia* have four disulfide bridges similar to cereal thionins. It contains 14–21 amino acids, and shares eight consecutive amino acid residues similar to crambin. A different group of thionins was isolated from barley leaves. Thionins were identified from its cDNA sequences and mature thionins were detected from barley leaves. The thionins isolated from these two fractions inhibit the growth of phytopathogenic fungi. The main characteristic of thionins, except crambin, is that they have a toxic effect on different biological systems. Their antimicrobial activity has been tested against several microorganisms. Susceptibility was established in phytopathogenic bacteria like *Pseudomonas*, *Xanhtomonas*, *Erwinia*, *Corynebacterium*, in yeast and in phytopathogenic fungi such as *Thielaviopsis paradoxa* and *Drechslera teres*.[11,13,16] Insect larvae present sensitivity to different purothionins when injected in the haemocele, but not when integrated to nutrients. Besides toxicity on different organisms, cytotoxic activity on different cells has been recorded as well as inhibitory effects in *in vitro* systems. The sensitivity of several mammalian cell lines to thionins was detected, for example, mouse fibroblasts cell line (L929), Baby hamster kidney cell (BHK-21), human cell line (HeLa) cells, and monkey kidney cell line (CV10), show an increase in the cell membrane permeability. Viscotoxins exert strong immunomodulatory effects. Thionins inhibit protein synthesis in cell-free system, probably through direct interaction with mRNA or at the initial translation level. The ribonucleotide redutase enzyme activity is inhibited in the presence of purothionins, interfering in the DNA synthesis and irreversible inhibition of β-glucuronidase.[11,17]

1.2.2 PLANT DEFENSINS

Defensins are the most ancient eukaryotic antimicrobial peptides. It is a group of small basic proteins which was isolated from wheat, barley, sorghum, radishes, and other species of *Brassicaceae* family.[18,19] Amino acid sequence analysis proved that it is a new family of thionins, different from α and β-thionins, named γ-thionins.[11,20] It was later proved that the γ-thionins show less similarity to other genetic variants of wheat (α1, α2, and β) and barley (α and β) thionins, as well as thionins from *Pyrularia*, crambin, and viscotoxins. These groups present a series of structural characteristics that separate them from γ-thionins, like the basic amino acid distribution and the disulfide bridges arrangement. In 1995, they were considered as a super family of antimicrobial peptides, which were present in vertebrates, invertebrates, and plants, they were renamed as plant defensins because of the high degree of similarities between plants and animal defensins. The *in vivo* activity of plant defensins is so far not understood. It has been observed that γ-hordothionins inhibit *in vitro* protein synthesis.[21] Three isoinhibitors of insects, α-amilases (SIα1, SIα2, and SIα3) were isolated. They have a high sequential similarity to γ-purothionins. Many plant defensins have a powerful fungicidal action, indicating that they can act as defense proteins.[22] Although there are some differences among their primary structures, they all have very similar tertiary structures, which greatly differ from those observed for α1-purothionin and crambin, having high structural similarity to scorpion neurotoxins. The three-dimensional structure for SIα1 is similar to charybdotoxin (2CRD), a small basic protein with 37 residues and 3 disulfide bridges, isolated from *Leiurus quinquestriatus hebraeus* venom, which binds specifically to potassium channels for blocking. The two γ-thionins that have been isolated from maize seeds are extremely similar to sorghum SIα1 and SIα3 isoinhibitors.[11]

1.2.3   *LIPID TRANSFER PROTEINS*

In various monocotyledonous and dicotyledonous plants, the nonspecific small lipid transfer proteins (ns-LTPs) are present that are capable of exchanging lipids between membranes *in vitro*. The ns-LTPs participate in the regulation of intracellular fatty acid pools, membrane biogenesis, involved in the defense reactions against phytopathogens, cutin formation, embryogenesis, symbiosis, and the adaptation of plants to various environmental conditions.[7] The exploration for lipid carrier proteins led to the discovery of lipid transfer proteins (LTPs) in various plant species. The assays used for the detection of lipid transfer activity are one of the most important steps of the LTP characterization. These assays examine the labeled lipid transfers from a donor membrane to an acceptor membrane. The donor membranes are liposomes, while the acceptors are cytoplasmic organelles, such as chloroplasts and mitochondria. The lipids to be transferred can be radiation labeled or, instead, be fluorescent lipids. When the membrane fractions are incubated in the presence of LTPs and later separated by centrifugation, the detection of the labeled lipids in the acceptor fraction indicates the effective transfer of lipids previously inserted in the donor membranes. These biological assays help to identify various LTPs from different parts of the plants.[23–26] The plant LTPs are divided into two subfamilies, their molecular weights are 9 kDa (LTP1s) and 7 kDa (LTP2s). They are able to transfer different types of polar lipids, like phosphatidylethanolamine, phosphatidylinositol, and phosphatidylcholine, besides galactolipids. Because of their low specificity, plant LTPs are named as nonspecific lipid transfer proteins. Techniques such as infrared spectrometry, Ramachandran plot, crystallography, and NMR are used to study the three-dimensional structure of plant LTPs.[11,27] The plant LTPs consist mainly of α-helix connected by disulfide bridges. They have four α-helixes and a C-terminal end lacking an arranged structure. The hydrophobic internal cavity, forming an adequate site for the interaction between the aliphatic chain of lipids and the hydrophobic amino acids is exposed in the cavity. This family of proteins is characterized by facilitating the lipid transfer among natural or artificial membranes and by binding to fatty acids *in vitro*. The isolation of cDNAs and genes encoding LTPs demonstrated the presence of a signal peptide, indicating that the LTPs would be secreted. Plant LTPs are important antimicrobial peptides involved in plant defense against phytopathogens. They possess potent antimicrobial activities, and are distributed in sufficient concentrations in the plant tissues and also have an increase in their gene expression levels soon after infection. Barley LTP2 expression in tobacco and *Arabidopsis* transgenic plants enhanced the reduction in necrotic diseases that caused by *Pseudomonas*, suggesting that LTPs are involved in plant defense mechanisms against plant phytopathogens. The antifungal and antibacterial properties of LTP110, a lipid transfer protein from rice were tested *in vitro* against rice pathogens, *Xanthomonas oryzae* and *Pyricularia oryzae*. LTP110 was able to inhibit the germination of *P. oryzae* spores, but only slightly inhibited the growth of *Xanthomonas*.[11,28]

1.2.4 SNAKINS

Snakin peptides have been isolated from potato tubers. They are a part of the cell wall-associated peptides snakin-1 (StSN1) and snakin-2 (StSN2). They are antimicrobial peptides with 63 amino acid residues. All snakins have 12 conserved cysteine amino acids and six disulfide bonds.[7,29,30] The mechanism of action of snakins is not yet understood. They do not inter-relate with artificial lipid membranes. The snakin peptides are basic and rich in cysteine amino acids, which form six disulfide bridges that give stability to their structure. StSN1 amino acid sequence shows similarity with members of the tomato GAST family and Arabidopsis GASA family and classified as a member of snakin GASA family.[31] Snakin GASA genes have small proteins where three distinct domains are present. A putative signal peptide of 18–29 residues, a variable region displaying high divergence between family members, both in amino acid composition and sequence length, and a C-terminal region of approximately 60 amino acids containing 12 cysteine amino acids are conserved positions of the GASA domain.[29,32]

1.2.5   CYCLOTIDES

Cyclotides are an important class of plant antimicrobial peptides. They are globular disulfide rich peptides, ranging in size from about 28 to 37 amino acids. Cyclotides have unique structural features, with a head to tail cyclized backbone and a knotted arrangement of three disulfide bonds, referred to as a cyclic cystine knot (CCK) motif. This motif gives the cyclotides a high resistance to chemical, thermal, and enzymatic degradation. This class of peptides are mostly found in the botanical families Apocynaceae, Rubiaceae, Curcubitaceae, Violaceae, and Poaceae.[33,34] Cyclotides are desirable targets for the pharmaceutical and agrochemical industries, because of their unique cyclic structural scaffold, biological activities and diversity of sequence. They are one of the most studied plant peptide families, and comprise a huge library of natural peptides that can be used in the search for new bactericides. Kalata B1 and circulin A show high activity against *Staphylococcus aureus*. But circulin B has a strong activity against *E. coli*, and is moderately active against *Staphylococcus aureus*.[20]

1.2.6   HEVEIN AND KNOTTIN-LIKE PEPTIDES

Hevein is a small cysteine rich, chitin binding peptide present in the latex of rubber trees. It inhibits the hyphal growth of fungi by binding to chitin. Other hevein-like proteins with antimicrobial activity have been isolated from different plants.[35,36] Hevein-like peptides are small, containing 43 amino acid chitin binding peptides. The hevein like AMPs differ in the number of disulfide bonds. Most of them possess eight cysteine residues forming four disulfide bonds, for example, hevein isolated from the seeds of *Pharabitis nil L.* and *Avena sativa*. The hevein-like proteins from *Amaranthus caudatus* seeds contains six disulfide bonds linked to the cysteine amino acids. Two AMPs from seeds of *Pharbitis nil L.* (Pn-AMP1 and Pn-AMP2) have potent antifungal activities against both chitin containing and non chitin containing fungi. The Pn-AMPs enter rapidly into fungal hyphae and destroy the hyphal tips, leading to the disruption of the fungal membrane, and linkage of cyto-plasmic materials.[5,37]

The knottin type of antifungal peptides have been isolated from *Mirabilis Jalapa L.* (Mj-AMP1) and from *Phytolacca Americana*. The structure of Phytolacca antifungal protein (PAFP-S) consists of a triple stranded, anti-parallel beta sheet with a long loop region connecting β-strands 1 and 2. This peptide which was isolated from garden pea (PA1b) has an insecticidal activity through the inhibition of vacuolar ATPase.[38–40]

1.3 MECHANISM OF ACTION

The amphipathicity of AMPs is enhanced by the induction of secondary structures, such as α-helices, β-sheets, or extended polyproline like helices. This amphipathicity has an important role in their antimicrobial action.[8] The mechanism of action of an AMP depends on a number of physicochemical properties: the amino acid sequence, amphipathicity, hydrophobicity, net charge, structural folding which includes secondary structure, dynamics and orientation in the membranes, oligomerization, peptide concentration, and membrane composition. The mechanism of action of the plant AMPs is of two types: (1) Membrane disruptive method (2) Intracellular targets.[9] There are several mechanisms of membrane disruption which have been proposed to explain the activity of the AMPs. Some of the models used to explain the membrane disrupting process are carpet model, barrel stave model, toroidal pore or wormhole model (Figure 1-1), and detergent type membrane lytic mechanism.[41,42] In the carpet model, the AMPs gather on the surface of the membrane to disrupt the membrane through barrel-stave, toroidal-pore, or detergent-type model. In the barrel-stave model, the AMPs insert with a transmembrane orientation in the membranes and aggregate to form an ion channel pore. In the toroidal pore model,[8,43] the peptides are located closer to the head group region with an initial orientation parallel to the lipid bilayer surface. In this orientation, the hydrophilic side of the helix is exposed to the hydrophilic lipid head groups and the water phase outside the bilayers, while the hydrophobic face of the helix is dived in the hydro-phobic core of the membrane to minimize the net free energy of the folding process. Aggregation of the peptides to appropriate concentration increases the twist strain on the membrane surface to a level such that toroidal pores will be formed. In the detergent type of mechanism, the peptides, carpet the surface of the lipid bilayer first, as in the toroidal pore model. The peptide aggregation to high concentration and the amphipathic nature of the peptide makes it to act as a detergent and break the lipid membrane into smaller fragments. These fragments can be like micelles. There are other models such as the sinking raft model[44] or the molecular electroporation, which is now neglected, but could be useful to explain the antimicrobial activity of certain AMPs.[8]

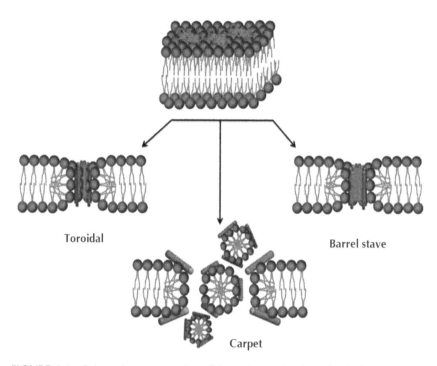

Toroidal   Barrel stave

Carpet

FIGURE 1-1   Schematic representation of the action mechanism of antimicrobial peptides. Barrel stave: peptides come close to the membrane target causing a pore in the membrane; In the toroidal model, lipids and peptides are over lapped; In the carper model, the hydrophilic regions of the peptides are exposed to solvents and hydrophobic region to the membrane. At a certain threshold concentration of peptides, the permeability of the membrane increases which facilitates pore formation.

In the sinking raft model, the AMPs bound to the cell membrane introduce a large membrane warp because of a mass imbalance, which makes the AMPs sink and generate pores in the membranes. While in the molecular electroporation model, the AMPs create a difference in the electrical potential across the membrane leading to the formation of pores through electroporation. Bacterial membranes are negatively charged with lipids such as phosphatidylglycerol (PG), cardiolipin (CL), or phosphatidylserine (PS). Electrostatic interaction between these negatively charged lipids and the positively charged AMPs enables the cationic peptides to bind to the bacterial membranes. The outer membrane of Gram-negative bacteria is negatively charged as it contains anionic lipopolysaccharides (LPS). LPS is normally stabilized by the divalent cations like $Ca2+$ and $Mg2+$, but the AMPs displace them to interact with the outer membrane.

The mammalian cell membranes contain zwitterionic phospholipids which are neutral in charge such as phosphatidylethanolamine (PE), phosphatidylcholine (PC), or sphingomyelin (SM) and as a result, they are less attractive to the cationic AMPs. In addition, the cholesterol present in the mammalian cell membranes will not allow the AMPs to interrupt the lipid bilayer structures of the mammalian cells. Hence, the AMPs are selectively toxic to bacteria. There are AMPs that act on intracellular targets within the bacteria without disrupting the cell membrane.[8,45] These AMPs inhibit the synthesis of protein and activities of certain enzymes and interfere with the metabolic processes of microbes, or interact with the DNA or RNA. Certain AMPs have been shown to have antiviral activities. They inhibit the replication of enveloped viruses such as influenza A virus, vesicular stomatitis virus (VSV) and human immunodeficiency virus (HIV-1).[46] The viral AMPs interact directly with the envelope of the virus, leading to permeation of the envelope and eventually, lysis of the virus particle, similar to the pore formation mechanism mentioned earlier for antibacterial activity of AMPs. Other mechanisms for antiviral activities like T22, an 18 amino acids peptide analogue of polyphemusin II, which is AMP isolated from the hemocyte of *Limulus polyphemus*, that specifically inhibit the ability of T cell line HIV-1 to induce cell fusion. On the other hand, lactoferricin, an N-terminal fragment of lactoferrin, inhibits the binding and uptake of human *papilloma virus*, human *cytomegalovirus*, and *herpes simplex virus* into human cells.[47,48,49] Changes in the membrane of a cell have important implications in the development of cancer, as they play an important role in the response of the cell to its surrounding environment. The cell membrane of a malignant tumor might grow even in the absence of signals promoting growth and it attaches and responds to neighboring cells differently as well. Cancer cell membranes usually carry a net negative charge because of the higher expression of anionic molecules like PS which account for 9% of the total phospholipids of the membranes[8,34] and glycosylated mucins. The net negative charge of the membrane of the cancer cells may also add to the selective cytotoxic activity of the anticancer peptides (ACPs). On the other hand, electrostatic interactions between the ACPs and normal cells are not preferential because of the neutral charge present on the healthy cells by the zwitterionic nature of their major membrane components, like SM, PE, and PC. The ACPs could kill cancer cells by one of the above mentioned mechanisms to disrupt the cell membrane. The other way is the initiation of apoptosis in cancer cells via mitochondrial membrane disruption following ACP uptake into the cytoplasm. Although the entire mechanism of action of plant AMPs is not yet understood, they were found to cause leakage of the cell constituents of *Neurospora crassa*. Plant AMPs are found to mediate a sustained Ca^{+2} influx and K^+ efflux, when they were added to the fungus *Neurospora crassa* that differs from thionins which cause permeabilization to isoaminobutyric acid. Defensins group I causes morphological distortion of the hyphae and also inhibits Gram-positive bacteria, while Defensins group II is the vice versa of group I. Defensins are not toxic to animal or plant cells. Some plant lipid transfer peptides were much more active against bacterial pathogens and their activities differed from one another. The plant lipid transfer peptides might complement each other when simultaneously present in a tissue.[5,8]

1.4   APPLICATION OF PLANT ANTIMICROBIAL PEPTIDES

1.4.1   PLANT AMPs AND PHYTOPATHOGENS

All plant AMPs differ in their way of attacking phytopathogens, they even could exhibit different cytotoxic effect against bacteria and fungi.

The major characteristic of thionins is the toxic effect on different biological systems. Thionins have activities against several microorganisms. Susceptibility was demonstrated in phytopathogenic bacteria such as *Pseudomonas*, *Xanhtomonas*, *Erwinia*, and *Corynebacterium*, yeast and in phytopathogenic fungi like *Thielaviopsis paradoxa* and *Drechslera teres*. Insect larvae are sensitive to different purothionins when injected in the haemocele, but not when incorporated to nutrients. Different thionins cause an increase in the cell membrane permeability, which was proved by experiments carried out on several mammal cell lineages such as mouse fibroblasts L929, baby hamster kidney cell line BHK-21, human cell line HeLa, and monkey cell lines. Viscotoxins exert strong immunomodulatory effects. Thionins are also able to inhibit protein synthesis in cell-free system, probably through direct interaction with mRNA or at the initial translation level. The activity of the enzyme ribonucleotide redutase is inhibited in the presence of purothionins, interfering in the DNA synthesis, and irreversible inhibition of β-glucuronidase. Thionins could act as thiol secondary messengers in the redox regulation of enzymes and also act as regulatory proteins.[11] The thionins found in seeds, such as viscotoxins, purothionins, and hordothionins, function as storage proteins, especially as sources of sulfur. For example, viscotoxins, whose levels fall dramatically in senescing leaves, are defense proteins with general effect. Thionins have been utilized to produce transgenic plants resistant to phytopathogenic fungi, like the α-hordothionin gene expression in tobacco, under the control of Cauliflower mosaic virus promoter (CaMV35S), increasing the resistance against *Pseudomonas syringae*. Viscotoxin expression in *Arabidopsis thaliana* conferring resistance to *Plamodiophora brassicae*, and the expression of an oat thionin in transgenic rice seedlings lead to protection against the phytopathogenic bacteria *Burkholderia plantarii* and *B. glumae*. The plant defensins, β-zeathionins, as well as scorpion toxins exhibit neurotoxic activity. The β-zeathionins are able to completely inhibit the sodium channels in voltage dependent ion channels.[11,50] Plant defensins have antifungal and antibacterial activities. They also have proteinase or amylase inhibitory effect. Plant defensins inhibit the growth of a wide range of fungi and are less toxic to mammalian cells or plants.[51]

1.4.2 AGRICULTURAL APPLICATIONS

Barley LTP2 expression in tobacco and *Arabidopsis* transgenic plants promoted greatly reduced necrotic infection caused by *Pseudomonas*. This was a proof that the LTPs are involved in plant defense mechanisms against microbial and phytopathogens, as well as confirming their biotechnological potential.

The role of LTPs in plant protection has been studied. The antifungal and antibacterial properties of LTP110, a lipid transfer protein which was extracted from rice were studied. This protein was cloned and the expressed protein was purified and tested *in vitro* against rice pathogens *Pyricularia oryzae* and *Xanthomonas oryzae*. LTP110 inhibited the germination of *Pyricularia oryzae* spores, but partially inhibited the growth of *Xanthomonas*.[28]

1.4.3 PLANT ANTIMICROBIALS FOR FOOD BIOPRESERVATION

Plants can be a good source of antifungal proteins and peptides including chitinases, glucanases, thaumatin-like proteins, thionins, and cyclophilin-like proteins. Some of them could possibly be exploited for food biopreservation and maintenance of seeds for longer time. Peptides could also be used for the preservation of canned food.[52] Thionins found in seeds like viscotoxins, purothionins, and hordothionins can also function as storage proteins, especially as sources of sulfur. Evidences of their role were reported for viscotoxins.

1.4.4 DEVELOPMENT OF TRANSGENIC PLANTS EXPRESSING ANTIMICROBIAL PEPTIDES

Gene constructions including sequence coding for AMPs have been expressed on crop plants providing different degrees of protection against phytopathogens. Animal defensin genes have been expressed in several plants. Cecropins A and B expressed in rice, confer protection against *Magneporthe grisea* and *Xanthomonas oryzae*, magainin expressed in tobacco confers protection against several fungi and bacteria,[53,54] and tachyplesin from crab expressed in potato was effective against infections by *Erwinia carotovora*. The insect defensins heliomicin and drosomycin expressed in tobacco confer protection against *B. cinerea* and the sarco-toxin from fruit fly expressed in tobacco protected against *Pseudomonas syringae* and *Erwinia carotovora ssp.*[55,56]

Plant defensins have been expressed in plants. The radish (Raphanus sativus antifungal peptide) defensin Rs-AFP2 was expressed in tobacco and tomato and confers protection against *Alternaria longipes*; Alfalfa defensin expressed in potato protects against *V. dahlia*; spruce (Picea abies) defensing SPI1 expressed in tobacco protects against *Heterobasidium annosum*;[19,40,56] pea defensing gene (DRR206) expressed in canola and tobacco protects against *Leptosphaeria maculans*; cabbage (Brassica stamen) plant defensing BSD1 expressed in tobacco protects against *Phytophthora parasitica*, wasabi defensing WT1 expressed in rice gives protection against *M. grisea*;[57–59] Dahlia merckii defensin (Dm-AMP1) expressed in eggplant gives protection against *Botrytis cinerea* and *Verticillium alboatrum*; and jalapa defensin Mj-AMP1 expressed in tomato protects it against the plant pathogen *Alternaria solani*[60,61] The hevein Pn-AMP expressed in tobacco gives protection against *Phytophthora parasitica* and barley hordothionin expression in tobacco confers protec-tion against *Pseudomonas syringae*.[1,62]

1.4.5   AMPs AS BIOCONTROL AGENTS

The production of AMPs has been endorsed to the inhibitory activity of several microorganisms that act against fungal and bacterial plant pathogens. Cyclopeptides which are produced by several plants and soil bacteria, have antifungal and antibacterial, cytotoxic, or surfactant properties. They contain a large number of natural peptides which can be utilized for the production of new bactericides.[1,33,34] Cyclic lipopeptides have been implicated to give *Bacillus amyloliquefaciens* FZB42 the ability to control *F. oxysporum* by using structural and functional characterizations of gene clusters involved in their synthesis and the defective mutants are unable to produce bacillomycin D and fengycin.[63]

Mycosubtilin produced by *Bacillus subtilis* mutants were more effective than the parent peptides in controlling *Phytium* damping off on tomato. The production of bacillomycin, fengycin, or iturin by several strains of *Bacillus subtilis* was used to control powdery mildew on melon caused by *Podosphaera fusca*. There is less evidence for the inference of cyclopeptides in biocontrol which has been provided by the in situ detection of compounds on host plant tissues.[64,65]

1.4.6   PLANT AMPs FOR CONTROLLING PLANT DISEASES

The mechanism of action against the target microorganism makes the plant AMPs attractive antimicrobial compounds for plant disease control. Most AMPs are cationic and bind to the surface of microorganisms through receptor-mediated interaction and insert into the cytoplasmic membrane. Several plant AMPs are membrane disruptive, but others are non membrane disruptive and cross the cell membrane to interact with intracellular targets and inhibit nucleic acid or protein synthesis, or enzymatic activity. Several AMPs have been used as the basis for the development of shorter and less toxic analogues by synthetic procedures. Some AMPs play a role in biocontrol agents of plant diseases. The animal and plant AMPs, which are produced by ribosomal synthesis, have provided tools for developing transgenic plants expressing gene coding for the synthesis of these antimicrobial peptides conferring partial or total resistance to plant pathogens.[1,76]

1.4.7   ANTICANCER

Three plant AMP families have cytotoxic and anticancer properties. These are defensins, thionins, and cyclotides. The method of cell lysis resulting from targeting the membrane shows a potential for synergy with the present cancer treatments. ACPs have shown additional killing of cancer cells when tested in concert with the antineoplastic drug treatments. AMPs that show cytotoxic activity against cancer cells have the ability to overcome the common problems with multiple drug resistant (MDR) proteins. The thionin from mistletoe exhibits an anticancer activity against cervical cancer cells (HeLa) and mouse melanoma cells. Another group of thionins that have anticancer and cytotoxic activity are the viscotoxins from *Viscum* spp. Viscotoxin B2 has anticancer activity against rat osteoblast-like sarcoma.[66] Plant defensin Sesquin from *Vigna sesquipedalis* was the first peptide reported with anticancer activity which inhibits the proliferation of Michigan Cancer Foundation MCF-7 and leukemia (Mus musculus, myeloblast mouse cell line 1) M1 cells. A defensin from *Phaseolus limensis* differentially inhibited the proliferation of leukemia cells, reaching 60% inhibition for M1 and 30% inhibition for Mouse lymphocytic leukemia cells L1210 cells. However, its effect against normal cells was not evaluated.[52,67] A defensin from the purple pole bean *Phaseolus vulgaris*, inhibited the proliferation of cancer cell lines (liver hepatocellular carcinoma cells) HepG2, MCF-7, but did not affect human embryonic liver cells or human erythrocytes under the same conditions. Cycloviolacin from *Viola odorata* is a promising peptide because of its selective toxicity to cancer cell lines compared with normal cells, which indicates the possibility of its use as an anticancer agent. Synthetic Cyclotides (MCo-PMI) showed the activity *in vivo* in a murine xenograft model with prostate cancer cell significantly suppressed tumor growth.[68–70]

The linear peptide *Cn*-AMP1, isolated from *Cocos nucifera*, was tested against MCF-7, Human cancer cell lines HCT-116, Human colorectal carcinoma Caco-2, murine macrophage cell line RAW264 cells, and human erythrocytes. It showed a decrease of cell viability in cancer cells without causing hemolysis. The cyclic heptapeptide cherimolacyclopeptide C, obtained from a methanol extract of the seeds of *Annona cherimola*, exhibited significant *in vitro* cytotoxicity against oral cancer cell lines KB cells. Small antitumor cyclic peptides RA-XVII and RAXVIII from the roots of *Rubia cordifolia* have cytotoxicity against P-388 cells. However, it was not determined whether these peptides are effective against normal cells.[71,72]

1.4.8 PHARMACY AND BIOTECHNOLOGY

AMPs are encoded by small genes with conserved sequences. Therefore, gene amplification and transgenesis are among the feasible ways to increase production and enhance specific activity of selected peptides. AMPs are also widely applied in the development of synthetic and genetically modified peptides. Expression of plant defensins in transgenic plants leads to the protection of vegetative tissues against pathogen bacteria. The radish defensin Rs-AFP2 was expressed in tobacco and tomato and confers protection against *Alternaria longipes* and the Jalapa defensin expressed in tomato protects against *Alternaria solani* . The hevein Pn-AMP expressed in tobacco protects against *P. parasitica*, and the expression of an alfalfa defensin in potato provided good resistance against the fungus *V. dahliae* under field conditions. The antimicrobial activity of defensins can be enhanced *in vivo* because of the synergistic interaction with other defense components.[3,40] Thionins are important tools for the genetic improvement and development of transgenic plants producing higher levels of thionins, increasing the pathogenic resistance and reducing crop losses in farming.[73] Another example is of the cyclotides, which have potential applications in both pharmaceutical and agricultural industries.

Cyclotides can be used for the development of novel antibiotics and bioinsecticides, like kalata B1, where polar and/or charged residues were modified.[74] Cell-penetrating peptides (CPPs) are promising candidates for intracellular drug delivery, RNA, DNA, and nanoparticles in a nondestructive manner. CPPs have been shown to facilitate delivering a wide variety of biomolecules across the skin. The enormous potential of this technology resides in the high efficiency and relatively low toxicity of CPPs conjugated to bioactive cargoes. Different CPPs can be successfully used for the delivery of high molecular weight drugs into cells as well as for vaccine development. The application of CPPs in pharmaceutical preparations is a promising field with a great potential in transdermal drug delivery systems.[75,76]

1.5 CONCLUSION

AMPs offer a good alternative for treating infections in relation to conventional antibiotics based on their broad spectrum activity and efficiency. In pharmaceutical areas, AMPs play a strong role in agriculture as plant protection products. Successful use of AMPs has been achieved through the commercial development of AMPs as biopesticides. Although numerous transgenic plants expressing AMPs that confer different degrees of protection against diseases have been developed, commercial cultivars have not been marketed because of regulatory limitations and social concerns. Synthetic approaches to obtain AMPs guided by combinatorial chemical methods provide powerful tools to optimize molecules derived from natural compounds with improved activity against selected target pathogens, including decreased cytotoxicity and increased protease stability. With a rapid development in proteomics, bioinformatics, peptide libraries, and modification strategies, the plant AMPs have emerged as novel promising anticancer drugs in future clinical applications. A majority of AMPs with potential uses have been studied at the *in vitro* level, fewer compounds have been tested on plant pathogens, and only a few are in the commercial market. Presently, very few reports are available on newly discovered AMPs.

Therefore, the future area of interest consists of developing less toxic and more stable AMPs as well as decreasing production and manufacturing cost by enhancing synthesis and biotechnological procedures using microbial system.

KEYWORDS

- **Antimicrobial activity**
- **Application**
- **Classification**
- **Mechanism of action**
- **Plant peptides**

REFERENCES

1. Montesinos, E. Antimicrobial peptides and plant disease control. *FEMS Microbiol. Lett.* **2007,** *270*, 1–11.
2. Guzmán-Rodríguez, J. J.; Ochoa-Zarzosa, A.; López-Gómez, R.; and Lopez-Meza, J. E. Plant antimicrobial peptides as potential anticancer agents. *BioMed. Res. Int.* **2015**, 1–11, DOI: 10.1155/2015/735087.
3. Thomma, B. P.; Cammue, B. P.; and Thevissen, K. Plant defensins. *Planta* **2002,** *216* (2), 193–202.
4. Lay, F. T; and Anderson, M. A. Defensins—components of the innate immune system in plants. *Cur. Prot. Pep. Sci.* **2005,** *6* (1), 85–101.

5. Garcia-Olmedo, F.; Rodriguez-Palenzuela, P.; and Molina, A. Antibiotic activities of peptides, hydrogen peroxide and peroxynitrite in plant defence. *FEBS Let.* **2001,** *498* (2–3), 219–222.

6. Pelegrini, P. B; del Sarto, R. P; Silva, O. N; Franco, O. L; and Grossi-De-Sa, M. F. Antibacterial peptides from plants: What they are and how they probably work. *Biochem. Res. Int.* **2011,** 1–9, DOI: 10.1155/2011/250349.

7. Nawrot, R.; Barylski, J.; Nowicki, G.; Broniarczyk, J.; Buchwald, W.; and Goździcka-Jozefiak, A. Plant antimicrobial peptides. *Folia Microbiol.* **2013,** *59,* 181–196.

8. Hoskin, D. W.; and Ramamoorthy, Y. Studies on anticancer activities of antimicrobial peptides. *Biochem. Biophy. Acta.* **2008,** *1778* (2), 357–375.

9. Giuliani, A.; Pirri, G.; and Nicoletto, S. F. Antimicrobial peptides: An overview of a promising class of therapeutics. *CEJB.* **2007,** *2* (1), 1–33.

10. Stotz, H. U.; Waller, F.; and Wang, K. Innate immunity in plants: The role of antimicrobial peptides. In *Antimicrobial Peptides and Innate Immunity.* Hiemstra, P. S., Zaat, S. A. J., Eds.; Springer Science & Business Media: Springer Basel, **2013,** 29–51. DOI: 10.1007/978-3-0348-0541-4_2.

11. Mariana, S.; and Fontes, W. Plant defense and antimicrobial peptides. *Prot. Pep. Let.* **2005,** *12,* 11–16.

12. Bohlmann, H.; Clausen, S.; Behnke, S.; Giese, H.; Hiller, C.; Reimann-Philipp, U.; Schrader, G.; Barkholt, V.; and Apel, K. Leaf-specific thionins of barley-a novel class of cell wall proteins toxic to plant-pathogenic fungi and possibly involved in the defense mechanism of plants. *EMBO J.* **1988,** *7,* 1559–1565.

13. Balls, A. K.; Hale, W. S.; Harris, T. H.; Balls, A. K.; Hale, W. S.; and Harris, T. H. A crystalline protein obtained from a lipoprotein of wheat flour. *Cereal Chem.* **1942,** *19,* 279–288.

14. Samuelsson, G. Mistletoe toxins. *System. Zool.* **1973,** *22,* 566–569.

15. Li, S. S.; Gullbo, J.; Lindholm, P.; Larsson, R.; Thunberg, E.; Samuelsson, G.; Bohlin, L.; Claeson, P.; and Ligatoxin, B. A new cytotoxic protein with a novel helix-turn-helix DNA-binding domain from the mistletoe Phoradendron liga. *Biochem. J.* **2002,** *366,* 405–413.

16. Kramer, K. J.; Klassen, L. W.; Jones, B. L.; Speirs, R. D.; and Krammer, A. E. Toxicity of purothionin and its homologues to the tobacco hornworm, Manduca sexta (L.) (Lepidoptera:Sphingidae). *Toxicol. Appl. Pharmacol.* **1979,** *48,* 179–183.

17. Stein, G. M.; Schaller, G.; Pfuller, U.; Wanger, M.; Wanger, B.; Schietzel, M.; and Bussing, A. Thionins from Viscum album L., influence of viscotoxins on the activation of granulocytes. *Biochem. Biophy. Acta.* **1999,** *1426,* 80–90.

18. Collila, F. J.; Rocher, A.; and Mendez, E. gamma-Purothionins: Amino acid sequence of two polypeptides of a new family of thionins from wheat endosperm. *FEBS Lett.* **1990,** *270,* 191–194.

19. Terras, F. R.; Eggermont, K.; and Kovaleva, V. Small cysteine-rich antifungal proteins from radish: their role in host defense. *Plant Cell.* **1995,** *7,* 573–588.

20. Candido, E. F.; Porto, W. F.; Amaro, D. S.; Viana1, J. C.; Dias, S. C.; and Franco, O. L. Structural and functional insights into plant bactericidal peptides. *Formatex* **2011,** *2,* 951–960.

21. Broekaert, W. F.; Terras, F. R. G.; Gammue, B. P. A.; and Osborn, R. W. Plant defensins: Novel antimicrobial peptides as components of the host defense system. *Plant Physiol.* **1995,** *108,* 1353–1358.

22. Thevissen, K.; Terras, F. R.; and Broekaert, W. F. Permeabilization of fungal membranes by plant defensins inhibits fungal growth. *App. Env. Microbiol.* **1999**, *65*, 5451–5458.
23. Nielsen, K. K.; Nielsen, J. E.; Madrid, S. M.; and Mikkelsen, J. D. New antifungal proteins from sugar beet (Beta vulgaris L.) showing homology to non-specific lipid transfer proteins. *Plant Mol. Biol.* **1996**, *31*, 539–552.
24. Carmen Ramirez-Medeles, M.; Aguilar, M. B.; Miguel, R. N.; Bolanos-Garcia, V. M.; Garcia-Hernandez, E.; and Soriano-Garcia, M. Amino acid sequence, biochemical characterization, and comparative modeling of a nonspecific lipid transfer protein from Amaranthus hypochondriacus. *Arch. Biochem. Biophys.* **2003**, *415*, 24–33.
25. Liu, Y. J.; Samuel, D.; Lin, C. H.; and Lyu, P. C. Purification and characterization of a novel 7-kDa non-specific lipid transfer protein-2 from rice (Oryza sativa). *Biochem. Biophys. Res. Commun.* **2002**, *294*, 535–540.
26. Castro, M. S.; Gerhardt, I. R.; Orru, S.; Pucci, P.; and Bloch, C. Purification and characterization of a small (7.3 kDa) putative lipid transfer protein from maize seeds. *J. Chromatogr. B Analyt. Technol. Biomed. Life Sci.* **2003**, *794*, 109–114.
27. Douliez, J. P.; Michon, T.; Elmorjani, K.; and Marion. D. Structure, biological and functions of lipids transfer proteins, the major lipid binding protein from cereal kernels. *J. Cereal Sci.* **2000**, *32*, 1–20.
28. Ge, X.; Chen, J.; Li, N.; Lin, Y.; Sun, C.; and Cao, K. Resistance function of rice lipid transfer protein LTP110. *J. Biochem. Mol. Biol.* **2003**, *36* (6), 603–7.
29. Segura, A.; Moreno, M.; Madueno, F.; Molina, A.; and García-Olmedo, F. Snakin-1, a peptide from potato that is active against plant pathogens. *Mol. Plant Microbe Interact.* **1999**, *12*, 16–23.
30. Berrocal-Lobo, M.; Segura, A.; Moreno, M.; López, G.; Garcia-Olmedo, F.; and Molina, A. Snakin-2, an antimicrobial peptide from potato whose gene is locally induced by wounding and responds to pathogen infection. *Plant Physiol.* **2002**, *128*, 951–961.
31. Almasia, N. I.; Narhiriak, V.; Hopp, E. H.; and Vazquez-Rovere, C. Isolation and characterization of the tissue and developmental specific potato snaking-1 promoter inducible by temperature and wounding. *Electr. J. Plant Biotech.* **2010**, *13* (5), fulltext-12.
32. Nahirñak, V.; Almasia, N. I.; Hopp, H. E.; and Vazquez-Rovere, C. Snakin/GASA proteins involvement in hormone crosstalk and redox homeostasis. *Plant Signal. Behav.* **2012**, *7*, 1004–1008.
33. Daly, N. L.; Rosengren, K. J.; and Craik, D. J. Discovery, structure and biological activities of cyclotides. *Adv. Drug Deliv. Rev.* **2009**, *61*, 918–930.
34. Gruber, C. W. Global cyclotide adventure: A journey dedicated to the discovery of circular peptides from flowering plants. *Biopolymers* **2010**, *94*, 565–572.
35. Van Parijs, J.; Broekaert, W. F.; Goldstein, I. J.; and Peumans, W. J. Hevein an antifungal protein from rubber-tree (Hevea braziliensis) latex. *Planta* **1991**, *183*, 258–264
36. Kiba, A.; Saitoh, H.; Nishihara, M.; Omiya, K.; and Yamamura, S. C-terminal domain of a hevein-like protein from Wasabia japonica has potent antimicrobial activity. *Plant Cell Physiol.* **2003**, *44*, 296–303.
37. Li, S. S.; and Claeson, P. Cys/Gly-rich proteins with a putative single chitin-binding domain from oat (Avena sativa) seeds. *Phytochemistry* **2003**, *63*, 249–255.
38. Chouabe, C.; Eyraud, V.; DaSilva, P.; Rahioui, I.; Royer, C.; Soulage, C.; Bonvallet, R.; Huss, M.; and Gressent, F. New mode of action for a knottin protein bioinsecticide: Pea albumin 1 subunit b (PA1b) is the first peptidic inhibitor of V-ATPase. *J. Biol. Chem.* **2011**, *286*, 36291–36296.

39. Cammue, B. P.; De Bolle, M. F.; Terras, F. R.; Proost, P.; Van Damme, J.; Rees, S. B.; Vanderleyden, J.; and Broekaert, W. F. Isolation and characterization of a novel class of plant antimicrobial peptides form Mirabilis jalapa L. seeds. *J. Biol. Chem.* **1992**, *267*, 2228–2233.

40. Gao, G. H.; Liu, W.; Dai, J. X; Wang, J. F.; Hu, Z.; Zhang, Y.; and Wang, D. C. Solution structure of PAFP-S: A new knottin-type antifungal peptide from the seeds of Phytolacca americana. *Biochemistry* **2001**, *40*, 10973–10978.

41. Shai, Y. Mechanism of the binding, insertion and destabilization of phospholipid bilayer membranes by alpha-helical antimicrobial and cell non-selective membrane-lytic peptides. *Biochim. Biophys. Acta.* **1999**, *1462*, 55–70.

42. Shai, Y. Mode of action of membrane active antimicrobial peptides. *Biopolymers* **2002**, *66*, 236–248.

43. Hallock, K. J.; Lee, D. K.; and Ramamoorthy, A. MSI-78, an analogue of the magainin antimicrobial peptides, disrupts lipid bilayer structure via positive curvature strain. *Biophys. J.* **2003**, *84*, 3052–60.

44. Pkorny, A.; and Almeida, P. F. Kinetics of dye efflux and lipid flip-flop induced by delta-lysin in phosphatidylcholine vesicle and the mechanism of graded release by amphipathic, alpha-helical peptides. *Biochemistry* **2004**, *43*, 8846–8857.

45. Hwang, P. M.; and Vogel, H. J. Structure-function relationships of antimicrobial peptides. *Biochem. Cell Biol.* **1998**, *76* (2–3), 235–46.

46. Morimoto, M.; Mori, H.; Otake, T.; Ueba, N.; Kunita, N.; Niwa, M.; Murakami, T.; and Iwanaga, S. Inhibitory effect of tachyplesin I on the proliferation of human immunodeficiency virus in vitro. *Chemotherapy* **1991**, *37*, 206–211.

47. Andersen, J. H.; Osbakk, S. A.; Vorland, L. H.; Traavik, T.; and Gutteberg, T. J. Lactoferrin and cyclic lactoferricin inhibit the entry of human cytomegalovirus into human fibroblasts. *Antiviral Res.* **2001**, *51*, 141–149.

48. Mistry, N.; Drobi, P.; Näslund, J.; Sunkari, V. G.; Jenssen, H.; and Evander, M. The anti-papillomavirus activity of human and bovine lactoferricin. *Antiviral Res.* **2007**, *75*, 258–265.

49. Utsugi, T.; Schroit, A. J.; Connor, J.; Bucana, C. D.; and Fidler, I. J. Elevated expression of phosphatidylserine in the outer leaflet of human tumor cells and recognition by activated human blood monocytes. *Cancer Res.* **1991**, *51*, 3062 3066.

50. Holtorf, S.; Ludwig-Muller, J.; Apel, K.; and Bohlmann, H. High-level expression of a viscotoxin in Arabidopsis thaliana gives enhanced resistance against Plasmodiophora brassicae. *Plant Mol. Biol.* **1998**, *36*, 673–680.

51. Vriens, K.; Cammue, B. P.; and Thevissen, K. Antifungal plant defensins: Mechanisms of action and production. *Molecules* **2014**, *19* (8), 12280–12303.

52. Ng, T. B. Antifungal proteins and peptides of leguminous and non-leguminous origins. *Peptides* **2004**, *25*, 1215–1222.

53. Sharma, A.; Sharma, R.; Imamura, M.; Yamakawa, M.; and Machii, H. Transgenic expression of cecropin B, an antibacterial peptide from Bombyx mori, confers enhanced resistance to bacterial leaf blight in rice. *FEBS Lett.* **2000**, *484* (1), 7–11.

54. DeGray, G.; Rajasekaran, K.; Smith, F.; Sanford, J.; and Daniell, H. Expression of an antimicrobial peptide via the chloroplast genome to control phytopathogenic bacteria and fungi. *Plant Physiol.* **2001**, *127*, 852–862.

55. Allefs, J. H. M.; DeJong, E. R.; Florak, D. E. A.; Hoogendoorn, J.; and Stiekema, W. J. Erwinia soft rot resistance of potato cultivars expressing antimicrobial peptide tachyplesin I. *Mol. Breed.* **1996**, *2* (2), 97–105.

56. Ohshima, M.; Mitsuhara, I.; Okamoto, M.; Sawano, S.; Nishiyama, K.; Kaku, H.; Natori, S.; and Ohashi, Y. Enhanced resistance to bacterial disease of transgenic tobacco plants over expressing sarco-toxin IA, a bactericidal peptide of insect. *J. Biochem.* **1999**, *125*, 431–435.

57. Wang, Y.; Nowak, G.; Culley, D.; Hadwiger, L. A.; and Fristensky, B. Constitutive expression of pea defense gene DRR206 confers resistance to blackleg (Leptosphaeria maculans) disease in transgenic canola (*Brassica napus*). *Mol. Plant-Microbe Interact.* **1999**, *12*, 410–418.

58. Park, C. H.; Kang, Y. H.; Chun, H. J.; Koo J. C.; Cheong, Y. H.; Kim C. Y.; Kim, M. C.; Chung, W. S.; Kim, J. C.; Yoo, J. H.; Koo, Y. D.; Koo, S. C.; Lim, C. O.; Lee, S. Y.; and Cho, M. J. Characterization of a stamen-specific cDNA encoding a novel plant defensin in Chinese cabbage. *Plant Mol. Biol.* **2002**, *50*, 59–69.

59. Kanzaki, H.; Nirasawa, S.; Saitoh, H.; Ito, M.; Nishihara, M.; Terauchi, R.; and Nakamura, I. Overexpression of the wasabi defensin gene confers enhanced resistance to blast fungus (Magnaporthe grisea) in transgenic rice. *Theor. Appl. Genet.* **2002**, *105*, 809–814.

60. Turrini, A.; Sbrana, C.; Pitto, L.; Ruffini Castiglione, M.; Giorgetti, L.; Briganti, R.; Bracci, T.; Evangelista, M.; Nuti, M. P.; and Giovannetti, M. The antifungal Dm-AMP1 protein from Dahlia merckii expressed in Solanum melongena is released in root exudates and differentially affects pathogenic fungi and mycorrhizal symbiosis. *New Phytol.* **2004**, *163* (2), 393–403.

61. Schaefer, S. C.; Gasic, K.; Cammue, B.; Broekaert, W.; van Damme, E. J.; Peumans, W. J.; and Korban, S. S. Enhanced resistance to early blight in transgenic tomato lines expressing heterologous plant defense genes. *Planta* **2005**, *222*, 858–866.

62. Carmona, M. J.; Molina, A.; Fernandez, J. A.; Lopez-Fando, J. J.; and Garcia-Olmedo, F. Expression of the alpha-thionin gene from barley in tobacco confers enhanced resistance to bacterial pathogens. *Plant J.* **1993**, *3* (3), 457–462.

63. Koumoutsi, A.; Chen, X. H; Henne, A.; Liesegang, H.; Hitzeroth, G.; Franke, P.; Vater, J.; and Borriss, R. Structural and functional characterization of gene clusters directing nonribosomal synthesis of bioactive cyclic lipopeptide in Bacillus amyloliquefaciens strain FZB42. *J. Bacteriol.* **2004**, *186* (4), 1084–96.

64. Leclere, V.; Bechet, M.; and Adam, A. Mycosubtilin overproduction by Bacillus subtilis BBG100 enhances the organism's antagonistic and biocontrol activities. *Appl. Environ. Microbiol.* **2005**, *71*, 4577–4584.

65. Romero, D.; Vicente, A.; Rakotaoly, R. H.; Dufour, S. E.; Veening, J. W.; Arrebola, E.; Cazorla, F.; Kuipers, O. P.; Paquot, M.; and Perez-Garcia, A. The iturin and fengycin families of lipopeptides are key factors in antagonism of Bacillus subtilis towards Podosphaera fusca. *Mol. Plant-Microbe Interact.* **2007**, *20* (4), 430–440.

66. Kong, J. L.; Du, X. B.; Fan, C. X.; Xu, J. F.; and Zheng, X. J. Determination of primary structure of a novel peptide from mistletoe and its antitumor activity. *Acta Pharmaceutica Sinica.* **2004**, *39* (10), 813–817.

67. Wong, J. H.; and Ng, T. B. Limenin, a defensin-like peptide with multiple exploitable activities from shelf beans. *J. Peptide Sci.* **2006**, *12* (5), 341–346.

68. Lin, P.; Wong, J. H.; and Ng, T. B. A defensin with highly potent antipathogenic activities from the seeds of purple pole bean. *Biosci. Rep.* **2010**, *30* (2), 101–109.

69. Gerlach, S. L.; Rathinakumar, R.; Chakravarty, G.; Göransson, U.; Wimley, W. C.; Darwin, S. P.; and Mondal, D. Anticancer and chemosensitizing abilities of cycloviolacin

O2 from *Viola odorata* and psyle cyclotides from *Psychotria leptothyrsa*. *Biopolymers* **2010,** *94* (5), 617–625.

70. Ji, Y.; Majumder, S.; Millard, M.; Borra, R.; Bi, T.; Elnagar, A. Y.; Neamati, N.; Shekhtman, A.; and Camarero, J. A. *In vivo* activation of the p53 tumor suppressor pathway by an engineered cyclotide. *J. Am. Chem. Soc.* **2013,** *135* (31), 11623–11633.

71. Silva, O. N.; Porto, W. F.; Migliolo, L.; Mandal, S. M.; Gomes, D. G.; Holanda, H. H.; Silva, R. S.; Dias, S. C.; Costa, M. P.; Costa, C. R.; Silva, M. R.; Rezende, T. M.; and Franco, O. L. Cn-AMP1: A new promiscuous peptide with potential for microbial infections treatment. *Biopolymers* **2012,** *98* (4), 322–331.

72. Wele, A.; Zhang, Y.; Ndoye, I.; Brouard, J. P.; Pousset, J. L.; and Bodo, B. A cytotoxic cyclic heptapeptide from the seeds of *Annona cherimola. J. Nat. Prod.* **2004,** *67* (9), 1577–1579.

73. Pelegrini, P. B.; and Franco, O. L. Plant gamma-thionins: Novel insights on the mechanism of action of a multi-functional class of defense proteins. *Int. J. Biochem. Cell Biol.* **2005,** *37*, 2239–2253.

74. Pelegrini, P. B.; Quirino, B. F.; and Franco, O. L. Plant cyclotides: An unusual class of defense compounds. *Peptides* **2007,** *28*, 1475–1481.

75. Nasrollahi, S. A.; Taghibiglou, C.; Azizi, E.; and Farboud, E. S. Cell penetrating peptides as a novel transdermal drug delivery system. *Chem. Biol. Drug Des.* **2012,** *80*, 639–646.

76. Brogden, K. A. Antimicrobial peptides: Pore formers or metabolic inhibitors in bacteria? *Nat. Rev. Microbiol.* **2005,** *3*, 238–250.

CHAPTER 2

DEVELOPMENT OF ANTITUBERCULAR AGENTS THROUGH HYBRIDIZATION STRATEGIES, FUTURE CHALLENGES AND PERSPECTIVES

SRINIVAS KANTEVARI[1,3*], ANVESH JALLAPALLY[1],
ANJANA DEVI TANGUTUR[2,3*], and DINESH KUMAR[2]

[1]*C P C Division (Organic Chemistry Division-II), CSIR-Indian Institute of Chemical Technology, Hyderabad, Telangana – 500007, India.*

[2]*Centre for Chemical Biology, CSIR-Indian Institute of Chemical Technology, Hyderabad, Telangana – 500007, India.*

[3]*Academy of Scientific and Innovative Research, CSIR-Indian Institute of Chemical Technology, Hyderabad, Telangana – 500007, India.*

*Corresponding authors: Srinivas Kantevari, Anjana Devi Tangutur. E-mail: kantevari@yahoo.com, anjdevi@gmail.com

CONTENTS

Abstract 29
2.1 Introduction..29
2.2 Strategies in the Development of New Antitubercular Agents 32
2.3 Molecular Modification 34
2.4 Molecular Hybrids as Antitubercular Agents....................................37
2.5 Examples of Antitubercular Agents Developed
     Via Hybridization Strategies 41
2.6 Future Challenges and Procepts in the Usage of Molecular
     Hybrids for the Development of TB Therapeutics............................52

2.7   Prospects of Molecular Hybridization Approach in the
       Development of Next Generation Antitubercular Drugs—
       "Doom's Day" Ahead 53
2.8   Conclusion 54
Keywords 54
References 55

ABSTRACT

Tuberculosis (TB) is one the most prevalent endemic lung infection caused by the pathogen *Mycobacterium tuberculosis* (*Mtb*). In spite of the availability of various combinations of antitubercular drugs with a six-month treatment regimen, the global TB situation is worsened by increased morbidity and mortality. Further, the control of TB has been exacerbated by the emergence of multidrug resistant, extremely drug resistant TB strains, and latent infection. TB drugs currently in use were discovered much before in the seventies and studies indicate that TB drug research remained silent for more than 40 years. To counteract this crisis, there is an urgent need to develop novel alternative antitubercular agents which can target processes that are critical for the growth and survival of the bacterium. Development of antitubercular agents *via* molecular hybridization strategies can therefore be imagined as a future toward a new dawn where the classical/conventional molecules usage will diminish and molecular hybrids emerge to bring over a dramatic improvement in the mechanism of action against *Mycobacterium tuberculosis* (*Mtb*). We herein review strategies for antitubercular drug discovery with an emphasis on molecular hybridization and provide examples of hybrid molecules as highly potential antitubercular agents for clinical use.

2.1 INTRODUCTION

Tuberculosis (TB) is one of the most ancient contagious diseases, caused by pathogen *Mycobacterium tuberculosis* which continued to remain as the world's second leading killer after Human Immunodeficiency Virus (HIV) with a molecular evidence going back to over 100 decades.[1,2,3] According to the latest World Health Organization (WHO) report, TB remains to be the most prevalent disease in the world today, with millions of new cases of infected individuals and millions of deaths being notified each year globally.[4-6] The WHO estimates that almost one third of the population is infected with *Mycobacterium tuberculosis* (*Mtb*). It is anticipated that by 2020, around one billion people will be infected newly, 125 million people will get sick, and over 30 million will die of TB if its control is not further strengthened.[7,8] In economically developing countries, over 95 percent of deaths occur every minute. In the global control of this infectious disease, there are major obstacles which include the difficulties to detect and cure sufficient number of cases to interrupt transmission.[9,10] In spite of this, there was no specific treatment until the 20th century, which saw the development of several antimicrobials leading to the modern day highly effective six-month regimen (i.e,) isoniazid (H), rifampicin (R), ethambutol (E), pyrazinamide (Z) daily for two months, followed by four months of isoniazid (H) and rifampicin (R) given three times a week (2HREZ/4HR$_3$),[10,11] . In case of resistance to first line therapy, second line drugs i.e., amikacin, kanamycin, capreomycin, enviomycin, viomycin, ciprofloxacin, levofloxacin, moxiflaxacin, ethionamide, prothinamide, closerin, and terizidone known as reserved therapy are used[12] against extensively drug-resistant tuberculosis (XDR-TB) or multidrug-resistant tuberculosis (MDR-TB).[13–15] Recommended treatment regimen (i.e., a cocktail of 6–8 drugs) works effectively with low rates of severe adverse reactions. TB is also responsible for opportunistic secondary infections and co-infections in immunocompromised individuals.[16] It is also reported that if left untreated, each person with active TB will infect (on an average) between 10 and 15 people every year. An estimated 5 percent of TB patients are co-infected with HIV.[17]

TB has also been described as "an orphan giant" or "white plague" that claims millions of lives every year. It is often described as the disease of the poor and the pharmaceutical industries tend to bypass drug discovery for TB owing to the lack of profitable markets.[18,19] However, concurrent with these hurdles, several drugs have been developed for controlling, curing, and preventing further transmission of the disease.[20] The Directly Observed Treatment Short-course (DOTS), a multidrug therapy program developed by the World Health Organization (WHO), is one among the most competent weapons against this global epidemic.[21,22] Chemotherapy remains the mainstay of its control strategy. Adding to that, the global TB situation is being further worsened by the emergence of multidrug resistant and extremely drug resistant TB strains. The long and complex TB regimen and lack of appropriate treatments further pose a major threat to the existing strategies for TB control.[23–27] This crisis can be counteracted by developing novel alternative antitubercular agents which can target processes that are critical for the growth and survival of this bacterium.[28] The design and development of such novel, promising, extraordinarily better drugs acting against the desired targets (Figure 2-1)[27–29] with effective pharmacologic potency and selective toxicity therefore remains the biggest challenge of the antitubercular drug discovery process.[30] TB drugs currently in use were discovered much before in the seventies and studies indicate that for 40 years TB drug research was still. With gained momentum in antitubercular discovery research, about seven compounds are presently in advanced stages of ongoing clinical trials (Figure 2-2) and several of them are under preclinical evaluations.[28,31–33]

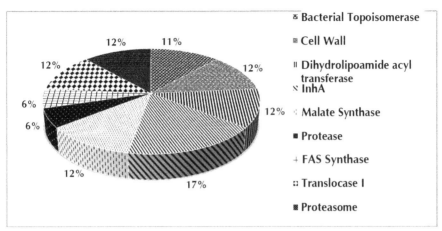

FIGURE 2-1 Different targets studied for the discovery of antitubercular agents.

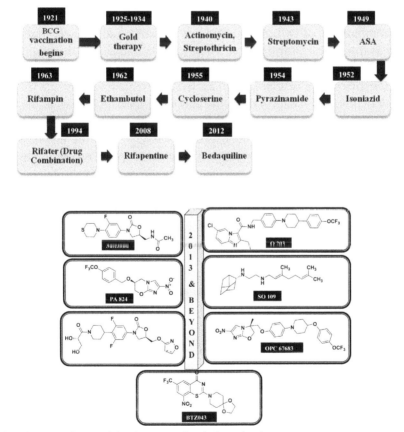

FIGURE 2-2 History of development of antitubercular drugs.

With regard to this, it is extremely noteworthy that the Council of Scientific and Industrial Research (CSIR), India has initiated a major programme on affordable healthcare named Open Source model for Drug Discovery (OSDD) in 2008. The OSDD is a CSIR led team India Consortium with global partnership, with a vision to provide affordable healthcare to the developing world by discovering novel therapies for neglected tropical diseases like Malaria, Tuberculosis, Leshmaniasis, etc.[34] This has more than 7300 registered users from more than 130 countries of the world, and has emerged as the major collaborative effort in drug discovery. This will provide a unique opportunity for students, researchers, doctors, scientists, technocrats, and others with diverse expertise to work for a common cause.[35] It is a community driven open innovation platform launched on the three cardinal principals of collaborate, discover, and share to collaboratively aggregate the genetic, biological, and chemical information available to hasten the discovery of drugs.[36,37]

A recent emerging strategy within medicinal chemistry and drug discovery is the combination of two pharmacophoric subunits into a single molecule, named as molecular hybridization.[38] The adequate fusion of these subunits leads to the design of a new hybrid architecture that maintains preselected characteristics of the original template with preserved activity.[39,40] Current research in this field seems to therefore endorse hybrid molecules as the next generation antitubercular agents.[41] This review therefore focuses on hybridization strategies for antitubercular drug discovery, with the emphasis on molecular hybrids. Therefore, herein we provide an update on several hybrid molecules that have been developed and have demonstrated ample potential for clinical use. We also describe the research work in the field of molecular hybridization currently being addressed in our laboratory which may contribute toward the design and development of chemotherapeutic agents targeting drug susceptible and resistant types (MDR and XDR tuberculosis).

2.2 STRATEGIES IN THE DEVELOPMENT OF NEW ANTITUBERCULAR AGENTS

2.2.1 HIGH THROUGHPUT SCREENING (HTS) TECHNIQUE

The existing antitubercular drugs have been developed using whole cell based high throughput screening approach, wherein various high throughput screening techniques using whole cells are applied to large libraries of compounds resulting in the identification of a set of potentially active agents.[42-44] One such recent example identified from a *Mycobacterium tuberculosis* (*Mtb*) whole cell high through-put screening (HTS) campaign is tetrahydropyrazolo[1,5-*a*]pyrimidine scaffold.[45] A series of derivatives of this class were synthesized to evaluate their structure-activity relationship (SAR) and structure-property relationship (SPR). Among these, Compound **1** displayed promising DMPK profile *in vivo* in mouse and exhibited potent *in vivo* activity in a mouse efficacy model, achieving a reduction of 3.5 log CFU of *Mtb* after oral administration to infected mice once a day, at 100 mg/kg for 28 days. Compound 1 (Figure 2-3) is therefore considered as a potential candidate for inclusion in combination therapies for both drug-sensitive and drug-resistant TB.[46]

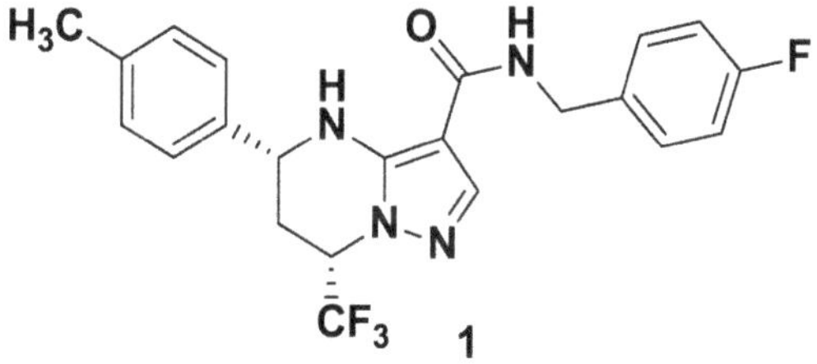

FIGURE 2-3   Compound 1.

The promising compounds are further tested for potency, mechanism of action, pharmacokinetics, safety, and efficacy *in vitro* and *in vivo*. The various strategies employed in the TB drug discovery process are therefore as mentioned below:

2.2.2   ANTITUBERCULAR EVALUATION OF COMPOUNDS ORIGINALLY DEVELOPED AGAINST OTHER DISEASES

Mycobacterium tuberculosis shares several targets for antimicrobial activity with other bacteria which offer a brilliant opportunity to explore the biological evaluation of a range of antimicrobials as antimycobacterials.[47] Some of the significant antimicrobial targets include DNA gyrase, topoisomerase, and beta lactamase for which fluoroquinolones, 8-methoxyquinolones, moxifloxacin, and gatifloxacin have been identified.[48-50] In some cases, though the particular class of compounds exhibit *in vitro* activity and possible *in vivo* efficacy, substantial neurotoxicity and haematological toxicity were seen during extended therapy. These drugs were mainly discovered from chance observations or from screening of huge libraries of compounds.

2.2.3 TARGET BASED DRUG DISCOVERY

Target based drug discovery[51] is a strategy which exploits the huge growing database of mycobacterial enzymes and other critical biological targets[30,52] such as *icl (isocitrate lyase), pcaA (a cyclopropane synthase), dnaE2 (Error-prone DNA polymerase), relA (GTP pyrophosphokinase and Guanosine-3',5'-bis(diphosphate)3'-pyrophosphohydrolase)* which are implicated to have a possible role in processes required for mycobacterial persistence instead of screening for whole organism killing. Active agents can further be "custom designed" for their interaction with target molecules and interference with the function of the target may form the basis of an *in vitro* study. The lead compounds can then be synthetically altered to maximize potency or other desired properties.[30,53]

2.3 MOLECULAR MODIFICATION

Molecular modification is a chemical alteration of a molecule, which could be a lead compound or a drug with the aim to enhance its pharmacokinetic, pharmaceutical, or pharmacodynamic parameters.[54–56] For several years, this strategy has been used as a medicinal chemist tool to modify natural products into useful synthetic drugs. In the development of antitubercular agents, strategies of molecular modification were effectively utilized in generating lead derivatives. In general, molecular modifications are of three types, namely, prodrug approach, bioisosterism, and molecular hybridization.

2.3.1 PRODRUG APPROACH

Prodrug strategy[57] has arguably been successful for a number of clinically used therapeutic agents. However, prodrug research also encounters various challenges and additional work is necessary in preclinical and clinical settings. Much of these findings can be attributed to understanding the bioconversion mechanisms of prodrugs.[58]

In antitubercular research, prodrug approach is used in several instances for developing improved lead candidates.[30] One such recent example is the synthesis of high-affinity reversible competitive inhibitors of *Mycobacterium tuberculosis* type II dehydroquinase.[59] Here, chemical modifications of the reported acid derivatives were carried out to improve internalization into *Mycobacterium tuberculosis* through an ester prodrug approach. The propyl esters appended are proved to be most efficient in achieving optimal *in vitro* antitubercular activities (Figure 2-4).

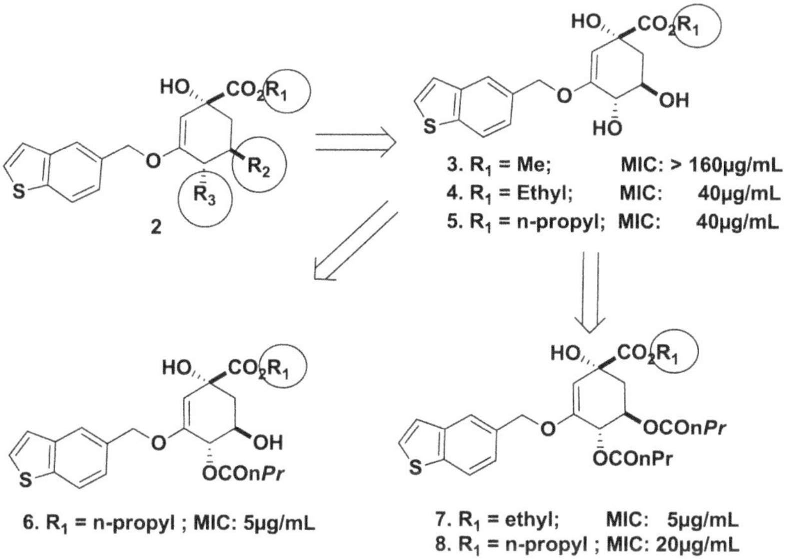

3. R_1 = Me;   MIC: > 160μg/mL
4. R_1 = Ethyl;   MIC: 40μg/mL
5. R_1 = n-propyl; MIC: 40μg/mL

6. R_1 = n-propyl ; MIC: 5μg/mL

7. R_1 = ethyl;   MIC: 5μg/mL
8. R_1 = n-propyl ; MIC: 20μg/mL

FIGURE 2-4   3-Methoxybenzothiophenyl prodrug derivatives.

Co-infection of TB associated with HIV weakens the immune system in allowing reactivation of latent TB and makes the patients more susceptible to drug-resistant strains.[15–17] In this context, antimycobacterial agents (such as isoniazid, norfloxacin, and ciprofloxacin) and HIV nucleoside analogs (such as zidovudine, stavudine, and lamivudine) were combined using prodrug strategy. The lamivudine prodrugs (Figure 2-5) with *in vitro* antiretroviral activities equipotent to lamivudine with EC_{50} of 0.0742 ± 0.04 μM exhibited inhibition of *Mycobacterium tuberculosis* virulent variant strain (H37Rv) at a concentration of 6.25 μg/mL.[60]

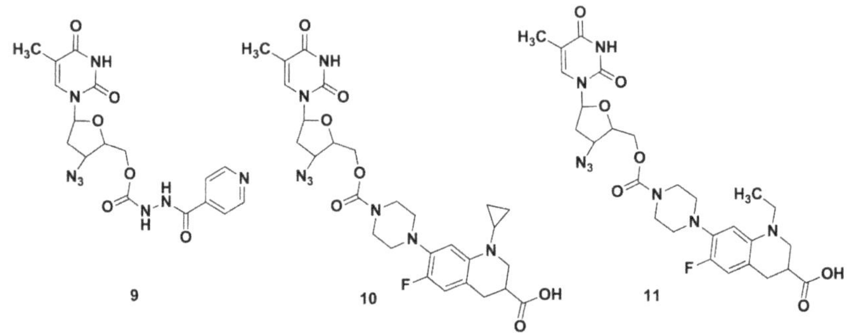

FIGURE 2-5   Lamivudine antitubercular prodrugs.

2.3.2   BIOISOSTERISM

The bioisosterism[61] approach is yet another important molecular modification tool to identify new compounds to treat tuberculosis. The term "bioisostere" originally contemplated by James Moir in 1909, was introduced in 1950, who defined it as compounds eliciting a similar biological effect. On the basis of this concept, several drugs were developed.[62]

Recently, Prithwiraj et al.[63] synthesized a library of thioester, amide, hydrazide, and triazole-phthalazine derivatives as isosteres of 4-alkoxy cinnamic acid. Many of these bioisosteres exhibited submicromolar minimum inhibitory concentrations against Mycobacterium tuberculosis virulent variant strain (H37Rv)and good cytotoxicity range toward THP-1 cells and possess satisfactory druggability (Polar Surface Area (PSA) = 50–80 Å²). Interestingly, 12-(4-isopentenyloxycinnamyl) triazolophthalazine derivative is extremely active against two isoniazid (INH)-resistant strains (Figure 2-6). This study also revealed that 12 does not interfere with mycolic acid biosynthesis, thereby indicating a different mode of action and representing an attractive lead compound for the development of new antitubercular agents.

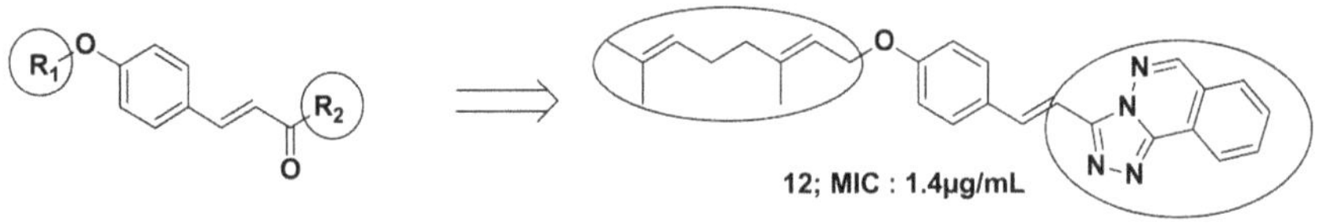

FIGURE 2-6   Bioisosteres of 4-alkoxy cinnamic acid derivatives.

Linezolid 13,[64] discovered by researchers at the Upjohn Company in Kalamazoo, Michigan, which belongs to the oxazolidinone class of antibacterial agents (Figure 2-7), was used in the treatment of nosocomial pneumonia, skin, and soft tissue infections. Later developments on the evaluation of this drug led to its use in the treatment of MDR tuberculosis.[65–66] On the basis of this interesting result, Sutezolid 14, a bioisosteric analog of linezolid was developed as an inhibitor of MDR tuberculosis and is currently in clinical trials. Following this success, other linezolid derivatives such as radezolid 16 and torezolid 15 were obtained by isosteric replacement and are being evaluated for treating MDR tuberculosis.[67]

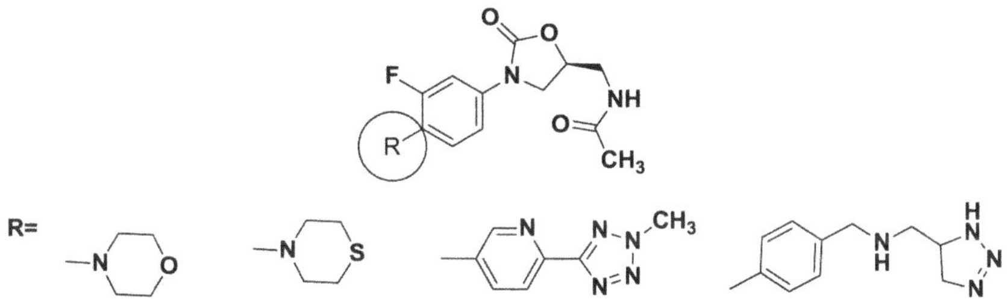

13 linezolid | 14 sutezolid | 15 torezolid | 16 radezolid

FIGURE 2-7 Oxazolidinone class of antibacterial agents.

2.3.3 MOLECULAR HYBRIDIZATION

As discussed earlier, molecular hybridization[38,68] plays a key role in rational drug design approach. Single hybrid molecules with improved efficacy have been developed as novel antitubercular drugs. However, on the basis of wide interest in the hybrid molecules as well as numerous encouraging efficacy and toxicity reports, the next generation antitubercular agents may as well be hybrid drugs as opposed to multicomponent ones.

2.4 MOLECULAR HYBRIDS AS ANTITUBERCULAR AGENTS

There are numerous advantages of employing hybrid molecules over multicomponent drugs in TB therapy. Compared with the latter, hybrid drugs may be less expensive because, in principle, the risks and costs involved may not be different from any other single entity. Another advantage is that of the lower risk of drug adverse interactions compared with multicomponent drugs. The downside however is that it is more difficult to adjust the ratio of activities at the different targets. Various hybrid molecules developed mainly for tropical diseases such as malaria and TB[69] are classified into three categories based on their chemical structure alteration. They are (1) Fused hybrids (2) Carbohybrids (3) Fragment based hybrids.

2.4.1   FUSED HYBRIDS

These are hybrid molecules designed *via* molecular hybridization of two or more bioactive compounds (at least one could be an antitubercular agent) and synthesized by fusing them together through a covalent bond. We discuss here some recent examples of fused hybrids as antitubercular agents.

Trans-cinnamic acid and its derivatives are well known in history as antituberculosis agents.[70] However, the potentiality of these compounds remained underutilized till the recent years. In an attempt to identify potential new agents active against tuberculosis, *trans*-phenylacrylamide (cinnamide) derivatives were designed using molecular hybridization approach and synthesized by fusing *trans*-cinnamic acids with guanylhydrazones.[70] Twenty fused phenylacrylamide hybrids were evaluated using resazurin microtitre plate assay (REMA) against *Mycobacterium tuberculosis* virulent variant strain (H37Rv). (2*E*)-*N*-((-2-(3,4-di methoxybenzylidene) hydrazinyl)(imino)methyl)-3-(4-methoxyphenyl)acrylamide **17** (MIC, 6.49 µM) (Figure 2-8) with a good safety profile of > 50-fold in VERO cell line was identified as the most active antitubercular agent.

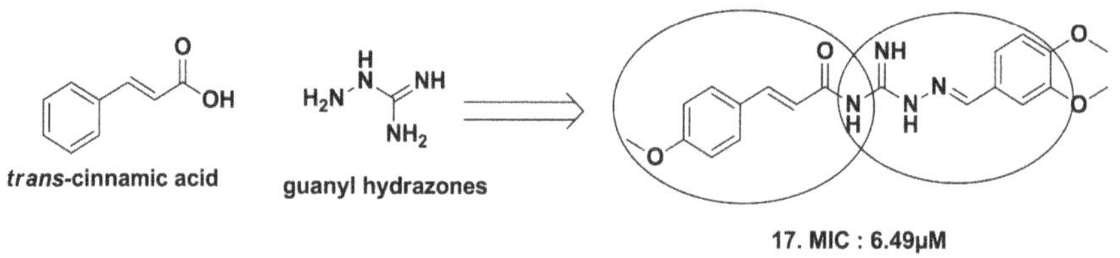

17. MIC : 6.49µM

FIGURE 2-8   Fused phenylacrylamide hybrids.

Further, this approach was employed to synthesize substituted 2-(2-(4-aryl oxybenzylidene)hydrazinyl)benzothiazole derivatives (Figure 2-9) with 2-hydrazinobenzothiazole and 4-(alicyclic/aryl/biaryl/heteroaryloxy)benzaldehyde as new antitubercular agents.[71] The synthesized compounds, when tested against *Mycobacterum tuberculosis* virulent variant strain (H37Rv), resulted in 6-chloro-2-(2-(4-(pyridin-4-yloxy) benzylidene) hydrazinyl) benzo[*d*] thiazole 18 (MIC, 1.35 µg/mL) as a potential lead.

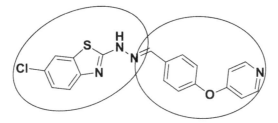

18. MIC : 1.35ug/mL

FIGURE 2-9   Benzothiazole derivatives as antitubercular agents.

2.4.2   CARBOHYBRIDS

Hindsgaul, a carbohydrate chemist at the University of Alberta in Edmonton, coined the name "carbohybrids". These are molecules containing a sugar portion linked to other organic groups that are simple to make and typically bind more tightly to protein drug targets. Hence, this strategy shows a great promise for a future in high-speed "combinatorial" technique for making hundreds of related carbohybrids.[72]

In mycobacteria, the cell wall structure consists of a dense network of cross-linked sugar residues esterified with mycolic acid at the ends. This understanding has prompted the investigation of various sugar prototypes with distinct characteristics as potential antitubercular agents.[73] A variety of carbohydrate derivatives are also known to interfere with the cell wall biosynthesis of *M. tuberculosis.*[74]

In this context, research work was done on the synthesis of nine new fluoroquinolone derivatives based on the modifications with sugar unit at the C-7 position of the known fluoroquinolones, ciprofloxacin, gatifloxcin, and moxifloxacin, as well as their antitubercular evaluation.[75] The synthesis of these new analogues was improved using microwave irradiation, providing several advantages such as better yields and shorter reaction times, in comparison with classical reaction conditions. Derivatives **19, 20** and **21** exhibited promising antitubercular activities (Figure 2-10).

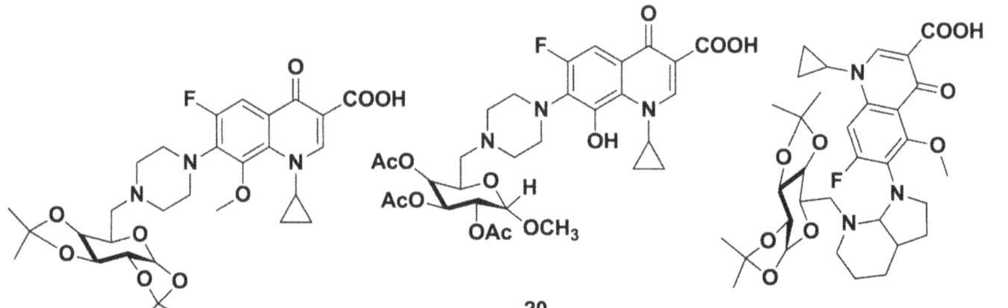

FIGURE 2-10   Carbohybrids as antitubercular agents.

Later, Sanki *et al.* synthesized a library of arabinose and trehalose-based molecules elaborated with esters, α-ketoesters, and α-ketoamides.[76] The compounds were again tested against antigen (Ag85C) *in vitro*. One of the compounds, a methyl ester **22**, showed modest inhibition of antigen (Ag85C) in the mm range but did not exhibit the inhibition of growth of *Mycobacterium smegmatis*. Significant improvement of the inhibitory activity *in vitro* will be necessary before any growth inhibition is observed (Figure 2-11).

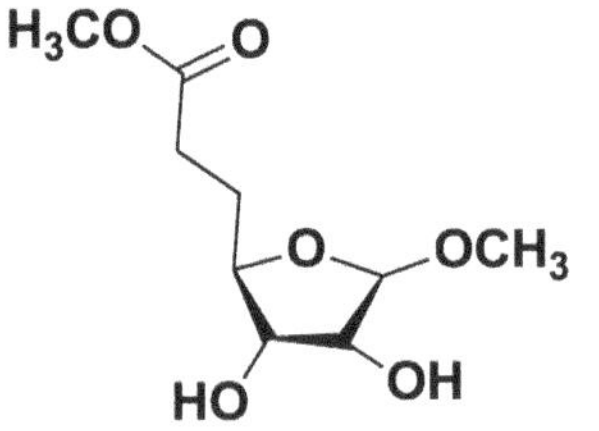

22 ag85C, IC$_{50}$ = 25mM

FIGURE 2-11   α-ketoesters compounds of arabinose.

2.4.3   FRAGMENT BASED HYBRIDS

One of the most exciting new strategies for lead generation is fragment based discovery.[53,77] Fragments are small, low molecular weight molecules that would usually form part of a clinical compound. Once bound to the active site of a target protein or enzyme, these fragments can be used for hybridization to develop highly selective and potent drug candidates. Joining of fragments in mix and match fashion will lead to new hybrids with envisaged biological activity profile. This strategy has been successfully used in tuber-culosis drug discovery in identifying inhibitors of the protein tyrosine phos-phatases PtpA and PtpB. Compounds **23** and **24** were found to inhibit PtpA with Ki values of 1.4 and 1.6 µM, respectively.[78,79,81] Compound **25** was identified using substrate activity screening and was found to bind to PtpB with a Ki of 0.22 µM[80,81] (Figure 2-12).

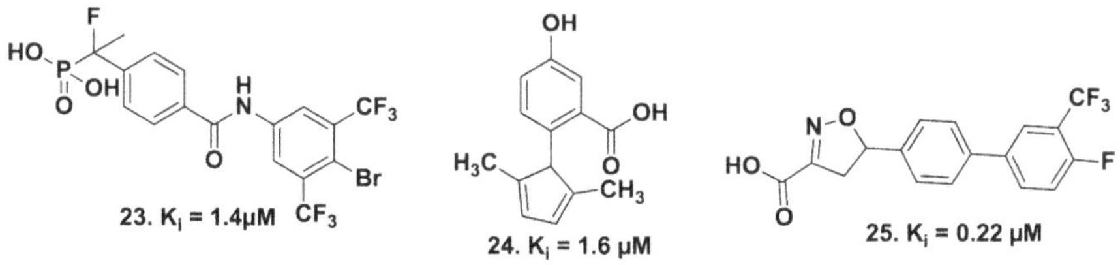

FIGURE 2-12   Fragment based protein tyrosine phosphatases PtpA and PtpB inhibitors.

2.5   EXAMPLES OF ANTITUBERCULAR AGENTS DEVELOPED VIA HYBRIDIZATION STRATEGIES

2.5.1   QUINOLINE-OXAZOLIDINONE HYBRIDS

A series of quinoline derivatives carrying oxazolidinone ring designed by hybridization strategy[82] were screened for their antimycobacterial proper-ties. Amongst the compounds tested, three analogues were active (at 0.65 µg/mL) against *Mycobacterum tuberculosis* virulent variant strain (H37Rv). The mode of action of these active compounds was assessed by docking of the receptor enoyl-acyl carrier protein (enoyl-ACP) reductase. From all the *in vitro* and docking studies, compound **26** was considered as the best anti-tubercular agent.

In yet another modification, Rane et al.[83] designed and synthesized a series of twenty marine bromopyrrole alkaloids containing 1,3,4-oxadiazole scaffold. Inhibition of *Mycobacterium tuberculosis* at concentrations as low as 1.6 and 1.5 µg/mL by compounds **27** and **28**, respectively, indicates that bromopyrrole-1,3,4-oxadiazole hybrids (Figure 2-13) can act as leads for the development of newer antitubercular compounds.

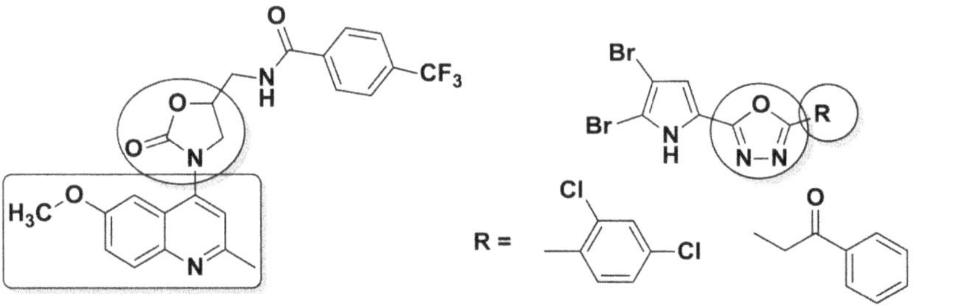

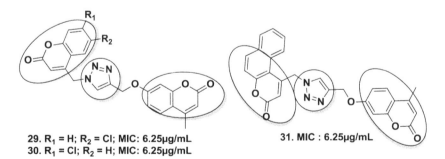

26. MIC : 0.625µg/mL   **27. MIC : 1.6 µg/mL**   **28. MIC : 1.5 µg/mL**

FIGURE 2-13   Quinoline-oxazolidinone hybrids.

2.5.2   TRIAZOLE BASED HYBRIDS

Triazoles with three nitrogen atoms in the five-membered aromatic ring are readily able to bind with a variety of enzymes and receptors *via* diverse noncovalent interactions, and display versatile biological activities.[84] In particular, substituted 1,2,3- and 1,2,4-triazoles hybridized with other bioactive cyclic/heterocyclic and aromatic counterparts are among the recent potential antitubercular agents.[85] Some of such triazole based hybrid antimycobacterial agents are exemplified here.

Naik et al.[86] synthesized a series of bis-chromenyl triazole hybrids under click reaction conditions and evaluated for their antimycobacterial activity against *M. tuberculosis* using MABA assay. The screening data revealed that **29**, **30**, and **31** (Figure 2-14) derivatives were as highly active as streptomycin with MIC of 6.25 µg/mL.

29. R₁ = H; R₂ = Cl; MIC: 6.25µg/mL   **31. MIC : 6.25µg/mL**
30. R₁ = Cl; R₂ = H; MIC: 6.25µg/mL

FIGURE 2-14   Bis-chromenyl triazole hybrids.

Further, Jardosh and Patel[87] designed 24 biquinolone-isoniazid hybrids *via* molecular hybridization technique and synthesized using multicomponent cyclocondensation (MCC) reaction. Screening of all synthesized hybrids using brine shrimp bioassay method resulted in the identification of **32** and **33** (Figure 2-15) as the most potent compounds with 99 percent inhibition against *Mycobacterium tuberculosis* bacteria with LC_{50} values of 35.39 and 34.59 mg/mL, respectively.

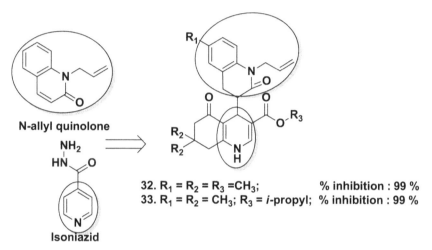

N-allyl quinolone

Isoniazid

32. $R_1 = R_2 = R_3 = CH_3$;   % inhibition : 99 %
33. $R_1 = R_2 = CH_3$; R_3 = *i*-propyl; % inhibition : 99 %

FIGURE 2-15   Biquinolone-isoniazid hybrids.

These studies[88] describe the biological activity attributed to factors like length of the linker and the substituents on the arylidene moiety concurrent with rhodanine. Compounds **34, 35,** and **36** of 4-aminoquinoline rhodanine series (Figure 2-16) designed through molecular hybridization approach, exhibit potent antitubercular activity with low cytotoxicity.

In recent years, the natural product inspired, structurally diverse new molecules play a major role in drug design and discovery process.[89] In the field of antitubercular agents, the lichen derived secondary metabolite usnic acid has been shown to display an interesting biological activity.[90,91] Unfortunately, its rapid metabolism and toxicity prevents its exploration as an antitubercular agent. In this context, the development of usnic acid derived hybrid analogs including dibenzofuran as a basic core combined with other bioactive heterocycles would be of interest for evaluation as antimycobacterial agents.

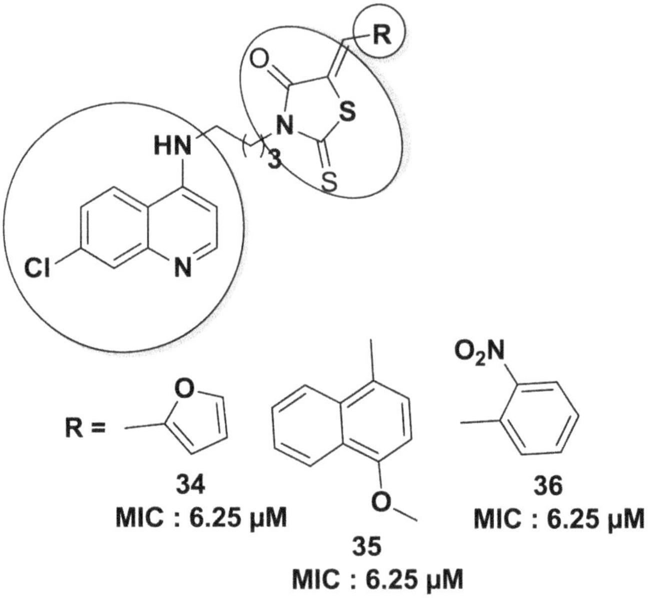

FIGURE 2-16   4-aminoquinoline rhodanine hybrids

Prado et al.[92] synthesized a series of natural product-like usnic acid hybrids, by angular fusion of the pyran ring onto the dibenzofuran system. Thus, a series of benzofuro [3,2-f][1]benzopyrans synthesized from 2-hydroxydibenzofuran displayed significant activities when tested against *Mycobacterium tuberculosis* virulent variant strain (H37Rv) and *Beijing* strains, with MIC$_{99}$ in the range of 1–10 µg/ml. Among them, 3,3-dimethyl-3*H*benzofuro[3,2-f][1]benzopyran **37** as well as the dihydro derivative **38** have a good inhibition activity on the growth of *M. tuberculosis* with an MIC$_{99}$ (less than 5 µg/ml) that is comparable with that of the first-line anti-tubercular drugs. Later, Tetmentzi et al.[93] introduced various functional modifications on antitubercular active 3,3-dimethyl-3*H*benzofuro[3,2-f][1] benzopyran for attaining better activity against *Mycobacterium bovis* BCG and the virulent strain *Mycobacterium tuberculosis* virulent variant strain (H37Rv) H37Rv. A linear 9-methoxy-2,2-dimethyl-2*H*-benzofuro[2,3-g][1] benzopyran **39** exhibiting a good antimycobacterial activity (MIC = 89 µM), better cytotoxicity profile, and selective index (SI= 83) resulted as the most promising antitubercular agent (Figure 2-17).

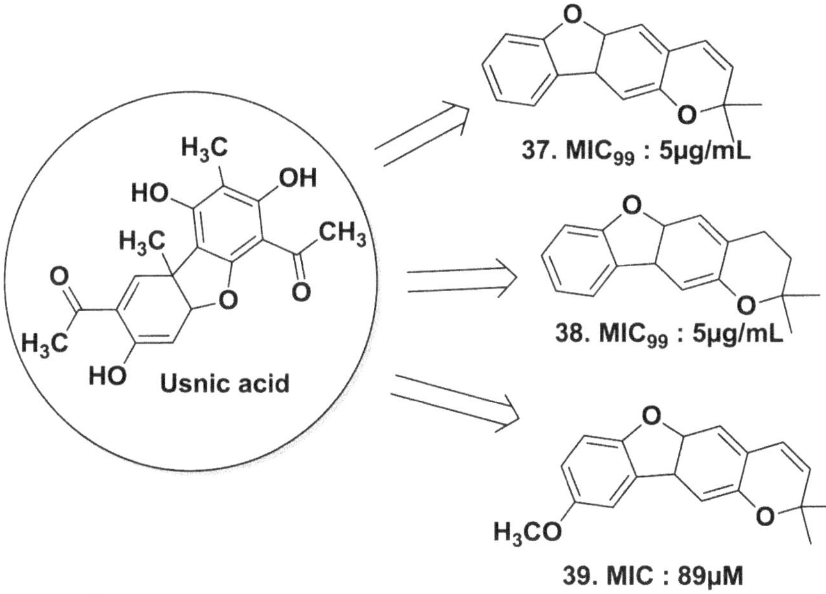

H3C
HO
OH
H3C
CH3
O
O
O
H3C
HO
Usnic acid
O
O
37. MIC99 : 5µg/mL
O
O
38. MIC99 : 5µg/mL
O
O
H3CO
39. MIC : 89µM

FIGURE 2-17 Antitubercular hybrids derived from usnic acid.

On the basis of these encouraging results, we envisaged that molecular hybridization of benzofurobenzopyran nucleus with bioactive heterocyclic units in one molecular frame could result in newer antitubercular molecules. Benzofurobenzopyran nucleus was initially hybridized with 1,3-oxazinone unit of antitubercular drugs linezolid and PNU-100480,[94] to deliver dibenzofuran derived hybrid heterocycles tethered with amidoalcohols or fused with 1,3-oxazinone unit.[95] Among all the 16 compounds screened for *in vitro* activity against *Mycobacterium tuberculosis* virulent variant strain (H37Rv), **40, 41, 42** (MIC 3.13 µg/mL) and **43, 44** (MIC 1.56 µg/mL) were found to be the most active agents (Figure 2-18).

Homoisoflavonoids constitute a class of natural products isolated from the bulbs, rhizomes, or roots of several genera of Hyacinthaceae and Caesal pinioideae.[96] Several natural and synthetic homoisoflavonoids were found to possess a wide range of biological properties including antitubercular activity. It is therefore of interest to hybridize bioactive dibenzofuran derivatives with natural bioactive homoisoflavonoids in one molecular frame for antimycobacterial evaluation. A series of natural products like dibenzofuran embodied homoisoflavonoids [(*E*)-3-(dibenzo[*b,d*]furan-2-ylmethylene) chroman-4-ones] were synthesized in very good yields using a base catalyzed Baylis–Hillmann (BH) reaction.[97] Out of the 11 compounds screened for *in vitro* antimycobacterial activity against *Mycobacterium tuberculosis* virulent variant strain (H37Rv), (*E*)-3-(dibenzo[*b,d*]furan-2-ylmethylene)-6-fluorochroman-4-one **45** and (*E*)-3-(dibenzo[*b,d*]furan-2-ylmethylene)-6-fluorochroman-4-one **46** were found to be active with an MIC of 12.5 µg/mL (Figure 2-18).

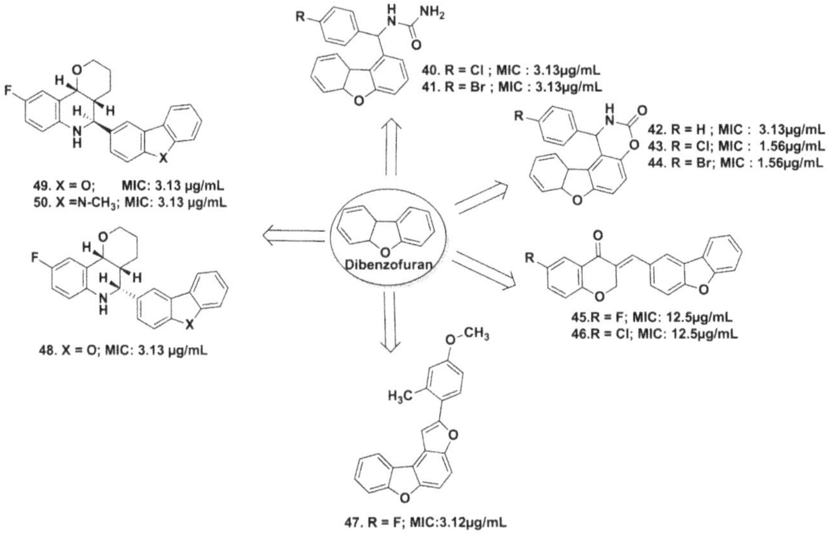

FIGURE 2-18   Dibenzofuran architectures as antitubercular agents.

Continued efforts on the modification of benzofuro[3,2-*f*][1]benzopyran by hybridizing with various pharmacophoric heterocycle fragments, led to the examination of a series of benzofuro[3,2-*e*]benzofurans as antimycobacterial agents.[98] Among the 10 compounds screened for *in vitro* antimycobacterial activity against *Mycobacterium tuberculosis* virulent variant strain (H37Rv), 2-(4-methoxy-2-methylphenyl)-3*H*-benzofuro[3,2-*e*]benzofuran **47** was found to be most active with MIC 3.12 µg/mL and has shown lower cytotoxicity with good selectivity index (SI > 10).

The FDA approval of Bedaquiline (TMC 207 or R207910) as the first antitubercular drug ended 40 years of drought for new drugs[99] to treat patients with multidrug resistant tuberculosis and raised interest in quinoline based drug candidates. It is therefore of interest to combine tetrahydroquinoline nucleus with bioactive tricyclic systems to examine their pharmacological properties. A series of hexahydro-2*H*-pyrano [3,2-*c*]quinoline hybrids derived from dibenzo[*b,d*]furan and 9-methyl-9*H*-carbazoles synthesized *via* $SnCl_2 \cdot 2H_2O$ catalyzed one-pot Povarov reaction (imino-Diels–Alder reaction) were examined for their antimycobacterial activity.[100] Among the 23 compounds screened, 5-(dibenzo[b,d]furan-2-yl)-9-fluoro-3,4, 4a,5,6,10b-hexahydro-2H-pyrano[3,2-c]quinoline **48**, 5-(dibenzo[b,d]furan-2-yl)-9-fluoro-3,4,4a, 5,6,10b-hexahydro-2H-pyrano[3,2-c]quinoline **49**, and 9-fluoro-5-(9-methyl-9H-carbazol-3-yl)-3,4,4a,5,6,10b-hexahydro-2H-pyrano[3,2-c] quinoline **50** (MIC 3.13 µg/mL) resulted as the most active antitubercular agents (Figure 2-18).

A wide range of heterocyclic moieties conjugated with 1,2,3-triazoles were reported to exhibit potent antitubercular activity.[86] Among them, benzofuran salicylic acid derivative 6 (I-A09, Figure 2-3) is a lead antitubercular agent currently in clinical evaluations.[101] Here, 1,2,3-triazole-based *Mycobacterium tuberculosis* inhibitors and synthetic and natural product-based tricyclic (carbazole, dibenzo[b,d]furan, and dibenzo[b,d]thiophene) antimycobacterial agents were integrated in one molecular platform to prepare various novel clubbed 1,2,3-triazole hybrids using click chemistry.[102] Several derivatives displayed MIC values below 6.25 µg/mL. From the point of view of the establishment of structure-activity correlations (SAR), the order of the *M. tuberculosis* inhibitory activity of the compounds is dibenzo[b,d]thiophene series > dibenzo[b,d]furan series > 9-methyl-9H-carbazole series.[103] Out of the library of all compounds tested, two compounds, **51** and **52**, possess the maximum *M. tuberculosis* inhibitory activity with MIC = 1.9 µM (0.78 µg/mL) (Figure 2-19) and is 26 times more active than pyrazinamide and four times more active than ethambutol and has low toxicity profile (SI in the range of 55−255 for four different cell lines). These results together provided the potential importance of molecular hybridization and the development of triazole clubbed dibenzo[b,d]thiophene-based lead candidates to treat mycobacterial infections. Pharmacological evaluations and *in vivo* efficacy of some of these potent hybrids are in progress.

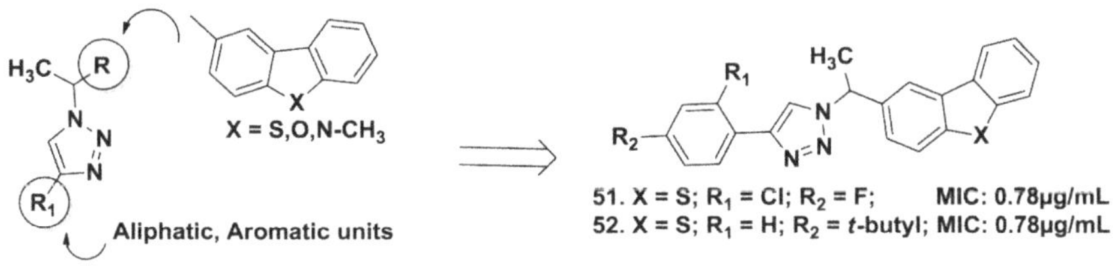

51. X = S; R₁ = Cl; R₂ = F;   MIC: 0.78µg/mL
52. X = S; R₁ = H; R₂ = *t*-butyl; MIC: 0.78µg/mL

FIGURE 2-19   1,2,3 triazole antitubercular conjugates.

In recent days, it is found interesting that some of the phenothiazine based successful drug candidates for treating neurodegenerative disorders were also found effective in inhibiting *M. tuberculosis*.[104] Chlorpromazine, Trifluoperazine (TPZ), and Thioridazine are a few with phenathiazine architecture (Figure 2-20) found to act in synergy with *M. tuberculosis* susceptible to regular antibiotics rifampicin and streptomycin.[105] Among them, TPZ is comparatively less toxic and displays good antitubercular activity.[106–108] Herein, we integrate both 2-(trifluoromethyl)-10H-phenothiazine and triazole pharmacophoric unit in one molecular platform to generate a newer scaffold 2-(trifluoromethyl) phenothiazine-[1,2,3]triazole hybrids.[109] Among all the newly synthesized compounds, three compounds **56, 57,** and **58** (MIC, 6.25 µg/mL) emerged as the most potent antitubercular agents with lower toxicity (selectivity index >10) against *Mycobacterium tuberculosis* virulent variant strain (H37Rv) (Figure 2-21). In continuation of our efforts, we herein demonstrated the potential utility of new hybrid analogues of adamantane ("lipophilic bullet") with appended 1,2,3-triazole fragment as potent antitubercular agents.[110] Evaluation of all the newly synthesized hybrids against *Mycobacterium tuberculosis* virulent variant strain (H37Rv) revealed that four compounds **59, 60, 61** with MIC 6.25 µg/mL, and **62** with MIC 3.13 µg/mL were the best active antitubercular agents with a selectivity index >15 (Figure 2-22).[111]

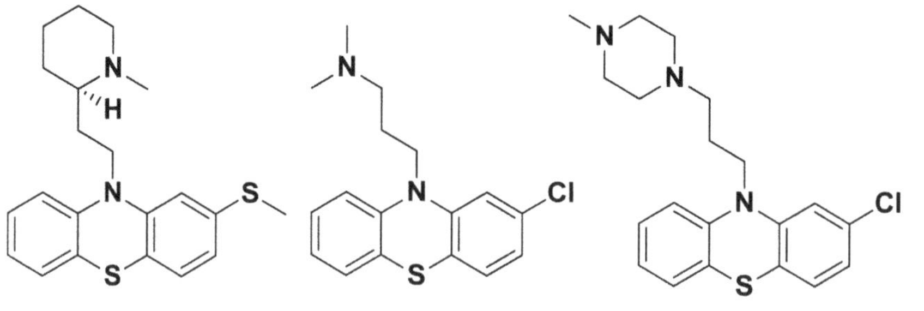

FIGURE 2-20   Phenothiazine based antitubercular drug candidates.

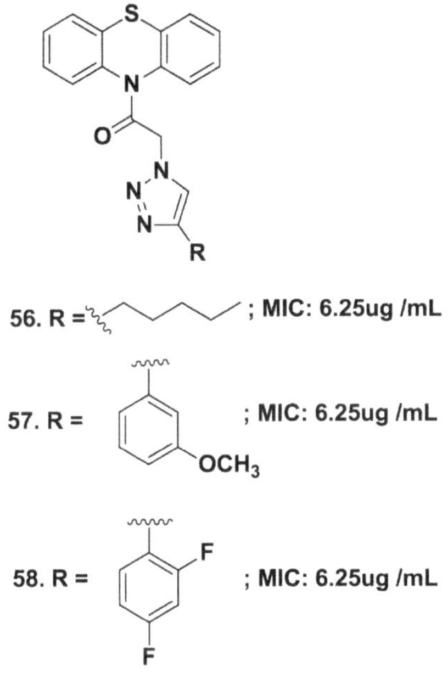

56. R = [structure] ; **MIC: 6.25ug /mL**

57. R = [structure] OCH$_3$; **MIC: 6.25ug /mL**

58. R = [structure] F, F ; **MIC: 6.25ug /mL**

FIGURE 2-21 Phenothiazine appended 1,2,3-triazoles.

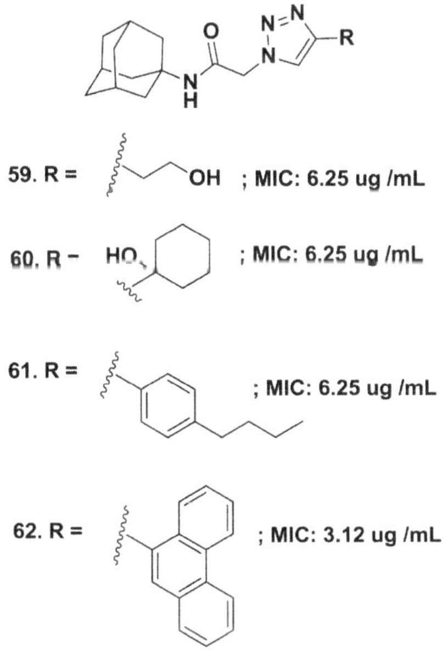

59. R = [structure] OH ; **MIC: 6.25 ug /mL**

60. R – HO, [structure] ; **MIC: 6.25 ug /mL**

61. R = [structure] ; **MIC: 6.25 ug /mL**

62. R = [structure] ; **MIC: 3.12 ug /mL**

FIGURE 2-22 Adamantane 1,2,3-triazole hybrids.

2.5.3   CARBAZOLE TETHERED PYRROLE HYBRIDS

Carbazole alkaloids (Figure 2-23) are playing an important role in discovering new antitubercular leads against nonresistant and multidrug resistant strains of *M. tuberculosis*. It is observed that Anti-TB activity is highly sensitive to subtle changes in substitution around the carbazole ring system. Our efforts on coupling of carbazole fragment with substituted pyrroles[112] resulted in one compound **67** as the promising lead analogue with MIC 3.13 μg/mL against *Mycobacterium tuberculosis* virulent variant strain (H37Rv) with low cytotoxicity profile (Figure 2-24).

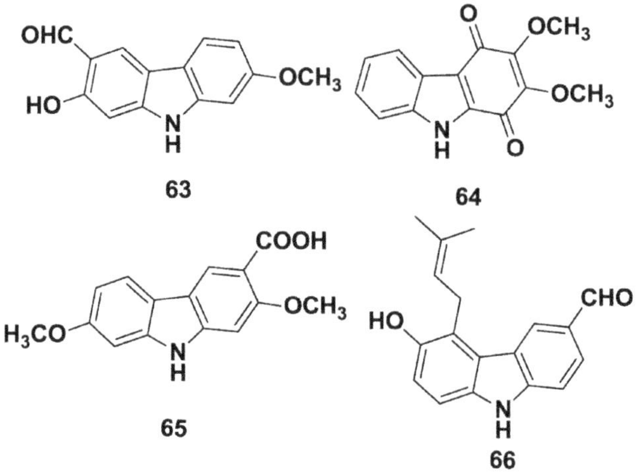

FIGURE 2-23   Carbazole alkaloids as antitubercular agents.

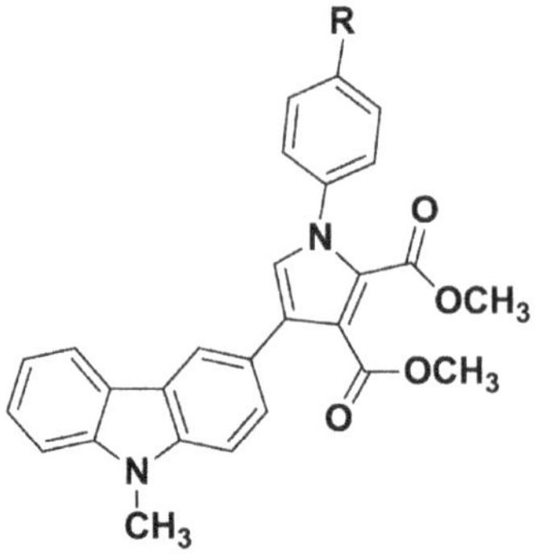

67. R = F; MIC: 3.12 ug/mL

FIGURE 2-24   Carbazole tethered pyrrole hybrids.

2.5.4   PIPERAZINE-THIOSEMICARBAZONE HYBRIDS

Piperazine is a nitrogen containing heterocycle present in more than 300 clinical drugs.[113] Thiosemicarbazone, a class of sulfur-containing ligands with metal-chelating abilities and reductive capacity, exhibit pronounced biological antitubercular activity (i.e., thiacetazone, **68** an antitubercular agent) (Figure 2-25).[114] Continuing our efforts for the development of novel antitubercular agents, we propose to combine three bioactive fragments such as imidazole-piperazine-thiosemicarbazone in one molecular architecture (Figure 2-26), and screened all the new compounds for *in vitro* antitubercular activity against *Mycobacterium tuberculosis* virulent variant strain (H37Rv) which resulted two compounds **69** and **70** with MIC, 3.13 µg/mL as the most potent analogs with a lower toxicity profile.[115]

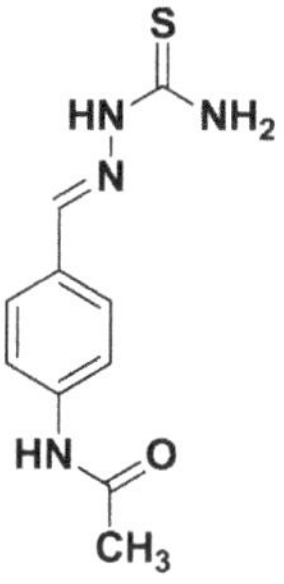

68

FIGURE 2-25   Thiacetazone, an antitubercular agent.

69. R =   ; MIC: 3.13 ug/mL

70. R =   ; MIC: 3.13 ug/mL

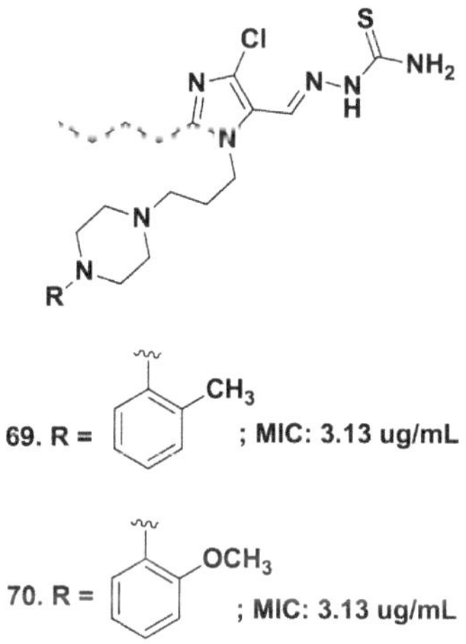

FIGURE 2-26   Piperazine-thiosemicarbazone hybrids.

2.6 FUTURE CHALLENGES AND PROCEPTS IN THE USAGE OF MOLECULAR HYBRIDS FOR THE DEVELOPMENT OF TB THERAPEUTICS

Hybrid antitubercular drugs are of great therapeutic interest as they can potentially overcome most of the pharmacokinetic drawbacks encountered when using conventional antitubercular drugs.[116] In fact, the future of hybrid antitubercular drugs is very bright for the discovery of highly potent and selective molecules which trigger two or more pharmacological mechanisms of action acting in synergy to inhibit *Mycobacterium tuberculosis*. However, there are certain inherent challenges which need to be met by these promising molecular hybrids.[30,53] They are:

2.6.1 PROMISED EFFICACY

Most of the TB drugs do not follow Lipinski's rule of 5,[30] which defines the optimal drug-like properties; whereas, pharmaceutical compound collection is biased towards these properties. This is one of the reasons linked to poor efficiency of new TB drugs identified through screening pharmaceutical library collection.[51] The key question asked is how effective are these hybrid molecules in balancing drug like properties without losing efficiency in inhibiting *Mtb*. Although some of the hybrid molecules described earlier look promising,[120] there is a lot needed to address the problem while converting identified hybrid chemical entities to performing as antitubercular drugs.

2.6.2 DEVELOPMENT OF DRUG RESISTANCE

Another challenge is to design hybrid antitubercular agents to control MDR-TB and XDR-TB. Treatment of TB has become more complicated by the emergence of MDR and XDR forms of TB.[2] Non judicious and inadequate use of antitubercular drugs has led to the emergence of these resistant strains. Identification of new compounds with activity against *M. tuberculosis* is again a difficult challenge. Drugs should have both bactericidal and sterilizing activity i.e., regimens against TB should kill rapidly growing mycobacteria and the persisting mycobacteria in the lesions.[117] Molecular hybrids may provide exclusive advantage in the treatment of the resistant strains.

2.6.3 RE-ENGINEERING OLD DRUGS VIA HYBRIDIZATION STRATEGIES

The approach for designing drugs for the purpose of therapy is "one-drug-one-target" since a long time and many diseases remains inadequately treated today, including TB. The hybridization strategies between molecules improve the activity entity by combining multiple target actions. Single drug with multiple actions is an interesting strategy in tuberculosis treatment.[118] Molecular hybridization of isoniazid and one quinolone derivative,[119] fluorquinole derivative and pyrazinamide through Mannich,[120] quinoxaline-1,4-di-N-oxide derivates[121] by molecular hybridization increases the antitubercular activity. Combined scaffold of three antitubercular drugs, isoniazid (first line), pyrazinamide (first line), and ciprofloxacin (second line) through molecular hybridization and resulting compounds had great activity against *Mtb*.[122] Conversion of clinically failed antitubercular drug candidates by re-engineering and hybridizing with more biologically compatible fragments and re-positioning them as newer analogues through dual/novel mode of action is yet another challenge in treating tuberculosis. The success of antimalarial hybrid drug trioxaquines,[123,124] promoted a renewed hope for the development of hybrids as antitubercular drugs.

2.7 PROSPECTS OF MOLECULAR HYBRIDIZATION APPROACH IN THE DEVELOPMENT OF NEXT GENERATION ANTITUBERCULAR DRUGS—"DOOM'S DAY" AHEAD

The field of molecular modification holds great expectations for rapid expansion into areas of drug development for diverse diseases including malaria, TB, inflammation, etc.[125] The expansion process is already underway as observed from the reported review articles and also indicated by studies carried out so far. The future holds great promise for continued advances in this area.[126] The greatest potential relies on our ability to translate this new emerging approach into completely novel molecules which can be applied not only for the treatment of drug resistant TB, but also for other chronic and/or orphaned diseases.[127] We, therefore, imagine a future where the classical/conventional molecules usage will diminish, holding a great upsurge in the usage of molecular hybrids by further exploitation and exploration of novel pathways and/or receptors/other target identification.

2.8 CONCLUSION

This review shows the various approaches by which hybrid antitubercular agents can be designed. Antitubercular molecules designed by molecular hybridization represents the most valid rational approach wherein, apart from having the dual pharmacophoric functionality, the drug holds great potential as a future lead to overcome the problem of drug resistance, effect on latent TB infection, and as a well enhanced patient compliance in a cost effective manner. The current global scenario demonstrates that many new hybrid molecules are being developed with promising *in vitro* results and some of them are reaching clinical trials with a high potential in clinical application. In addition, many known effective anti-TB molecules, which because of the development of resistance were not successfully used, can form part of the hybrids with a high anti-TB potential. It is also clear that the hybrids themselves, with their entire structure, can provide additional biological properties which only the combination of drugs can do, and which cannot be achieved by single entities. In this direction, it is note-worthy that many novel pathways for *Mtb* are explored to strengthen the TB pipeline research. This will offer molecular modification strategy, a brilliant opportunity to identify novel leads for antitubercular therapy. Continuous advances in the field of molecular hybridization allows the future to hold great promise in the quest of novel hybrid molecules which can offer more effective, safe, and shorter duration regimens for TB finally identifying a "molecular hybrid" as an effective lead in the "TB drug space". To conclude, imagination and creativity are other key elements to construct a successful hybrid antitubercular agent.

KEYWORDS

- Classical/conventional molecules
- Hybridization strategies
- Pharmacophoric behavior
- Tuberculosis

REFERENCES

1. Sharma, S. K.; and Mohan, A. Tuberculosis: From an incurable scourge to a curable disease—journey over a millennium. *Ind. J. Med. Res.* **2013**, *137*, 455–493.
2. Boire, N. A.; Riedel, V. A. A.; Parrish, N. M.; and Riedel, S. Tuberculosis: From an untreatable disease in antiquity to an untreatable disease in modern times? *J. Anc. Dis. Prev. Rem.* **2013**, *1* (2), 106.
3. Ananad, G. Global TB fight hits a wall. *Wall Street J.* **2013**.
4. Zumla, A.; Raviglione, M.; Hafner, R.; and von Reyn, C. F. Tuberculosis. *New Engl. J. Med.* **2013**, *368*, 745–755.
5. Dowdy, D. W.; and Chaisson, R. E. Post-2015 tuberculosis strategies in a pre-2015 world. *Int. J. Tuberc. Lung Dis.* **2013**, *17*, 143–143.
6. Diel, R.; Loddenkemper, R.; Zellweger, J. P.; Sotgiu, G.; D'Ambrosio, L.; Centis, R.; van der Werf, M. J.; Dara, M.; Detjen, A.; Gondrie, P.; Reichman, L.; Blasi, F.; and Migliori, G. B. Old ideas to innovate tuberculosis control, preventive treatment to achieve elimination. *Eur. Respir. J.* **2013**, *42*, 785–801.
7. Global tuberculosis control, WHO report 2011(WHO/HTM/TB/2011.16), World Health Organization, Geneva (Switzerland), **2011**. ISBN: 978-92-4-1546438-0; http.//www.who.int/tb/publications/global_report/2011/en/index.html.
8. Glaziou, P.; Floyd, K.; and Raviglione, M. Global burden and epidermology of tuberculosis. *Clin. Chest. Med.* **2009**, *30*, 621–636.
9. Dye, C.; and Williams, B. G. The population dynamics and control of tuberculosis. *Science* **2010**, *328*, 856–861.
10. Barry, C. E.; Boshoff, H. I.; Dartois, V.; Dick, T.; Ehrt, S.; Flynn, J.; Schnappinger, D.; Wilkinson, R. J.; and Young, D. The spectrum of latent tuberculosis: Rethinking the biology and intervention strategies. *Nat. Rev. Microbiol.* **2009**, *7*, 845–855.
11. Burman, W. J. Rip van wrinkle wakes up, development of tuberculosis treatment in the 21st Century. *Clin. Infect. Dis.* **2010**, *50*, S165–S172.
12. Marais, B. J.; and Zumla, A. History of tuberculosis and drug resistance. *N. Engl. J. Med.* **2013**, *368*, 88–88.
13. Lawn, S. D.; and Zumla, A. Advances in tuberculosis diagnostics, the Xpert MTB/RIF assay and future prospects for a point-of-care test. *Lancet Infect. Dis.* **2013**, *13*, 349–361.
14. Prabowo, S. A.; Gröschel, M. I.; Schmidt, E. D.; Skrahina, A.; Mihaescu, T.; Hastürk, S.; Mitrofanov, R.; Pimkina, E.; Visontai, I.; de Jong, B.; Stanford, J. L.; Cardona, P. J.; Kaufmann, S. H.; and van der Werf, T. S. Targeting multidrug-resistant tuberculosis (MDR-TB) by therapeutic vaccines. *Med. Microbiol. Immunol.* **2013**, *202*, 95–104.
15. Abubakar, I.; Zignol, M.; Falzon, D.; Raviglione, M.; Ditiu, L.; Masham, S.; Adetifa, L.; Ford, N.; Cox, H.; Lawn, S. D.; Marais, B.; McHugh, T. D.; Mwaba, P.; Bates, M.; Lipman, M.; Zijenah, L.; Logan, S.; McNerney, R.; Zumla, A.; Sarda, K.; Nahid, P.; Hoelscher, M.; Pletschette, M.; Memish, Z.; Kim, P.; Hafner, R.; Cole, S.; Migliori, G.; Maeurer, M.; Schito, M.; and Zumla, A. Drug-resistant tuberculosis, time for visionary political leadership. *Lancet Infect. Dis.* **2013**, *13*, 529–539.
16. Dierberg, K. L.; and Chaisson, R. E. Human immunodeficiency virus-associated tuberculosis: Update on prevention and treatment. *Clin. Chest Med.* **2013**, *34*, 217–228. DOI: 10.1016/j.ccm.2013.02.003.
17. Satti, H.; McLaughlin, M. M.; and Seung, K. J. Short report, Drug-resistant tuberculosis treatment complicated by antiretroviral resistance in HIV co-infected patients: A

report of six cases in Lesotho. *Am. J. Trop. Med. Hyg.* **2013,** *89,* 174–177. DOI: 10.4269/ajtmh.13-0046.

18. Royce, S.; Falzon, D.; van Weezenbeek, C.; Dara, M.; Hyder, K.; Hopewell, P.; Richardson, M. D.; and Zignol, M. Multidrug resistance in new tuberculosis patients, burden and implications. *Int. J. Tuberc. Lung Dis.* **2013,** *17,* 511–513.

19. Trouiller, P.; Olliaro, P.; Torreele, E.; Orbinski, J.; Laing, R.; and Ford, N. Drug development for neglected diseases, a deficient market and a public-health policy failure. *Lancet* **2002,** *359,* 2188–2194.

20. Beena, and Rawat, D. S. Antituberculosis drug research, A critical overview. *Med. Res. Rev.* **2013,** *33,* 693–764.

21. Gabriel, A. P.; and Mercado, C. P. Evaluation of task shifting in community-based DOTS program as an effective control strategy for tuberculosis. *Sci. World J.* **2011,** *11,* 2178–2186.

22. Kaboru, B. B.; Uplekar, M.; and Lönnroth, K. Engaging informal providers in TB control, what is the potential in the implementation of the WHO Stop TB Strategy? A discussion paper. *World Health Popul.* **2011,** *12,* 5–13.

23. Wright, A.; Zignol, M.; Van Deun, A.; Falzon, D.; Gerdes, S. R.; Knut, F.; Hoffner, S.; Drobniewski, F.; Barrera, L.; Soolingen, D.; Boulabhal, F.; Paramasivan, C. N.; Kam, K. M.; Mitarai, S.; Nunn, P.; and Raviglione, M. Epidemiology of antituberculosis drug resistance 2002–07: An updated analysis of the Global Project on Anti-Tuberculosis Drug Resistance Surveillance. *Lancet* **2009,** *373,* 1861–1873.

24. Gandhi, N. R.; Nunn, P.; Dheda, K.; Schaaf, H. S.; Zignol, M.; Soolingen, D. V.; Jensen, P.; and Bayona, J. Multidrug-resistant and extensively drug-resistant tuberculosis, a threat to global control of tuberculosis. *Lancet* **2010,** *375,* 1830–1843.

25. Cegielski, J. P. Extensively drug–resistant tuberculosis, there must be some kind of way out of here. *Clin. Infect. Dis.* **2010,** *50,* S95–S200.

26. Berry, M.; and Kon, O. M. Multidrug and extensively drug-resistant tuberculosis, an emerging threat. *Eur. Respir. Rev.* **2009,** *18,* 195–197.

27. Zar, H. J.; and Udwadia, Z. F. Advances in tuberculosis 2011–2012. *Thorax* **2013,** *68,* 283–287.

28. Xiong, X.; Xu, Z.; Yang, Z.; Liu, Y.; Wang, D.; Dong, M.; Parker, E. J.; and Zhu, W. Key targets and relevant inhibitors for the drug discovery of tuberculosis. *Curr. Drug Targets* **2013,** *14,* 676–699.

29. Lamichhane, G. Novel targets in M. tuberculosis, search for new drugs. *Trends Mol. Med.* **2011,** *17,* 25–33.

30. Koul, A.; Arnoult, E.; Lounis, N.; Guillemont, J.; and Andries, K. The challenge of new drug discovery for tuberculosis. *Nature* **2011,** *469,* 483–490.

31. Ma, Z.; Lienhardt, C.; McIlleron, H.; Nunn, A. J.; and Wang, X. Global tuberculosis drug development pipeline, the need and the reality. *Lancet* **2010,** *375,* 2100–2109.

32. Asif, M. A review of antimycobacterial drugs in development. *Mini Rev. Med. Chem.* **2012,** *12,* 1404–1418.

33. Cole, S. T.; and Riccardi, G. New tuberculosis drugs on the horizon. *Curr. Opin. Microbiol.* **2011,** *14,* 570–576.

34. Bhardwaj, A.; Scaria, V.; Raghava, G. P.; Lynn, A. M.; Chandra, N.; Banerjee, S.; Raghunandanan, M. V.; Pandey, V.; Taneja, B.; Yadav, J.; Dash, D.; Bhattacharya, J.; Misra, A.; Kumar, A.; Ramachandran, S.; Thomas, Z.; and Brahmachari, S. K. Open source drug discovery—a new paradigm of collaborative research in tuberculosis drug development. *Tuberculosis* **2011,** *91,* 479–486.

35. The Open Source Drug Discovery Initiative (2011) Tuberculosis Pipeline. www.osdd. net/tb-pipeline.

36. Sugumaran, G. Open Source Drug Discovery—redefining IPR through open source innovations. *Curr. Sci.* **2012,** *102,* 1637–1639.

37. Årdal, C.; and Røttingen, J. A. Open source drug discovery in practice: A case study. *PLoS Negl. Trop. Dis.* **2012,** *6,* e1827. DOI: 10.1371/journal.pntd.0001827.

38. Viegas-Junior, C.; Danuello, A.; Bolzani, V. S.; Barreiro, E. J.; and Fraga, C. A. M. Molecular hybridization, A useful tool in the design of new drug prototypes. *Curr. Med. Chem.* **2007,** *7,* 1459–1477.

39. Fraga, C. A. Drug hybridization strategies: Before or after lead identification? *Expert. Opin. Drug Discov.* **2009,** *4,* 605–609.

40. Mueller-Schiffmann, A.; Sticht, H.; and Korth, C. Hybrid compounds from simple combinations to nanomachines. *BioDrugs* **2012,** *26,* 21–31.

41. Bhakta S. An integration of interdisciplinary translational research in anti-TB drug discovery, out of the university research laboratories to combat *Mycobacterium tuberculosis. Mol. Biol.* **2013,** *2,* e108. DOI: 10.4172/2168-9547.1000e108.

42. Ekins, S.; Reynolds, R. C.; Franzblau, S. G.; Wan, B.; Freundlich, J. S.; and Bunin, B. A. Enhancing hit identification in *Mycobacterium tuberculosis* drug discovery using validated dual-event Bayesian models. *PLoS One* **2013,** *8,* e63240. DOI: 10.1371/journal. pone.0063240.

43. Stanley, S. A.; Grant, S. S.; Kawate, T.; Iwase, N.; Shimizu, M.; Wivagg, C.; Silvis, M.; Kazyanskaya, E.; Aquadro, J.; Golas, A.; Fitzgerald, M.; Dai, H.; Zhang, L.; and Hung, D. T. Identification of novel inhibitors of M. tuberculosis growth using whole cell based high-throughput screening. *ACS Chem. Biol.* **2012,** *17,* 1377–1384.

44. Koch, O.; Jager, T.; Heller, K.; Khandavalli, P. C.; Pretzel, J.; Becker, K.; Flohe, L.; and Selzer, P. M. Identification of M. tuberculosis thioredoxin reductase inhibitors based on high-throughput docking using constraints. *J. Med. Chem.* **2013,** *56,* 4849–4859.

45. Remunan, M. J.; Pérez-Herrán, E.; Rullas, J.; Alemparte, C.; Martinez-Hoyos, M.; Dow, D. J.; Afari, J.; Mehta, N.; Esquivias, J.; Jimenez, E.; Ortega-Muro, F.; Fraile-Gabaldon, M. T.; Spivey, V. L.; Loman, N. J.; Pallen, M. J.; Constantinidou, C.; Minick, D. J.; Cacho M.; Rebollo-Lopez M. J.; Gonzalez, C.; Sousa, V.; Angulo-Barturen, I.; Mendoza-Losana, A.; Barros, D.; Besra G. S.; Ballell, L.; and Cammack, N. Tetrahydropyrazolo[1,5-*a*]pyrimidine-3-carboxamide and *N*-benzyl-6',7'-dihydrospiro[piperidine-4,4'-thieno[3,2-*c*] pyran] analogues with bactericidal efficacy against *Mycobacterium tuberculosis* Targeting MmpL3. *PLoS One* **2013,** *8,* e60933. DOI: 10.1371/journal.pone.0060933.

46. Yokokawa, F.; Wang, G.; Chan, W. L.; Ang, S. H.; Wong, J.; Ma, I.; Rao, S. P. S.; Manjunatha, U.; Lakshminarayana, S. B.; Herve, M.; Kounde, C.; Tan, B. H.; Thayalan, P.; Ng, S.H.; Nanjundappa, M.; Ravindran, S.; Gee, P.; Tan, M.; Wei, L.; Goh, A.; Chen, P. Y.; Lee, K. S.; Zhong, C.; Wagner, T.; Dix, I.; Chatterjee, A. K.; Pethe, K.; Kuhen, K.; Glynne, R.; Smith, P.; Bifani, P.; and Jiricek, J. Discovery of tetrahydropyrazolopyrimidine carboxamide derivatives as potent and orally active antitubercular agents. *ACS Med. Chem. Lett.* **2014,** *4,* 451–455.

47. Bueno, J. Antitubercular in vitro drug discovery: Tools for begin the search. In *Understanding Tuberculosis—New Approaches to Fighting Against Drug Resistance;* Cardona, P-J., Ed.; In Tech publications: Carodona, 2012; 47–168. ISBN: 978-953-307-948-6, DOI: 10.5772/29634.

48. Dalhoff, A. Global fluoroquinolone resistance epidemiology and implictions for clinical use. *Interdiscip. Perspect. Infect. Dis.* **2012,** *2012,* 1–37. DOI: 10.1155/2012/976273.

49. Chang, K. C.; and Dheda, K. Rationalizing use of fluoroquinolones and pyrazinamide in the battle against multidrug-resistant tuberculosis. *Am. J. Respir. Crit. Care Med.* **2013,** *188*, 10–11.

50. Asif, M.; Siddiqui, A. A.; and Husain, A. Quinolone derivatives as antitubercular drugs. *Med. Chem. Res.* **2013,** *22*, 1029–1042.

51. Miller, C. H.; and O'Toole, R. F. Navigating tuberculosis drug discovery with target-based screening. *Exp. Opin. Drug. Discov.* **2011,** *6*, 839–854.

52. Chung, B. K.; Dick, T.; and Lee, D. Y. In silico analyses for the discovery of tuberculosis drug targets. *J. Antimicrob. Chemother.* **2013,** *68*, 2701–2709.

53. Coxon, G. D.; Cooper, C. B.; Gillespie, S. H.; and McHugh, T. D. Strategies and challenges involved in the discovery of new chemical entities during early-stage tuberculosis drug discovery. *J. Infect. Dis.* **2012,** *15*, S258–S264. DOI: 10.1093/infdis/jis191.

54. Barot, K. P.; Nikolova, S.; Ivanov, I.; and Ghate, M. D. Antitubercular drug development: Current status and research strategies. *Mini Rev. Med. Chem.* **2013,** *13* (11), 1664–1684.

55. Dutra, L. A.; Ferreira de Melo, T. R.; Chin, C. M.; and Santos, J. L. D. Antitubercular drug discovery, the molecular modification as promise tool. *Int. Res. Pharm. Pharmacol.* **2012,** *2*, 001–009.

56. Santos, J. L. D.; Dutra, L. A.; Ferreira de Melo, T. R.; and Chin, C. M. New antitubercular drugs designed by molecular modification. Understading tuberculosis—New approaches to fighting against drug resistance. Chapter 7 (In Tech Publication) **2012,** 169–186.

57. Zawilska, J. B.; Wojcieszak, J.; and Olejniczak, A. B. Prodrugs, a challenge for the drug development. *Pharmacol. Rep.* **2013,** *65*, 1–14.

58. Huttunen, K. M.; Raunio, H.; and Rautio, J. Prodrugs--from serendipity to rational design. *Pharmacol Rev.* **2011,** *63*, 750–771. DOI: 10.1124/pr.110.003459.

59. Tizón, L.; Otero, J. M.; Prazeres, V. F.; Llamas, S. A.; Fox, G. C.; van Raaij, M. J.; Lamb, H.; Hawkins, A. R.; Ainsa, J. A.; Castedo, L.; and González-Bello, C. A prodrug approach for improving antituberculosis activity of potent *Mycobacterium tuberculosis* type II dehydroquinase inhibitors. *J. Med. Chem.* **2011,** *54*, 6063–6084.

60. Sriram, D.; Yogeeswari, P.; and Gopal, G. Synthesis, anti-HIV and antitubercular activities of lamivudine prodrugs. *Eur. J. Med. Chem.* **2005,** *40*, 1373–1376.

61. Thornber, C. W. Isosterism and molecular modification in drug design. *Chem. Soc. Rev.* **1979,** *8*, 563–580.

62. Lima, L. M.; and Barreiro, E. J. Bioisosterism, a useful strategy for molecular modification and drug design. *Curr. Med. Chem.* **2005,** *12*, 23–49.

63. Prithwiraj, D.; Koumba, Y. G.; Constant, P.; Bedos-Belval, F.; Duran, H.; Saffon, N.; Daffé, M.; and Baltas, M. Design, synthesis, and biological evaluation of new cinnamic derivatives as antituberculosis agents. *J. Med. Chem.* **2011,** *54*, 1449–1461.

64. Ford, C. W.; Zurenko, G. E.; and Barbachyn, M. R. The discovery of linezolid, the first oxazolidinone antibacterial agent. *Curr. Drug. Targets. Infect. Dis.* **2001,** *1*, 181–199.

65. Cox, H.; and Ford, N. Linezolid for the treatment of complicated drug-resistant tuberculosis, a systematic review and meta-analysis. *Int. J. Tuberc. Lung. Dis.* **2012,** *16*, 447–454.

66. Pinon, M.; Scolfaro, C.; Bignamini, E.; Cordola, G.; Esposito, I.; Milano, R.; Mignone, F.; Bertaina, C.; and Tovo, P. A. Two pediatric cases of multidrug-resistant tuberculosis treated with linezolid and moxifloxacin. *Pediatrics* **2010,** *126*, e1253–1256. DOI: 10.1542/peds.2009-2172.

67. Shaw, K. J.; Barbachyn, M. R. The oxazolidinones, past, present, and future. *Ann. N. Y. Acad. Sci.* **2011,** *1241*, 48–70. DOI: 10.1111/j.1749-6632.2011.06330.x.

68. Maia, C. R.; and Fraga, M. A. Discovery of dual chemotherapy drug candidates designed by molecular hybridization. *Curr. Enz. Inhib.* **2010**, *6*, 171–182.

69. Muregi, F. W.; and Ishih, A. Next-generation antimalarial drugs: Hybrid molecules as a new strategy in drug design. *Drug Dev. Res.* **2010**, *71*, 20–32.

70. Bairwa, R.; Kakwani, M.; Tawari, N. R.; Lalchandani, J.; Ray, M. K.; Rajan, M. G.; and Degani, M. S. Novel molecular hybrids of cinnamic acids and guanylhydrazones as potential antitubercular agents. *Bioorg. Med. Chem. Lett.* **2010**, *20*, 1623–1625.

71. Bairwa, V. K.; and Telvekar, V. N. Novel 2-(2-benzyli denehydrazinyl) benzo[d]thiazole as potential antitubercular agents. *Comb. Chem. High Throughput Screen.* **2013**, *16*, 244–247.

72. Robert, F. S. The best of both worlds? *Science* **2001**, *291*, 2342.

73. Maddry, J. A.; Suling, W. J.; and Reynolds, R. C. Glycosyltransferases as targets for inhibition of cell wall synthesis in *M. tuberculosis* and *M. avium*. *Res. Microbiol.* **2006**, *147*, 106–112.

74. Wen, X.; Crick, D. C.; Brennan, P. J.; and Hultin, P. G. Analogues of the mycobacterial arabinogalactan linkage disaccharide as cell wall biosynthesis inhibitors. *Bioorg. Med. Chem.* **2003**, *11*, 3579–3587.

75. Saraiva, M. F.; de Souza, M. V.; Tran, H. D. M.; Araújo, D. P.; de Carvalho, G. S.; and de Almeida, M. V. Synthesis and antitubercular evaluation of new fluoroquinolone derivatives coupled with carbohydrates. *Carbohydr. Res.* **2010**, *345*, 761–767.

76. Sanki, A. K.; Boucau, J.; Umesiri, F. E.; Ronning, D. R.; and Sucheck, S. J. Design, synthesis and biological evaluation of sugar-derived esters, alpha-ketoesters and alpha-ketoamides as inhibitors for *Mycobacterium tuberculosis* antigen 85C. *Mol. Biosyst.* **2009**, *5*, 945–56.

77. Erlanson, D. A. Fragment-based lead discovery: A chemical update. *Curr. Opin. Biotechnol.* **2006**, *17*, 643–652.

78. Rawls, K. A.; Lang, P. T.; Takeuchi, J.; Imamura, S.; Baguley, T. D.; Grundner, C.; Alber, T.; and Ellman, J. A. Fragment-based discovery of selective inhibitors of the *Mycobacterium tuberculosis* protein tyrosine phosphatase PtpA. *Bioorg. Med. Chem. Lett.* **2009**, *19*, 6851–6854.

79. Manger, M.; Scheck, M.; Prinz, H.; von Kries, J. P.; Langer, T.; Saxena, K.; Schwalbe, H.; Furstner, A.; Rademann, J.; and Waldmann, H. Discovery of *Mycobacterium tuberculosis* protein tyrosine phosphatase A (MptpA) inhibitors based on natural products and a fragment-based approach. *Chembiochem.* **2005**, *6*, 1749–1753.

80. Rawls, K. A.; Grundner, C.; and Ellman, J. A. Design and synthesis of nonpeptidic, small molecule inhibitors for the *Mycobacterium tuberculosis* protein tyrosine phosphatase PtpB. *Org. Biomol. Chem.* **2010**, *8*, 4066–4070.

81. Soellner, M. B.; Rawls, K. A.; Grundner, C.; Alber, T.; and Ellman, J. A. Fragment-based substrate activity screening method for the identification of potent inhibitors of the *Mycobacterium tuberculosis* phosphatase PtpB. *J. Am. Chem. Soc.* **2007**, *129*, 9613–9615.

82. Thomas, K. D.; Adhikari, A. V.; Chowdhury, I. H.; Sandeep, T.; Mahmood, R.; Bhattacharya, B.; and Sumesh, E. Design, synthesis and docking studies of quinoline-oxazolidinone hybrid molecules and their antitubercular properties. *Eur. J. Med. Chem.* **2011**, *46*, 4834–4845.

83. Rane, R. A.; Gutte, S. D.; and Sahu, N. U. Synthesis and evaluation of novel 1,3,4-oxadiazole derivatives of marine bromopyrrole alkaloids as antimicrobial agent. *Bioorg. Med. Chem. Lett.* **2012**, *22*, 6429–6432.

84. Zhou, C. H.; and Wang, Y. Recent researches in triazole compounds as medicinal drugs. *Curr. Med Chem.* **2012,** *19*, 239–280.
85. Ferreira, V. F.; da Rocha, D. R.; da Silva, F. C.; Ferreira, P. G.; Boechat, N. A.; and Magalhaes, J. L. Novel 1H-1,2,3-, 2H-1,2,3-, 1H-1,2,4- and 4H-1,2,4-triazole derivatives, a patent review (2008–2011). *Expert. Opin. Ther. Pat.* **2013,** *23*, 319–331.
86. Naik, R. J.; Kulkarni, M. V.; Sreedhara, R. P. K.; and Nayak, P. G. Click chemistry approach for bis-chromenyl triazole hybrids and their antitubercular activity. *Chem. Biol. Drug. Des.* **2012,** *80*, 516–523.
87. Jardosh, H. H.; and Patel, M. P. Design and synthesis of biquinolone-isoniazid hybrids as a new class of antitubercular and antimicrobial agents. *Eur. J .Med. Chem.* **2013,** *65*, 348–359.
88. Chauhan, K.; Sharma, M.; Saxena, J.; Singh, S. V.; Trivedi, P.; Srivastava, K.; Puri, S. K.; Saxena, J. K.; Chaturvedi, V.; and Chauhan, M. S. Synthesis and biological evaluation of a new class of 4-aminoquinoline-rhodanine hybrid as potent anti-infective agents. *Eur. J. Med. Chem.* **2013,** *62*, 693–704.
89. Garcia, A.; Bocanegra-Garcia, V.; Palma-Nicolas, J. P.; and Rivera, G. Recent advances in antitubercular natural products. *Eur. J. Med. Chem.* **2012,** *49*, 1–23.
90. Ingolfsdottir, K. Usnic acid. *Phytochemistry* **2002,** *61*, 729–736.
91. Ramos, D. F.; and Almeida da Silva, P. E. Antimycobacterial activity of usnic acid against resistant and susceptible strains of *Mycobacterium tuberculosis* and non-tuberculous mycobacteria. *Pharm. Biol.* **2010,** *48*, 260–263. DOI: 10.3109/13880200903085490.
92. Prado, S.; Ledeit, H.; Michel, S.; Koch, M.; Darbord, J. C.; Cole, S. T.; Tillequin, F.; Brodin, P. Benzofuro[3,2-f][1]benzopyrans, a new class of antitubercular agents. *Bioorg. Med. Chem.* **2006,** *14*, 5423–5428.
93. Termentzi, A.; Khouri, I.; Gaslonde, T.; Prado, S.; Saint-Joanis, B.; Bardou, F.; Amanatiadou, E. P.; Vizirianakis, I. S.; Kordulakova, J.; Jackson, M.; Brosch, R.; Janin, Y. L.; Daffe, M.; Tillequin, F.; and Michel, S. Synthesis, biological activity, and evaluation of the mode of action of novel antitubercular benzofurobenzopyrans substituted on A ring. *Eur. J. Med. Chem.* **2010,** *45*, 5833–5847.
94. Kantevari, S.; Yempala, T.; Sridhar, B.; Yogeeswari, P.; and Sriram, D. Synthesis and anti-tubercular evaluation of amidoalkyl dibenzofuranols and 1*H*-benzo[2,3]benzofuro[4,5-e] [1,3]oxazin-3(2*H*)-ones. *Bioorg. Med. Chem. Lett.* **2011a,** *21*, 4316–4319.
95. Kantevari, S. Yempala, T.; and Vuppalapati, S. V. N. A new one-pot three-component, solvent free synthesis of amidoalkyl dibenzofuranols and dibenzofuran-condensed 1,3-oxazin-3-ones. *Synthesis* **2010,** *6*, 959–966.
96. Jeong, H. J.; Kim, Y. M.; Kim, Y. H.; Kim, J. Y.; Park, J. Y.; Park, S. J.; Ryu, Y. B.; and Lee, W. S. Homoisoflavonoids from Caesalpinia sappan displaying viral neuraminidases inhibition. *Biol. Pharm. Bull.* **2012,** *35*, 786–789.
97. Yempala, T.; Sriram, D.; Yogeeswari, P.; and Kantevari, S. Molecular hybridization of bioactives, synthesis and antitubercular evaluation of novel dibenzofuran embodied homoisoflavonoids via Baylis-Hillman reaction. *Bioorg. Med. Chem. Lett.* **2012,** *22*, 7426–7430.
98. Yempala, T.; Sridevi, J. P.; Yogeeswari, P.; Sriram, D.; and Kantevari, S. Design, synthesis and antitubercular evaluation of novel 2-substituted-3H-benzofuro benzofurans via palladium-copper catalysed Sonagashira coupling reaction. *Bioorg. Med. Chem. Lett.* **2013,** *23*, 5393–5396.

99. Osborne, R. First novel anti-tuberculosis drug in 40 years. *Nat. Biotech.* **2013**, *31*, 89–91.

100. Kantevari, S.; Yempala, T.; Surineni, G.; Sridhar, B.; Yogeeswari, P.; and Sriram, D. Synthesis and antitubercular evaluation of novel dibenzo[b,d]furan and 9-methyl-9H-carbazole derived hexahydro-2H-pyrano[3,2-c]quinolones via Povarov reaction. *Eur. J. Med. Chem.* **2011b**, *46*, 4827–4833.

101. Zhou, B.; He, Y.; Zhang, X.; Xu, J.; Luo, Y.; Wang, Y.; Franzblau, S. G.; Yang, Z.; Chan, R. J.; Liu, Y.; Zheng, J.; and Zhang, Z. Y. Targeting mycobacterium protein tyrosine phosphatase B for antituberculosis agents. *Proc. Natl. Acad. Sci. U S A* **2010**, *107* (10), 4573–4578.

102. Thirumurugan, P.; Matosiuk, D.; and Jozwiak, K. Click chemistry for drug development and diverse chemical-biology applications. *Chem. Rev.* **2013**, *113*, 4905–4979.

103. Patpi, S. R.; Pulipati, L.; Yogeeswari, P.; Sriram, D.; Jain, N.; Sridhar, B.; Murthy, R.; Devi, T. A.; Kalivendi, S. V.; and Srinivas, K. Design, synthesis, and structure-activity correlations of novel dibenzo[b,d]furan, dibenzo[b,d]thiophene, and N-methylcarbazole clubbed 1,2,3-triazoles as potent inhibitors of *Mycobacterium tuberculosis*. *J. Med. Chem.* **2012**, *55*, 3911–3922.

104. Wainwright, M.; Amaral, L.; and Kristiansen, J. E. The evolution of antimycobacterial agents from non-antibiotics. *Open J. Pharmacol.* **2012**, *2*, 1–11.

105. Pluta, K.; Mlodawska, M. B.; and Jelen, M. Recent progress in biological activities of synthesized phenothiazines. *Eur. J. Med. Chem.* 2011, *46*, 3179–3189.

106. Amaral, L.; and Kristiansen, J. E. Phenothiazines: An alternative to conventional therapy for the initial management of suspected multidrug resistant tuberculosis. A call for studies. *Int. J. Antimicrob. Agents.* **2000**, *14*, 173–176.

107. Amaral, L.; Kristiansen, J. E.; and Lorian, V. J. Synergic effect of chlorpromazine on the activity of some antibiotics. *J. Antimicrob. Chemother.* **1992**, *30*, 556–558.

108. Amaral, L.; Kristiansen, J. E.; Viveiros, M.; and Atouguia, J. Activity of phenothiazines against antibiotic-resistant *Mycobacterium tuberculosis*, a review supporting further studies that may elucidate the potential use of thioridazine as anti-tuberculosis therapy. *J. Antimicrob. Chemother.* **2001**, *47*, 505–511.

109. Addla, D.; Jallapally, A.; Gurram, D.; Yogeeswari, P.; Sriram, D.; and Kantevari, S. Rational design, synthesis and antitubercular evaluation of novel 2-(trifluoromethyl) phenothiazine-[1,2,3]triazole hybrids. *Bioorg. Med. Chem. Lett.* **2014a**, *24*, 233–236.

110. Wanka, L.; Iqbal, K.; Schreiner, P. R. The lipophilic bullet hits the targets, medicinal chemistry of adamantane derivatives. *Chem. Rev.* **2013**, *113*, 3516–3604.

111. Addla, D.; Jallapally, A.; Gurram, D.; Yogeeswari, P.; Sriram, D.; and Kantevari, S. Design, synthesis and evaluation of 1,2,3-triazole-adamantylacetamide hybrids as potent inhibitors of *Mycobacterium tuberculosis*. *Bioorg. Med. Chem. Lett.* **2014b**, *24*, 1974–1979.

112. Surineni, G.; Yogeeswari, P.; Sriram, D.; and Kantevari, S. Design and synthesis of novel carbazole tethered pyrrole derivatives as potent inhibitors of *Mycobacterium tuberculosis*. *Bioorg. Med. Chem. Lett.* **2015**, *25*, 481–491.

113. Maia, R. D. C.; Tesch, R.; and Fraga, C. A. M. Phenylpiperazine derivatives: A patent review (2006–Present). *Expert. Opin. Ther. Pat.* **2012**, *22*, 1169–1178.

114. Coxon, G. D.; Craig, D.; Corrales, R. M.; Vialla, E.; Gannoun-Zaki, L.; and Kremer, L. Synthesis, antitubercular activity and mechanism of resistance of highly effective thiacetazone analogues. *PloS One* **2013**, *8*, e53162.

115. Jallapally, A.; Addla, D.; Yogeeswari, P.; Sriram, D.; and Kantevari, S. 2-Butyl-4-chloroimidazole based substituted piperazine-thiosemicarbazone hybrids as potent inhibitors of *Mycobacterium tuberculosis*. *Bioorg. Med. Chem. Lett.* **2014,** *24,* 5520–5524.

116. Mitchison, D. A. The search for new sterilizing anti-tuberculosis drugs. *Front. Biosci.* **2004,** *9,* 1059–1072.

117. Morphy, J. R.; and Rankovic, Z. The physicochemical challenges of designing multiple ligands. *J. Med. Chem.* **2006,** *49,* 4961–4970.

118. Shindikar, A. V.; and Viswanathan, C. L. Novel fluoroquinolones, design, synthesis, and in vivo activity in mice against *Mycobacterium tuberculosis* H37Rv. *Bioorg. Med. Chem. Lett.* **2005,** *15,* 1803–1806.

119. Sriram, D.; Yogeeswari, P.; and Reddy, S. P. Synthesis of pyrazinamide Mannich bases and its antitubercular properties. *Bioorg. Med. Chem. Lett.* **2006,** *16,* 2113–2116.

120. Ansizu, S.; Moreno, E.; Solano, B.; Villar, R.; Buerguete, A.; Torres, E.; Perez-Silanes, S.; Aladana, L.; and Monge, A. New 3-methylquinoxaline-2-carboxamide 1,4-di-*N*-oxide derivatives as anti-*Mycobacterium tuberculosis* agents. *Bioorg. Med. Chem.* **2010,** *18,* 2713–2719.

121. Imramovsky, A.; Slovenko, P.; Vins, J.; Kocevar, M.; Jampílek, J.; Reckova, Z.; and Kaustova, J. A new modification of anti-tubercular active molecules. *Bioorg. Med. Chem.* **2007,** *15,* 2551–2559.

122. Chauhan, S. S., Sharma, M., and Chauhan, P. M., Trioxaquines: Hybrid molecules for the treatment of malaria. *Drug News Perspect.* **2010,** *23,* 632–646.

123. Bellot, F.; Coslédan, F.; Vendier, L.; Brocard, J.; Meunier, B.; and Robert, A. Trioxaferroquines as new hybrid antimalarial drugs. *J. Med. Chem.* **2010,** *53,* 4103–4109.

124. Vermund, S. H.; and Hayes, R. J. Combination prevention, new hope for stopping the epidemic. *Curr. HIV/AIDS Rep.* **2013,** *10,* 169–186.

125. Grosset, J. H.; Singer, T. G.; and Bishai, W. R. New drugs for the treatment of tuberculosis: Hope and reality. *Int. J. Tuberc. Lung Dis.* **2012,** *16,* 1005–1014.

126. Beceiro, A.; Tomas, M.; and Bou, G. Antimicrobial resistance and virulence: A successful or deleterious association in the bacterial world? *Clin. Microbiol. Rev.* **2013,** *26,* 185–230.

127. Napier, R. J.; Shinnick, T. M.; and Kalman, D. Back to the future, host-targeted chemotherapeutics for drug-resistant TB. *Future Microbiol.* **2012,** *7,* 431–435.

CHAPTER 3

NOVEL BIOACTIVE COMPOUNDS FROM AN ENDOPHYTIC *CHAETOMIUM ARCUATUM* STRAIN SAF-2 ISOLATED FROM *SEMECARPUS ANACARDIUM*

C. GANESH KUMAR[1]*, NARENDER REDDY GODUMAGADDA[1], POORNIMA MONGOLLA[1], SUJITHA POMBALA[1], PREETHI BADRINARAYAN[2], CHERUKU RAVINDRA REDDY[3], BISWANATH DAS[3], K.V.S. RAMAKRISHNA[4], BALASUBRAMANIAN SRIDHAR[5], and AHMED KAMAL[1]

[1]*Medicinal Chemistry and Pharmacology Division, CSIR-Indian Institute of Chemical Technology, Uppal Road, Hyderabad 500007, Telangana, India.*

[2]*Center for Molecular Modelling, CSIR-Indian Institute of Chemical Technology, Uppal Road, Hyderabad 500007, Telangana, India.*

[3]*Natural Products Chemistry Division, CSIR-Indian Institute of Chemical Technology, Uppal Road, Hyderabad 500007, Telangana, India.*

[4]*Nuclear Magnetic Resonance Center, CSIR-Indian Institute of Chemical Technology, Uppal Road, Hyderabad 500007, Telangana, India.*

[5]*Laboratory of X-ray Crystallography, CSIR-Indian Institute of Chemical Technology, Uppal Road, Hyderabad 500007, Telangana, India.*

Corresponding author: C. Ganesh Kumar. E-mail: cgkumar@iict.res.in

CONTENTS

Abstract 65

3.1   Introduction..65

3.2   Materials and Methods...66

3.3   Results and Discussion 71

3.4   Conclusion 82

Acknowledgments...83

Keywords 83

References..83

ABSTRACT

In continuation to our search for novel bioactive compounds from fungi, we profiled bioactive secondary metabolites from an endophytic fungus, *Chaetomium arcuatum* strain (*Semecarpus anacardium* fruit, SAF) SAF-2, isolated from the fruit of *Semecarpus anacardium*. Bioactivity-guided fractionation of the methanolic extract resulted in two compounds which were structurally elucidated as eugenetin and 6-hydroxymethyleugenin based on ^{1}H and ^{13}C NMR, Fourier transform infrared (FT-IR), and mass spectroscopic techniques. The crystal structure of 6-hydroxymethyleugenin was also confirmed by X-ray crystallography. Cytotoxicity results *in vitro* revealed that 6-hydroxymethyleugenin exhibited promising cytotoxicity against MCF7, MDA-MD-231, A549, and COLO205, while HeLa and K562 did not show any cytotoxicity. Eugenetin showed moderate cytotoxicity against MCF-7, HeLa, MDA-MD-231, and A549, while COLO205 and K562 did not show any cytotoxicity. The IC$_{50}$ (50% inhibitory concentration) values of acetylcholine esterase (AChE) inhibitory activity determined by Ellman assay for eugenetin and 6-hydroxymethyleugenin were 28.40 µM and 9.83 µM, respectively; while galanthamine (standard) exhibited IC$_{50}$ value of 2.50 µM. To understand the binding mode predictions of acetylcholinesterase complexed with eugenetin and 6-hydroxymethyleugenin in comparison with galanthamine, *in silico* studies were carried out. This is a first report on endophytic *Chaetomium arcuatum* producing bioactive chromanone compounds like eugenetin and 6-hydroxymethyleugenin exhibiting promising acetylcholinesterase inhibitory and anti-cancer activities.

3.1 INTRODUCTION

Alzheimer's disease (AD) is a neurodegenerative disorder connected with specific reduction of cholinergic neurons in the brain.[1] The characteristic feature of this illness is that the patient suffers from significant learning and memory impairment imputable to the loss of cholinergic neurons and deposition of aggregated Aβ-peptides forming amyloidal plaques. According to cholinergic hypothesis, these amyloid plaques are responsible for the cholinergic dysfunction in the brains of the AD patients.[2] During the last few decades, several cholinergic drugs have been launched in the market, including synthetic compounds tacrine,[3] galanthamine,[4] donepezil,[5] and rivastigmine.[6] The mechanism of action of the drugs approved for the AD therapy act by counterbalancing the acetylcholine deficiency, that is, they try to enhance the acetylcholine level in the brain by inhibiting acetylcholine esterase enzyme which terminates the nerve impulse transmission by hydrolyzing neurotransmitter acetylcholine at cholinergic synapses.[7] It has been reported that use of these drugs produced substantial cognitive improvement.[8–10]

Fungi are still not well explored candidates with regard to their bioprospecting for new drugs for the AD treatment. It is estimated that up to 1.5 million different fungi (Eumycota) exist, of which, only approximately 7–8% are explored to date.[11] Bioprospecting of fungal endophytes is gaining considerable attention in the recent years as prolific sources of novel bioactive compounds with diverse structures and biological activities such as antibacterial, antifungal, immunosuppressant, and anti-cancer compounds which find a potential use in modern medicine.[12–14] There are several medicinal plants producing different bioactive compounds exhibiting medicinal properties with geographical distribution in the Indian subcontinent and several counties worldwide.[15] *Semecarpus anacardium* (family: Anacardiaceae) is a deciduous tree with a wide distribution in the sub-Himalayan tract and in the tropical parts of India.[16] Fruit and nut extracts of *S. anacardium* have shown multiple medicinal usages in *Ayurveda* and have been reported to possess various therapeutic properties such as anti-inflammatory,[17] antioxidant,[18] immunomodulatory,[19] and antiarthritisproperties.[20] Recently, acetylcholine esterase (AChE) inhibitory activities were reported from the stem barks[21] and fruit resin[22] of *Semecarpus anacardium*. Considering the biological importance of *Semecarpus anacardium*, we screened for endophytic fungi from the bark, stem, leaves, and fruits of this plant. The secondary metabolites were profiled from a newly isolated endophytic *Chaetomium arcuatum* strain SAF-2 and assessed for their cytotoxicity and AChE inhibition activities.

3.2   MATERIALS AND METHODS

3.2.1   CHEMICALS AND REAGENTS

Bovine serum albumin (BSA), acetylcholine esterase (AChE) (EC 3.1.1.7) type V-S, from electric eel (*Electrophorus electricus*), 5,5'-dithiobis-(2-nitrobenzoic acid) (DTNB), acetylthiocholine iodide (ATCI), dimethylsulfoxide (DMSO), Trizma hydrochloride and galanthamine hydrobromide from *Lycoris* sp., and GenElute gel elution kit were purchased from Sigma,

St. Louis, MO, USA. All the chemicals used in the study were of analytical reagent grade and used without further purification.

3.2.2 ISOLATION AND IDENTIFICATION OF ENDOPHYTIC FUNGI

Bark, stem, leaves, and fruits of *Semecarpus anacardium* L (Carl Linnaeus). were collected from a local farm located on the outskirts of the city of Hyderabad, Telangana, India. These plant samples were transported to the laboratory in an ice box and were subjected to screening for endophytic fungi within 24 h. A total of 36 endophytic fungi were isolated from plant tissue samples of *S. anacardium*. Initially, all the plant tissue samples were washed thoroughly in running tap water to eliminate the epiphytic microbes, and then sterilized by rinsing in 75% isopropanol for 2 min and allowed to dry, followed by 0.1% mercuric chloride (v/v) treatment for 10 min. Later, the samples were rinsed 4–5 times in sterile distilled water and then were cut into pieces of length 0.5 cm;, the leaves were cut into 0.5 cm^2 sections with a flame sterilized surgical blade, and then transferred to petri dishes (9 cm in diameter) containing potato dextrose agar (PDA) medium (amended with 60 μg ml^{-1} streptomycin and 100 μg ml^{-1} ampicillin). The petri dishes were sealed with parafilm incubated at 30 °C in a light chamber with light-dark regimes of 12 h each and monitored every day to check the growth of endophytic fungal hyphae emerging from the segments. Individual hyphal tips of various fungi were removed from the agar plates, placed on fresh PDA medium plates and incubated at 30 °C for at least 10 days. Each fungal culture was checked for purity and transferred to another PDA plate by the hyphal tip method.[23] To ensure that surface sterilization had removed all hyphae and chlamydospores externally adhering to the segments, they were placed in PDA agar plates and incubated at 30 °C. Only the segments that were negative in this test were used for the isolation of endophytes. The fungal isolates were numbered and stored on PDA plates at 4 °C.

The endophytic fungus that showed acetylcholine inhibitory activity was identified based on the analysis of the DNA sequences of the ITS1–5.8S–ITS2, ITS regions of their rRNA gene. Genomic DNA was extracted from the fungal mycelia grown with PDA using ZR fungal/bacterial DNA identification kit (Zymo Research, Irvine, CA, USA). Primers ITS1 (TCC GTA GGT GAA CCT GCG G) and ITS4 (TCCTCCGCTTATTGATATGC) were used to amplify the ITS regions from the DNA extract. The Polymerase Chain Reaction (PCR) reaction was performed on an Eppendorf Mastercycler epgradient S (Eppendorf AG, Hamburg, Germany) with the following cycles: (1) 94 °C for 5 min; (2) 45 cycles of 94 °C for 30 sec, 56 °C for 30 sec, and 72 °C for 30 sec; and (3) 72 °C for 7 min. The PCR product spanning approximately 500–600 bp was checked on one percent agarose electrophoresis gel. It was then purified using quick spin column and buffers (washing buffer and elution buffer) according to the manufacturer's protocol of GenElute gel extraction kit (Sigma-Aldrich, St. Louis, MO, USA). DNA sequencing was done using ABI 3130 Genetic Analyzer system (Applied Biosystems, USA).

3.2.3   FUNGAL STRAIN AND FERMENTATION CONDITIONS

The pure culture of the fungal strain SAF-2 isolated from the *S. anacardium* fruit sample was identified as *Chaetomium arcuatum* and maintained in the inhouse culture repository of Council of Scientific and Industrial Research (CSIR)-Indian Institute of Chemical Technology, Hyderabad with the accession number ICTF-1152. The strain SAF-2 was precultured aerobically at 30 °C for 5 days in 4×1000 ml Erlenmeyer flasks containing 250 ml of potato dextrose broth medium (pH 6.0) and agitated at 150 rev min^{-1} in an orbital shaker for 120 h. The fermented medium was later filtered through a muslin cheese cloth to remove the fungal biomass and then the infiltrate was centrifuged at 2000 rpm to obtain a cell-free supernatant.

3.2.4   EXTRACTION, ANALYSIS, AND PURIFICATION OF METABOLITES

The compounds were extracted from the cell-free supernatant by absorption onto Diaion HP-20 (@3%, Supelco, Bellafonte, PA, USA) resin. The resin was washed with water and then extracted with methanol to obtain the crude extract which was further analyzed by thin-layer chromatography (TLC) on silica gel 60 plates (F_{254}, Merck). The plates were developed in ethyl acetate-hexane (60:40, v/v) solvent mixture and visualized under ultraviolet light at 254 nm which revealed the presence of two compounds. The crude extract was further concentrated under reduced pressure on a rotary vacuum evaporator (Rotavapor R-205, Büchi, Switzerland). The crude extract was further profiled on a silica gel (100–200 mesh) column (3×60 cm) with ethyl acetate-hexane solvent system as a mobile phase. Compound 1 was eluted in ethyl acetate-hexane mixture (15:85, v/v). The same solvent mixture was continued till the compound was completely eluted and after drying of the fractions, it gave a white crystalline solid. Compound 2 was eluted in ethyl acetate-hexane mixture (25:75, v/v), and drying of the fractions resulted in a white crystalline solid. Both the compounds were UV-active when visualized under ultraviolet light at 254 nm.

3.2.5　STRUCTURAL CHARACTERIZATION

The purified compounds were further subjected to ^{1}H and ^{13}C NMR, FT-IR, and high-resolution mass spectrometry (HR-MS) studies for elucidating the structures. The UV-visible spectra were measured by dissolving the samples in spectroscopic acetonitrile and recorded at 30 °C on a UV-visible double beam spectrophotometer (Lambda 25, Perkin-Elmer, Shelton, CT). The ^{1}H and ^{13}C NMR spectra were recorded on a Bruker Avance 500 MHz NMR spectrometer (Bruker, Switzerland) in deuterated DMSO at room temperature, and the chemical shifts were represented in δ values (ppm) with tetramethylsilane (TMS) as an internal standard. FT-IR spectra of the KBr pellets were collected at a resolution of 4 cm^{-1} in the wavenumber region of 400 to 4000 cm^{-1} on a Thermo-Nicolet Nexus 670 FT-IR spectrophotometer (Thermo Fisher Scientific Inc., Madison, WI, USA). The HR-MS spectra were recorded on a QSTAR XL Hybrid ESI-Q TOF mass spectrometer (Applied Biosystems Inc., Fosters City, CA, USA). Electrothermal Digital 9000 Series melting point apparatus (Model IA9200, Barnstead, UK) was used for determining the melting point of the purified compounds. The purified sample (1 mg) of each compound was placed in a glass capillary tube and this tube was placed in an aluminum heating block which was heated at a fast ramp rate of 10 °C min^{-1} from a temperature range of 25 °C to 300 °C for the measurement of melting point.

3.2.6　IN VITRO CYTOTOXICITY TESTING

Cytotoxicity of the compounds at various concentrations was assessed on the basis of the measurement of the *in vitro* growth in the 96 well plates by cell-mediated reduction of tetrazolium salt to water insoluble formazan crystals.[24] Cell lines for testing *in vitro* cytotoxicity included HeLa derived from human cervical cancer cells (ATCC No. CCL-2), A549 derived from human alveolar adenocarcinoma epithelial cells (ATCC No. CCL-185), MDA-MB-231 derived from human breast adenocarcinoma cells (ATCC

No. HTB-26), MCF7 derived from human breast adenocarcinoma cells (ATCC No. HTB-22), COLO 205 derived from human colon cancer cell line (ATCC No CCL-222), K562 derived from human chronic myelogenous leukemia cell line (ATCC No. CCL-243), and HEK 293 derived from human embryonic kidney cell line (ATCC No. CRL-1573) which were obtained from the American Type Culture Collection, Manassas, VA, USA. The effect of the different test compounds on the viability of tumor cell lines was measured at a wavelength of 540 nm on a multimode reader (Infinite® M200, Tecan, Switzerland). The IC_{50} values (50% inhibitory concentration) were calculated from the plotted absorbance data for the dose-response curves. The IC_{50} values (in μM) were expressed as the average of two independent experiments.

3.2.7 IN VITRO ACETYLCHOLINESTERASE ASSAY

The acetylcholinesterase (AChE) inhibitory activity for the crude extract and purified compounds was measured using a 96-well microtitre plate reader based on Ellman method.[25,26] The concentration of the compound that inhibited 50% of AChE activity (IC_{50}) was estimated by plotting the percentage activity and the percentage inhibition of AChE versus inhibitor concentrations on the same graph. The concentration at the intersection of these two curves was the IC_{50} value.

3.2.8 X-RAY CRYSTALLOGRAPHY

X-ray crystallography data for purified compound 2 showing promising acetylcholine esterase inhibitory activity was collected at room temperature using a Bruker Smart Apex CCD diffractometer with graphite monochromated MoKα radiation ($\lambda = 0.71073$Å) with ω-scan method[27]. Preliminary lattice parameters and orientation matrices were obtained from four sets of frames. Unit cell dimensions were determined from 3428 reflections for Compound 2. Integration and scaling of intensity data was accomplished using the SAINT program.[27] The structures were solved by direct methods using SHELXS97 and refinement was carried out by full-matrix least-squares technique using SHELXL97.[28] Anisotropic displacement parameters were included for all non-hydrogen atoms. The hydrogen atoms attached to the oxygen atoms were located in a difference density map and refined isotropically. The H atoms were positioned geometrically and treated as riding on their parent C atoms [C − H = 0.93 − 0.97 Å and U_{iso} (H) = $1.5U_{eq}$(C) for methyl H or $1.2U_{eq}$ (c) for other H atoms]. The methyl groups were allowed to rotate but not to tip.

3.2.9 MOLECULAR DOCKING STUDIES

The acetylcholine esterase complexed with galanthamine PDB: 1DX6 was downloaded from PDB. The binding modes and interactions of the AChE inhibitors from strain SAF-2 were determined using modelling protocols.[29] The docking protocol for mapping the inhibitor interactions were used which were based on our previous experiences.[30,31] Protein preparation wizard was used to prepare the protein after adding hydrogen. The side-chain residues were refined using Prime. Ligands were submitted to the LigPrep module to generate a range of ionization states populated at a given pH range of 7.4 ± 2 followed by an exhaustive conformational sampling with Confgen. The two modules of docking namely simple precision (SP) and extra precision (XP) were used for glide docking.[32] A 10 Å grid was generated by specifying the co-crystal galanthamine as the grid centee. The default SP docking settings and the conformations obtained from SP were used as an input for XP.

3.3   RESULTS AND DISCUSSION

The most significant therapeutic applications of AChE inhibitors are in the treatment of AD; various forms of dementia such as dementia with Lewy bodies, Down's syndrome, vascular dementia, and Parkinson's dementia; and also for the treatment of ataxia, migraine, myasthenia gravis, glaucoma, Postural Tachycardia Syndrome (PTS), and Korsakoff disease.[33] Natural sources like fungi can produce a variety of structurally distinct and biologically active secondary metabolites which can act as AChE inhibitors. Moreover, natural products have an edge over synthetic products in terms of safety concerns. In the recent past, several cholinergic drugs have been launched in the market. However, a major pullback on their widespread use as a general therapy is because of certain limitations such as short half life or side effects like hepatotoxicity exhibited by some AChE inhibitors like physostigmine and tacrine.[34] Synthetic AChE inhibitors like alkylpyridium polymers, dehydroevodiamine, and carbamate types have been reported to exhibit bioavailability problems and their possible side effects.[35] In view of these limitations, there is an urgent need to identify new AChE inhibitors with less toxicity and more potency. A majority of the AChE inhibitors examined till date were primarily from plants, and a few molecules were derived from marine sponges and microbes.

3.3.1   IDENTIFICATION OF THE ENDOPHYTIC FUNGUS

On the basis of 18S rRNA gene sequence homology analysis with Genbank sequence database, the fungal isolate showed 99% homology with different strains of *Chaetomium arcuatum* and was thus identified as *Chaetomium arcuatum*. The 18S rRNA gene sequence has been deposited in the National Centre for Biotechnology Information (NCBI) with Genbank number KF060290.

3.3.2   PHYLOGENETIC TREE ANALYSIS

The sequence obtained was annotated using the Chromas software and submitted to the NCBI GenBank database. It was aligned by Clustal W multiple sequence alignment program to other ITS rRNA gene sequences of endophytic fungal taxa obtained from GenBank database. All characters were equally weighted and unordered. The alignment gaps were treated as missing data. The evolutionary history was inferred using the Neighbor-Joining method.[36] The optimal tree with the sum of branch length = 0.93267357 was shown. The percentage of replicate trees in which the associated taxa clustered together in the bootstrap test (1000 replicates) was shown next to the branches.[37] The tree was drawn to scale, with branch lengths in the same units as those of the evolutionary distances used to infer the phylogenetic tree. The evolutionary distances were computed using the Jukes-Cantor method[38] and were in the units of the number of base substitutions per site. The analysis involved 17 nucleotide sequences. All positions containing gaps and missing data were eliminated. There were a total of 283 positions in the final dataset. Evolutionary analyses (Figure 3-1) were conducted using MEGA5 software.[39]

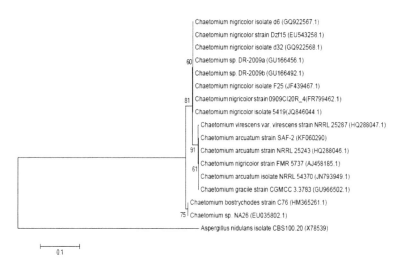

FIGURE 3-1 Evolutionary relationships of taxa were conducted using MEGA5 software.

3.3.3 PURIFICATION AND STRUCTURAL ELUCIDATION OF THE COMPOUNDS

The culture filtrate of strain SAF-2 revealed the presence of two spots in TLC with *Rf* values of 0.7 and 0.5 for compounds 1 and 2, respectively, which were separated in ethyl acetate-hexane solvent system. Compound 1 was isolated as a white crystalline solid with a yield of 20 mg l⁻¹; mp 161–163 °C; UV λ_{max} (EtOH) ($\log_e$): 225 (4.247). The ¹H NMR (DMSO-d₆ 500 MHz) spectrum (Figure 3-2) showed chemical shifts at δ = 2.38 ppm and δ = 1.98 ppm which corresponded to the methyl groups, δ = 3.89 ppm represented the proton of the methoxy group attached to the aromatic ring, δ = 6.23 and 6.68 ppm attributed to the aromatic protons, and δ = 12.96 ppm corresponded to the phenolic hydrogen. The ¹³C NMR (DMSO-d₆ 500 MHz) spectrum (Figure 3-3) showed that C11 carbon at δ = 55.82 ppm corresponded to the methoxy group which was attached to the C7 carbon of the benzene ring, C10 carbon at δ =13.76 ppm and C9 carbon at δ = 19.88 ppm attributed to the methyl groups attached to C8 of the benzene ring and C2 of the pyran ring. The other carbon signals with chemical shifts, δ = 89.33, 104.78, 108.15, 111.79, 157.56, 161.48, 165.35, 166.15, and 181.77 ppm matched with the 4H-chromen-4-one basic structure. The FT-IR spectrum (Figure 3-4) showed v = 3422.13 cm⁻¹ which corresponded to a broad stretching band indicating the presence of a phenolic –OH group,

$v = 2926.64$ cm^{-1} represented the stretching of the C=CH group, $v = 1660.14$ cm^{-1} attributed to the stretching of the C–O–C= group, $v = 1728.50$ cm^{-1} represented the stretching of the C=O group, and $v = 1626.24$ cm^{-1} corresponded to the stretching of C–O–CH$_3$. Molecular formula: C$_{12}$H$_{14}$O$_4$. MS (ESI) m/z: 221 [M+H]$^+$; HR-MS (ESI) m/z calculated for C$_{12}$H$_{13}$O$_4$ [M+H]$^+$: 221.08084, found: 221.08055 (Figure 3-5).

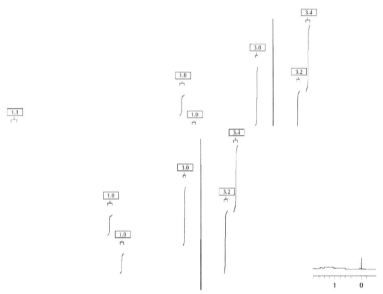

FIGURE 3-2 ^{1}H NMR spectrum of eugenetin.

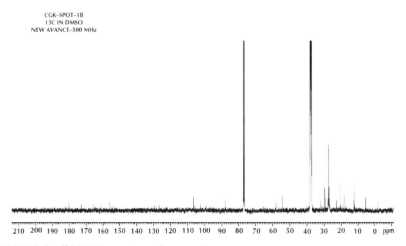

FIGURE 3-3 ^{13}C NMR spectrum of eugenetin.

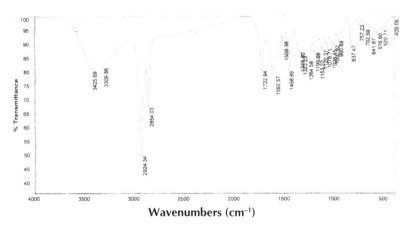

FIGURE 3-4   FT-IR spectrum of eugenetin.

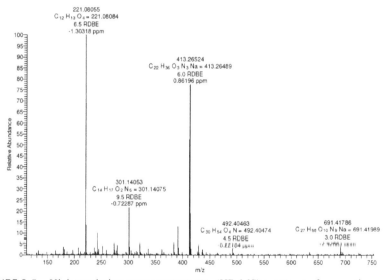

FIGURE 3-5   High-resolution mass spectrometry (HR-MS) spectrum of eugenetin.

Compound (2) was isolated as a white crystalline solid with a yield of 40 mg l⁻¹; mp. 191–193 °C; UV λ_{max} (EtOH) (log$_e$): 225 (4.247). A chemical shift was shown by ¹H NMR (DMSO-d$_6$ 500 MHz) spectrum (Figure 3-6) at δ = 2.39 ppm corresponding to the methyl group, δ = 4.1 ppm represented the aliphatic hydroxyl group, δ = 3.92 ppm attributed to the methoxy group attached to the benzene ring, δ = 4.62 ppm corresponded to the –CH$_2$ group, δ = 6.07 and 6.46 ppm corresponded to the aromatic protons of the benzene ring and pyran, respectively, and δ = 7.76 ppm represented the phenolic –OH group. The spectrum ^{13}C NMR (DMSO-d$_6$ 500 MHz) (Figure 3-7) showed that C11 carbon at δ = 55.84 ppm of methoxy group was attached to the C7 carbon of the benzene ring, C10 carbon at δ = 50.42 ppm of hydroxy methyl group was attached to the C6 of the benzene ring, and C9 carbon at δ = 19.76 ppm of the methyl group was attached to the C2 of the pyran ring. The other carbon signals corresponding to δ = 89.65, 104.03, 108.17, 111.77, 157.01, 158.57, 163.32, 167.04, and 181.74 ppm matched with the 4H-chromen-4-one basic structure. The ^{1}H and ^{13}C spectroscopic data of compound 2 were again similar to those of compound 1, suggesting that compound 2 was a derivative of compound 1. The FT-IR spectrum (Figure 3-8) showed ν = 3407.02 cm^{-1} broad stretching band indicating the presence of phenolic –OH group, ν = 3054.78 cm^{-1} attributed to stretching of the –CH$_2$ group, ν = 2924.57 cm^{-1} corresponded to aromatic stretching of the C=CH group, ν = 1658.02 cm^{-1} represented stretching of the C=O group, and ν = 1623.23 cm^{-1} corresponded to the C–O–C group. Molecular formula: C$_{12}$H$_{12}$O$_5$. MS (ESI) m/z: 219 [M+H–H$_2$O]$^+$; HR-MS m/z calculated for C$_{12}$H$_{12}$O$_5$ [M+H–H$_2$O]$^+$: 219.06519, found: 219.06488 (Figure 3-9). On the basis of these spectral data, compound 1 and 2 were identified as Eugenetin (IUPAC name: 5-Hydroxy-7-methoxy-2-6-dimethyl-4H-chromen-4-one) (Figure 3-10a) and 6-Hydroxymethyleugenin (IUPAC name: 5-Hydroxy-6-(hydroxymethyl)-7-methoxy-2-methyl-4H-chromen-4-one) (Figure 3-10b), respectively.

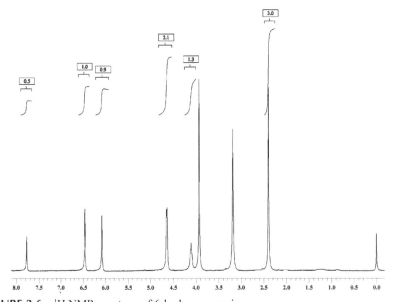

FIGURE 3-6 ^{1}H NMR spectrum of 6-hydroxyeugenin.

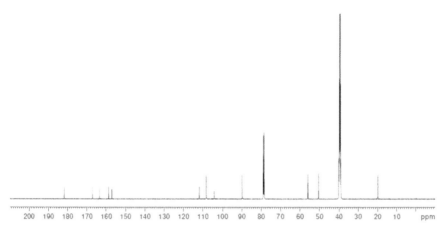

FIGURE 3-7 ^{13}C NMR spectrum of 6-hydroxyeugenin.

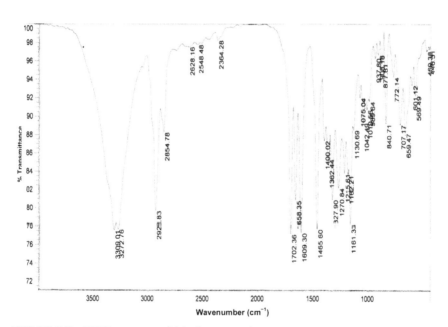

FIGURE 3-8 FT-IR spectrum of 6-hydroxyeugenin.

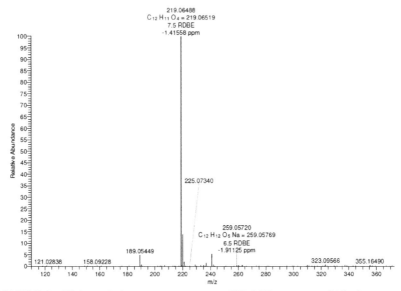

FIGURE 3-9   High-resolution mass spectrometry (HR-MS) spectrum of 6-hydroxyeugenin.

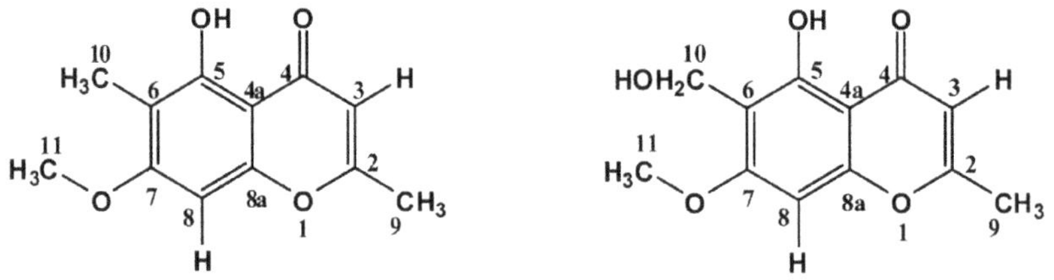

(A) Eugenetin   (B) 6-Hydroxymethyleugenin

FIGURE 3-10   The molecular structures of (A) Eugenetin and (B) 6-Hydroxymethyleugenin profiled from *Chaetomium arcuatum* strain SAF-2 as confirmed from the spectral data.

Fungi are the least explored microbial sources for AChE inhibitors and the different bioactive compounds produced by different fungal species include visoltricin from *Fusarium tricinctum*;[40] arisugacin from *Penicillium* sp. FO-4259;[41] quinolactacins A1 and A2 from *Penicillium citrinum*;[42] *p*-terphenyls from the mangrove endophytic fungus *Penicillium chermesinum* (ZH4-E2);[43] territrems and terreulactones A, B, C, and D from *Aspergillus* terreus;[44–46] sporotricolone from *Sporotrichum* sp.;[47] colletochlorin B and ilicicolin C and E from *Nectria galligena*;[48] huperzine A from an endophytic fungus *Acremonium* sp. 2F09P03B;[49] huperzine A which was isolated from two endophytic fungi, *Blastomyces* sp. and *Botrytis* sp.,[50] and from an endophytic fungus *Shiraia* sp. Slf14;[51] infractopicrin and 10-hydroxy-infractopicrin from the toadstool *Cortinarius infractus*;[52] xyloketal A from the mangrove fungus *Xylaria* sp. (no. 2508);[53] sporothrin A from the mangrove endophytic fungus *Sporothrix* sp. (#4335);[54] and lanostanoids from a mushroom called *Haddowia longipes*.[55] In the present study, we report the purification of two compounds from *Chaetomium arcuatum* strain SAF-2 and on the basis of the structure elucidation studies, they were identified as eugenetin and 6-hydroxymethyleugenin.

3.3.4 CRYSTAL STRUCTURE OF 6-HYDROXYMETHYLEUGENIN

The crystal structure of compound 2 (6-hydroxymethyleugenin, Figure 3-11) was also confirmed by X-ray crystallographic data analysis. Crystal data for 6-hydroxymethyleugenin: $C_{12}H_{12}O_5$, $M = 236.22$, colorless block, $0.19 \times 0.17 \times 0.11$ mm^3, triclinic, space group P-1 (No. 2), $a = 7.7589(9)$, $b = 8.2332(9)$, $c = 10.1352(11)$ Å, $\alpha = 107.040(2)$, $\beta = 97.156(2)$, $\gamma = 116.333(2)$ °, $V = 529.74(10)$ Å^3, $Z = 2$, $D_c = 1.481$ g/cm^3, $F_{000} = 248$, CCD area detector, MoKα radiation, $\lambda = 0.71073$ Å, $T = 294(2)$K, $2\theta_{max} = 50.0°$, 5107 reflections collected, 1857 unique (R$_{int} = 0.0171$). Final $GooF = 1.078$, $R1 = 0.0395$, $wR2 = 0.1125$, R indices based on 1659 reflections with I > 2σ(I) (refinement on F^2), 164 parameters, 0 restraints, μ = 0.116 mm^{-1}. The accession number CCDC 916685 contains supplementary crystallographic data for the structure. These data can be obtained free of charge at www.ccdc.cam.ac.uk/conts/retrieving.html [or from the Cambridge Crystallographic Data Centre (CCDC), 12 Union Road, Cambridge CB2 1EZ, UK; Fax: +44(0) 1223 336 033; Email: deposit@ccdc.cam.ac.uk].

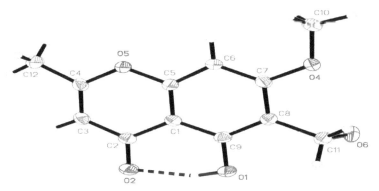

FIGURE 3-11   Crystal structure of (X) with the atom-numbering scheme. Displacement ellipsoids are drawn at the 30% probability level. Intramolecular hydrogen bond is shown as dashed lines.

3.3.5 IN VITRO CYTOTOXICITY STUDIES

On the basis of the cytotoxicity results (Table 3-1), 6-hydroxymethyleugenin exhibited higher cytotoxic potency against MCF7, MDA-MD-231, A549, and COLO205, while HeLa and K562 did not show any cytotoxicity. Eugenetin showed moderate cytotoxicity against MCF-7, HeLa, MDA-MD-231, and A549, while COLO205 and K562 did not show any cytotoxicity.

TABLE 3-1 *In Vitro* Cytotoxicity of Bioactive Compounds Produced by *Chaetomium arcuatum* Strain SAF-2

Test Cell Lines	IC50 Values (in μM)		
	Eugenetin	**6-hydroxymethyleugenin**	**Doxorubicin**
A549	96.14	17.89	1.21
HeLa	24.81	—[a]	0.45
MCF-7	11.89	2.79	1.05
MDA-MB-231	36.14	3.86	0.50
COLO 205	–	29.89	0.98
K562	–	–	0.32
HEK293 (normal)	–	–	–

[a] No activity.

3.3.6 ACETYLCHOLINE ESTERASE INHIBITION STUDIES

The acetylcholine esterase inhibitory activity of the crude extract from *Chaetomium arcuatum* strain SAF-2 showed 60% AChE inhibitory activity. The IC_{50} values for eugenetin and 6-hydroxymethyleugenin determined based on the Ellman method were 28.40 μM and 9.83 μM, respectively, while the IC_{50} value for galanthamine (standard inhibitor) was 2.50 μM. The results showed that the IC_{50} value for AChE inhibitory activity of 6-hydroxymethyleugenin was higher than that of eugenetin, which could be because of the lack of the hydroxyl group at C-6 in the former compound. On the basis of their chemical structures (Figure 3-4), there is a difference of only one functional group between the two compounds; 6-hydroxymethyleugenin has an extra hydroxyl group at the C-6 position when compared with eugenetin. The structure-activity relationship (SAR) can be explained by the fact that the presence of heteronuclear benzopyrone aromatic character in 4H-chromen-4-ones is one of the important structural feature essential for the potent interaction between 4H-chromen-4-ones and the peripheral anionic site (PAS) of AChE enzyme.[56]

The AChE inhibition and cytotoxicity screening assays revealed that eugenetin and 6-hydroxymethyleugenin have slightly different bioactivity potencies. Eugenetin and 6-hydroxymethyleugenin are 4H-chromen-4-ones. Chromone derivatives which are collectively known as chromanones have been isolated from a wide variety of natural sources like plants, fungi, and lichens[57] and possess a broad spectrum of bioactivities including the inhibition of gray hair formation by the promotion of melanin formation,[58] anti-cancer and immunostimulatory activity,[59] anti-oxidants[,60] and bradykinin B1 antagonists.[61]

3.3.7 MOLECULAR DOCKING STUDIES

3.3.7.1 ACTIVE SITE ANALYSIS

AChE is characterized by a deep and narrow "gorge" which contains the catalytic site at about 4 Å from its base. The active site is dissected into various sub-pockets to identify their role and specificity for various chemotypes of the inhibitor.[62] The catalytic active site of AChE consists of two sub-pockets, one the anionic site and the second esteratic sub-pocket. The positive quaternary amine of acetylcholine as well as other cationic substrates and inhibitors bind to the anionic sub-pocket. There are 14 conserved aromatic amino acids lining the active site which govern the binding of the cationic substrates. Among them, tryptophan (Trp) 84(86) is known to contribute to the efficacy in a major fashion. Substitution of this residue with alanine has shown to decrease the activity 3000-fold. The esteratic sub-pocket which is also the site for acetylcholine hydrolysis consists of three important catalytic residues Ser 200(203), His 440(447), and Glu 327(334) which are also called as the "catalytic-triad". Adjacent to this triad, lies the "acyl pocket" formed by residues Phe 288(295) and Phe 290(297). The last and important part is the "hydrophobic pocket" which is made up of Trp 84(86), Tyr 130(133), Tyr 330(337), and Phe 331(338). This pocket is known for its involvement in the stacking interactions wherein Trp 84(86) shows a tendency to facilitate cation-π interactions with charged species of the ligand of any of the −NH groups of the g-loop Gly 118(121), 119(122), and adjacent Ala 201(204) form the oxyanion hole.

3.3.7.2 INHIBITOR BINDING

All the three inhibitors show a tendency to bind to the esteratic sub-pocket of the receptor (Figure 3-12). Galanthamine scored best in Glide followed by eugenetin and 6-hydroxy methyl eugenetin. The cyclohexene ring forms a stacked-like arrangement with the indole ring of Trp 84(86), while no such tendency is observed in the case of eugenetin and 6-hydroxymethyleugenin. The 6-hydroxy methyl group forms close contacts with Phe 290(297). All the three inhibitors predominantly occupy the choline binding pocket, while the acyl binding pocket is occupied to a preliminary extent by the 6-hydroxy-methyl group of 6-hydroxymethyleugenin. The main H-bond interactions with the hydroxyl group of Ser 200(203) are displayed by all the three inhibitors. Galanthamine forms two additional hydrogen bonds wherein the hydroxyl group forms H-bond with OE1 of Glu 199(202) and the N-methyl group with OD2 of Asp 72(74). The Gly 118(121) and Gly 119(122) resi-dues of the oxyanion hole formed H-bonds with water indicating the possi-bility to form water bridged H-bonds with the inhibitor.

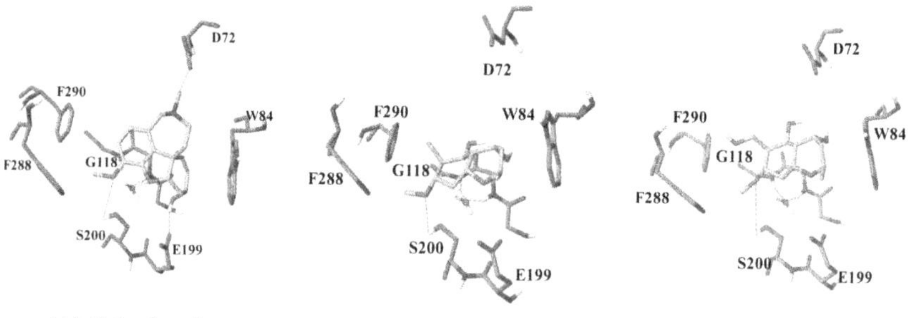

| (A) Galanthamine | (B) Eugenetin | (C) 6-Hydroxymethyleugenin |

*The amino acid residue numbers in the figure and the text refer to *Torpedo californica* ace-tylcholinesterase and the numbers in parentheses wherever specified refer to the positions of the analogous residues in *Homo sapiens* acetylcholinesterase.

FIGURE 3.12 Binding interactions of the inhibitors with acetylcholine esterase when docked with Glide 4.5.

3.4 CONCLUSION

The present study demonstrated that the newly isolated endophytic fungus *Chaetomium arcuatum* strain SAF-2 isolated from the fruit of *Semecarpus anacardium*, was an efficient producer of two known chromanone metab-olites, eugenetin and 6-hydroxymethyleugenin, exhibiting potential acetylcholinesterase inhibitory activity. These two metabolites have slightly different metabolic profiles. Both these compounds were active against the different cancer cell lines. Moreover, 6-hydroxymethyleugenin exhibited higher AChE inhibitory activity as compared with eugenetin. To the best of our knowledge, this is a first study on the chromanone metabolites from the endophytic fungus *Chaetomium arcuatum* strain SAF-2 exhibiting AChE inhibitory activities. In addition, X-ray crystallography data was also established for 6-hydroxymethyleugenin for the first time, as it showed promising AChE inhibitory activity as compared with eugenetin.

ACKNOWLEDGMENTS

The authors acknowledge the financial assistance provided to Godumagadda Narender Reddy, Pombala Sujitha and Cheruku Ravindra Reddy in the form of a Senior Research Fellowship by the Council of Scientific and Industrial Research (CSIR), New Delhi. Dr. G. Narahari Sastry, Head, Centre for Molecular Modeling, CSIR-Indian Institute of Chemical Technology is thanked for his scientific inputs. The Department of Science and Technology, New Delhi is thanked for the Women Scientist Fellowship to Preethi Badrinarayan . Further, we declare that there is no conflict of interest with any researcher or funding agency.

KEYWORDS

- 6-hydroxymethyleugenin
- Bioactives
- *Chaetomium arcuatum*
- Endophyte
- Eugenetin
- *Semecarpus anacardium*

REFERENCES

1. Davies, P.; and Maloney, A. J. F. Selective loss of central cholinergic neurons in Alzheimer's disease. *Lancet* **1976,** *2,* 1403–1403.

2. Bartus, R. T.; Dean, R. L. R.; Beer, B.; and Lippa, A. S. The cholinergic hypothesis of geriatric memory dysfunction. *Science* **1982**, *217*, 408–414.

3. Davis, K. L.; and Powchik, P. Tacrine. *Lancet* **1995**, *345*, 625–630.

4. Rainer, M. Galanthamine in Alzheimer's disease: A new alternative to tacrine? *CNS Drugs* **1997**, 7, 89–97.

5. Kawakami, Y.; Inoue, A.; Kawai, T.; Wakita, M.; Sugimoto, H.; and Hopfinger, A. J. The rationale for E2020 as a potent acetylcholinesterase inhibitor. *Bioorg. Med. Chem.* **1996**, *4*, 1429–1446.

6. Williams, B. R.; Nazarians, A.; and Gill, M. A. A review of rivastigmine: A reversible cholinesterase inhibitor. *Clin. Ther.* **2003**, *25*, 1634–1653.

7. Heinrich, M.; and Teoh, H. L. Galanthamine from snowdrop-the development of a modern drug against Alzheimer's disease from local Caucasian knowledge. *J. Ethnopharmacol.* **2004**, *92*, 147–162.

8. Bores, G. M.; Huger, F. P.; Petko, W.; Mutlib, A. E.; Camacho, F.; Rush, D. K.; Selk, D. E.; Wolf, V.; Kosley, R. W.; Davis, L.; and Vargas, H. M. Pharmacological evaluation of novel Alzheimer's disease therapeutics: Acetylcholinesterase inhibitors related to galanthamine. *J. Pharmacol. Exp. Ther.* **1996**, *277*, 728–738.

9. Johansson, M.; Hellström-Lindahl, E.; and Nordberg, A. Steady-state pharmacokinetics of tacrine in long-term treatment of Alzheimer patients. *Dementia* **1996**, 7, 111–117.

10. Wagstaff, A. J.; and McTavish, D. Tacrine. A review of its pharmacodynamic and pharmacokinetic properties, and therapeutic efficacy in Alzheimer's disease. *Drugs Aging* **1994**, *4*, 510–540.

11. Mueller, G. M.; and Schmit, J. P. Fungal biodiversity: What do we know? What can we predict? *Biodiversity Conservat.* **2007**, *16*, 1–5.

12. Strobel, G.; and Daisy, B. Bioprospecting for microbial endophytes and their natural products. *Microbiol. Mol. Biol. Rev.* **2003**, *67*, 491–502.

13. Zhang, H. W.; Song, Y. C.; and Tan, R. X. Biology and chemistry of endophytes. *Nat. Prod. Rep.* **2006**, *23*, 753–771.

14. Rodriguez, R. J.; White, J. F.; Arnold, A. E.; and Redman, R. S. Fungal endophytes: Diversity and functional roles. *New Phytol.* **2009**, *182*, 314–330.

15. Lawal, I. O.; Uzokwe, N. E.; Igboanugo, A. B. I.; Adio, A. F.; Awosan, E. A.; Nwogwugwu, J. O.; Faloye, B.; Olatunji, B. P.; and Adesoga, A. A. Ethnomedicinal information on collation and identification of some medicinal plants in research institutes of south-west Nigeria. *Afr. J. Pharm. Pharmacol.* **2010**, *4*, 1–7.

16. Kirithikar, K. R.; and Basu, B. D. *Indian Medicinal Plants*, 2nd eds; Basu L. M: Allahabad, 1933; pp 666–667.

17. Ramprasath, V. R.; Shanthi, P.; and Sachdanandam, P. Anti-inflammatory effect of *Semecarpus anacardium* Linn. Nut extract in acute and chronic inflammatory conditions. *Biol. Pharm. Bull.* **2004**, *27*, 2028–2031.

18. Ramprasath, V. R.; Shanthi, P.; and Sachdanandam, P. Evaluation of antioxidant effect of *Semecarpus anacardium* Linn. Nut extract on the components of immune system in adjuvant arthritis. *Vascul. Pharmacol.* **2005**, *42*, 179–186.

19. Ramprasath, V. R.; Shanthi, P.; and Sachdanandam, P. Immunomodulatory and anti-inflammatory effects of *Semecarpus anacardium* Linn. Nut milk extract in experimental inflammatory conditions. *Biol. Pharm. Bull.* **2006**, *29*, 693–700.

20. Vijayalakshmi, T.; Muthulakshmi, V.; and Sachdanandam, P. Effect of milk extract of *Semecarpus anacardium* nuts in glycohydrolases and lysosomal stability in adjuvant arthritis in rats. *J. Ethnopharmacol.* **1997**, *58*, 1–8.

21. Vinutha, B.; Prashanth, D.; Salma, K.; Sreeja, S. L.; Pratiti, D.; Padmaja, R.; Radhika, S.; Amita, A.; Venkateshwarlu, K.; and Deepak, M. Screening of selected Indian medicinal plants for acetylcholinesterase inhibitory activity. *J. Ethnopharmacol.* **2007**, *109*, 359–363.

22. Adhami, H. R.; Linder, T.; Kaehlig, H.; Schuster, D.; Zehl, M.; and Krenn, L. Catechol alkenyls from *Semecarpus anacardium*: Acetylcholinesterase inhibition and binding mode predictions. *J. Ethnopharmacol.* **2012**, *139*, 142–148.

23. Strobel, G.; Yang, X.; Sears, J.; Kramer, R.; Sidhu, R. S.; and Hess, W. M. Taxol from *Pestalotiopsis microspora*, an endophytic fungus of *Taxus wallachiana*. *Microbiology* **1996**, *142*, 435–440.

24. Mosmann, T. Rapid colorimetric assay for cellular growth and survival; application to proliferation and cytotoxicity assays. *J. Immunol. Methods* **1983**, *65*, 55–63.

25. Rhee, I. K.; van de Meent, M.; Ingkaninan, K.; and Verpoorte, R. Screening for acetylcholinesterase inhibitors from Amaryllidaceae using silica gel thin-layer chromatography in combination with bioactivity staining. *J. Chromatogr. A.* **1991**, *915*, 217–223.

26. Ellman, G. L.; Courtney, K. D.; Andres, V. Jr.; and Featherstone, R. M. A new and rapid colorimetric determination of acetylcholinesterase activity. *Biochem. Pharmacol.* **1961**, *7*, 88–95.

27. Anonymous. *SMART & SAINT Software Reference Manuals. Versions 6.28a and 5.625;* Bruker Analytical X-ray Systems Inc.: Madison, Wisconsin, USA, 2001.

28. Sheldrick, G. M. A short history of SHELX. *Acta Cryst. Sect. A Foundat. Cryst.* **2008**, *64*, 112–22.

29. Badrinarayan, P.; and Sastry, G. N. Virtual high-throughput screening in new lead identification. *Comb. Chem. High Throughput Screen.* **2011**, *14*, 840–860.

30. Srivastava, H. K.; Chourasia, M.; Kumar, D.; and Sastry, G. N. Comparison of computational methods to model DNA minor groove binders. *J. Chem. Inf. Model.* **2011**, *51*, 558–571.

31. Bohari, M. H.; and Sastry, G. N. FDA approved drugs complexed to their targets: Evaluating pose prediction accuracy of docking protocols. *J. Mol. Model.* **2012**, *18*, 4263–4274.

32. Friesner, R. A.; Banks, J. L.; Murphy, R. B.; Halgren, T. A.; Klicic, J. J.; Mainz, D. T.; Repasky, M. P.; Knoll, E. H.; Shelley, M.; Perry, J. K.; Shaw, D. E.; Francis, P.; and Shenkin, P. S. Glide: A new approach for rapid, accurate docking and scoring. 1. Method and assessment of docking accuracy. *J. Med. Chem.* **2004**, *47*, 1739–1749.

33. Giacobini, E. Cholinesterase inhibitors: New roles and therapeutic alternatives. *Pharmacol. Res.* **2004**, *50*, 433–440.

34. Knapp, M. J.; Knopman, D. S.; Solomon, P. R.; Pendlebury, W. W.; Davis, C. S.; and Gracon, S. I. A 30-week randomized controlled trial of high-dose tacrine in patients with Alzheimer's disease. The Tacrine study group. *J. Am. Med. Assoc.* **1994**, *271*, 985–991.

35. Schneider, L. S. New therapeutic approaches to Alzheimer's disease. *J. Clin. Psychiatry* **1996**, *57*, 30–36.

36. Saitou, N.; and Nei, M. The neighbor-joining method: A new method for reconstructing phylogenetic trees. *Mol. Biol. Evol.* **1987**, *4*, 406–425.

37. Felsenstein, J. Confidence limits on phylogenies: An approach using the bootstrap. *Evolution* **1985**, *39*, 783–791.

38. Jukes, T. H.; and Cantor, C. R. Evolution of protein molecules. In *Mammalian Protein Metabolism*; Munro, H. N., Ed.; Academic Press: New York, 1969; pp 21–132.

39. Tamura, K.; Peterson, D.; Peterson, N.; Stecher, G.; Nei, M.; and Kumar, S. MEGA5: Molecular evolutionary genetics analysis using maximum likelihood, evolutionary distance, and maximum parsimony methods. *Mol. Biol. Evol.* **2011,** *28,* 2731–2739.

40. Solfrizzo, M.; and Visconti, A. Anticholinesterase activity of the *Fusarium* metabolite visoltricin and its N-methyl derivative. *Toxicol. In Vitro* **1994,** *8,* 461–465.

41. Omura, S.; Kuno, F.; Otoguro, K.; Sunazuka, T.; Shiomi, K.; Masuma, R.; and Iwai, Y. Arisugacin, a novel and selective inhibitor of acetylcholinesterase from *Penicillium* sp. FO-4259. *J. Antibiot. (Tokyo)* **1995,** *48,* 745–746.

42. Kim, W. G.; Song, N. K.; and Yoo, I. D. Quinolactacins A1 and A2, new acetylcholinesterase inhibitors from *Penicillium citrinum. J. Antibiot. (Tokyo)* **2001,** 54, 831–835.

43. Huang, H.; Feng, X.; Xiao, Z.; Liu, L.; Li, H.; Ma, L.; Lu, Y.; Ju, J.; She, Z.; and Lin, Y. Azaphilones and p-terphenyls from the mangrove endophytic fungus *Penicillium chermesinum* (ZH4-E2) isolated from the South China Sea. *J. Nat. Prod.* **2011,** 74, 997–1002.

44. Chen, J. W.; and Ling, K. H. Territrems: Naturally occurring specific irreversible inhibitors of acetylcholinesterase. *J. Biomed. Sci.* **1996,** *3,* 54–58.

45. Cho, K. M.; Kim, W. G.; Lee, C. K.; and Yoo, I. D. Terreulactones A, B, C, and D: Novel acetylcholinesterase inhibitors produced by *Aspergillus terreus.* I. Taxonomy, fermentation, isolation and biological activities. *J. Antibiot. (Tokyo)* **2003,** 56, 344–350.

46. Kim, W. G.; Cho, K. M.; Lee, C. K.; and Yoo, I. D. Terreulactones A, B, C, and D: Novel acetylcholinesterase inhibitors produced by *Aspergillus terreus.* II. Physico-chemical properties and structure determination. *J. Antibiot. (Tokyo)* **2003,** 56, 351–357.

47. Shivanandappa, T.; Sattur, A. P.; Shereen; Divakar, S.; and Karanth, K. N. G. Compound as Cholinesterase Inhibitor from Fungus *Sporotrichum* Species. U.S. Patent 0,216,473, Nov 20, **2003.**

48. Gutierrez, M.; Theoduloz, C.; Rodríguez, J.; Lolas, M.; and Schmeda-Hirschmann, G. Bioactive metabolites from the fungus *Nectria galligena,* the main apple canker agent in Chile. *J. Agric. Food Chem.* **2005,** *53,* 7701–7708.

49. Li, W.; Zhou, J.; Lin, Z.; and Hu, Z. Study on fermentation for production of huperzine A from endophytic fungus 2F09P03B of *Huperzia serrata. Chin. Med. Biotechnol.* **2007,** *2,* 254–259.

50. Ju, Z.; Wang, J.; and Pan, S. Isolation and preliminary identification of the endophytic fungi which produce Huperzine A from four species in Hupziaceae and determination of Huperzine A by HPLC. *Fudan Univ. J. Med. Sci.* **2009,** *36,* 445–449.

51. Zhu, D.; Wang, J.; Zeng, Q.; Zhang, Z.; and Yan, R. A novel endophytic huperzine A-producing fungus, *Shiraia* sp. Slf14, isolated from *Huperzia serrata. J. Appl. Microbiol.* **2010,** *109,* 1469–1478.

52. Geissler, T.; Brandt, W.; Porzel, A.; Schlenzig, D.; Kehlen, A.; Wessjohann, L.; and Arnold, N. Acetylcholinesterase inhibitors from the toadstool *Cortinarius infractus. Bioorg. Med. Chem.* **2010,** *18,* 2173–2177.

53. Lin, Y.; Wu, X.; Feng, S.; Liang, G.; Luo, J.; Zhou, S.; Vrijmoed, L. L.; Jones, E. B.; Krohn, K.; Steingrover, K.; and Zsila, F. Five unique compounds: Xyloketales from mangrove fungus *Xylaria* sp. from the South China Sea coast. *J. Org. Chem.* **2001,** 66, 6252–6256.

54. Wen, L.; Cai, X.; Xu, F.; She, Z.; Chan, W. L.; Vrijmoed, L. L. P.; Jones, E. B. G.; and Lin, Y. Three metabolites from the mangrove endophytic fungus *Sporothrix* sp. (#4335) from the South China Sea. *J. Org. Chem.* **2009,** 74, 1093–1098.

55. Zhang, S. S.; Ma, Q. Y.; Huang, S. Z.; Dai, H. F.; Guo, Z. K.; Yu, Z. F.; and Zhao, Y. X. Lanostanoids with acetylcholinesterase inhibitory activity from the mushroom *Haddowia longipes*. *Phytochemistry* **2015**, *110*, 133–139.

56. Bartolini, M.; Bertucci, C.; Cavrini, V.; and Andrisano, V. Beta amyloid aggregation induced by human acetylcholinesterase inhibition studies. *Biochem. Pharmacol.* **2003**, *65*, 407–416.

57. Tsui, W.; and Brown, G. D. Chromones and chromanones from *Baeckea frutescens*. *Phytochemistry* **1996**, *43*, 871–876.

58. Okamoto, Y.; Kobayashi, T.; Imokawa, G.; Hori, T.; and Nishizawa, Y. Hair preparations containing chromones for inhibition of gray hair. *Jpn. Kokai Tokkyo Koho* **1997**, JP 09188608.

59. Gamal-Eldeena, A. M.; Djemgou, P. C.; Tchuendem, M.; Ngadjui, B. T.; Tane, P.; and Toshifumi, H. Anti-cancer and immunostimulatory activity of chromones and other constituents from *Cassia petersiana. J. Biosci.* **2007**, *62*, 331–338.

60. Dias, M. M.; Machado, N. F. L.; and Marques, M. P. M. Dietary chromones as antioxidant agents—the structural variable. *Food Funct.* **2011**, *2*, 595–602.

61. Bryan, M. C.; Biswas, K.; Peterkin, T. A.; Rzasa, R. M.; Arik, L.; Lehto, S. G.; Sun, H.; Hsieh, F. Y.; Xu, C.; Fremeau, R. T.; and Allen, J. R. Chromenones as potent bradykinin B1 antagonists. *Bioorg. Med. Chem. Lett.* **2012**, *22*, 619–622.

62. Badrinarayan, P.; and Sastry, G. N. Sequence, analysis of p38 MAP kinase: Exploiting DFG-out conformation as a strategy to design new type II leads. *J. Chem. Inf. Model.* **2011**, *51*, 115–129.

CHAPTER 4

UNDERSTANDING THE ROLE OF BIOMARKERS IN THE PATHOGENESIS OF SEPSIS

S. BURGULA[1*], M. SWATHI RAJU[1], A. NICHITA[1],
A. PAWAN KUMAR[1], and KARTHIK RAJKUMAR[1]

[1]Department of Microbiology, Osmania University, Hyderabad, Telangana – 500007, India.

*Corresponding author: S. Burgula. E-mail: s_burgula@osmania.ac.in

CONTENTS

Abstract 90
4.1 Introduction...91
4.2 Methods..92
4.3 Discussion 92
4.4 Conclusion 104
Keywords 104
References..104

ABSTRACT

Sepsis is a serious medical condition caused by a severe systemic infection leading to a systemic inflammatory response. The more critical subsets of sepsis include severe sepsis (sepsis with organ dysfunction) and septic shock (sepsis with refractory arterial hypotension). It is characterized by a systemic over-reaction of the inflammatory and the coagulatory system resulting from an infection, which can quickly lead to multiple organ failure and death.

It is known that the host response to Gram-negative bacterial LPS—an important component of sepsis pathology involves its interaction with Toll-like receptors (TLRs) and cluster differentiation 14 (CD14) receptors present on the monocytes and macrophages which initiate the production of proinflammatory mediators such as interleukin (IL-6), interleukin (IL-8), and tumor necrosis factor (TNF-α). LPS also induces the production of acute phase response proteins by the liver. Serum procalcitonin is currently the only USFDA (US Food and Drug Administration) approved biomarker for the diagnosis and monitoring of the progression of sepsis, though other acute phase reactants like CRP (C reactive protein) and Serum amyloid A protein are also in use. The diagnosis of sepsis and evaluation of its severity is complicated by the highly variable and nonspecific nature of the signs and symptoms of sepsis. Recently, genes involved in innate immune recognition have gained attention. In particular, variation in the genes encoding the so-called TLRs, Mannose-Binding Lectin (MBL), activated protein C (APC), etc., have been extensively studied. Although these studies have helped in devising methods for treatment, yet the incidence of sepsis is rising. Very little is known about the genetic susceptibility of an individual; the risk of sepsis increases because of the various immunosuppressive procedures given to patients like drugs, chemotherapy, radiotherapy, etc. Recently identified biomarkers slouable triggering receptor expressed on myeloid cells (sTREM), soluable form of urokinase type plasminogen receptor (suPAR), Nociceptin, Presepsin have shown significant results in the diagnosis of sepsis and further understanding on how they correlate with sepsis outcome can prove their potential as novel biomarkers. In this review, a panorama of sepsis biomarkers identified till date, their role in therapy, and the latest developments in the understanding of sepsis will be discussed.

4.1 INTRODUCTION

The systemic inflammatory response to a variety of severe clinical insults results in sepsis in susceptible individuals. In a survey conducted in eastern India, sepsis constituted 17% of all admissions to the ICU of which 45% admissions ended in fatality.[1] Severe sepsis takes more lives than breast, colon/rectal, head and neck, throat, and prostate cancer combined. According to recent reports, the incidence of sepsis is rising at the rates between 1.5% and 8% per year,[2] despite advancements in critical care support and equipments. Every hour, at least 25 people die of sepsis in USA alone. India with its much larger population compounded by poor medical and surgical infrastructure in non-urban areas, several small and mid-size hospitals in large cities with meager critical care infrastructure, has a bigger challenge of addressing the problem. The question of why the incidence is rising has been extensively discussed, but a final answer has not yet been found. Interestingly, the spectrum of responsible microorganisms and their sensitivity to various antibiotics appears to elude microbiologists which has made it harder for devising therapeutic strategies to combat sepsis in patients. While very little is known about the genetic susceptibility of an individual, the risk of sepsis increases because of the various immunosuppressive procedures given to patients, like drugs, chemotherapy, radiotherapy, etc. Sepsis seems to be a consequence predominantly of bacterial infection although increase in fungal infections leading to sepsis has been reported in recent times.

By definition, sepsis is not a true disease, but rather an innate physiological response by the immune system to infection, which may vary with patient, causative organism, site of infection and in some cases even with gender. The standard protocols employed for the diagnosis of sepsis mainly involve the use of microbial cultures, which are time consuming because of the slow growth or non cultivatable organisms and also shown to be insensitive in many cases where low quantities of the causative organism are present in the sample.[3] Basic antimicrobial therapy is the primary treatment given to control bacterial infection, but still, the patient is reported to enter septic shock, the major reason for the failure being antimicrobial treatment. In majority of the cases, it might be because of the inability to clear bacterial toxins or particles which act as mediators of inflammation, e.g., bacterial lipopolysaccharides of Gram-negative bacteria such as *E.coli, Klebsiella pneumoniae*, and lipoteichoic acid of Gram-positive bacteria such as *Streptococcus pneumoniae* and *Staphylococcus aureus.* These disadvantages can be overcome by alternative diagnostic methods using molecular-based tests like enzyme linked immunosorbent assay (ELISA), immunoluminometric assays, Polymerase

Chain Reaction (PCR), and magnetic probes for automated microbial detection. Another major focus in sepsis research apart from the development of faster and more sensitive detection methods of infectious microbes, is monitoring the serum protein profiles for the identification of protein biomarker meeting the criteria of an ideal biomarker best defined as "a characteristic that is objectively measured and evaluated as an indicator of normal biologic processes (or) pathogenic processes," as previously described by Doherty *et al.*[4] Though several biomarkers have been studied and identified in animal and human models of sepsis, an ideal biomarker which is specific, quick, and reliable for early diagnosis is yet to be identified.

4.2 METHODS

Electronic search tools as NCBI, MEDLINE, and Pub Med databases using the keywords "infection", "inflammation", "sepsis", "systemic inflammatory response syndrome (SIRs)", "septic shock", and "biomarker" were used to identify clinical and experimental studies which evaluated the diagnostic and prognostic significance of a biomarker in sepsis. The search retrieved references covering more than 170–178 different biomarkers of inflammation and infection. Text mining databases like information hyperlinked over protein (iHOP), Chilibot, Gene ontology, Uniport, etc., were searched for gene and protein information. Only few, highly significant biomarkers are discussed in detail in this review.

4.3 DISCUSSION

There are more than 170 biomarkers studied for their relevance in sepsis diagnosis, while only a few such as procalcitonin (PCT), C-reactive protein (CRP), and lactate are clinically approved and were proved to be of prognostic value. On other side, clinical trials for sepsis treatment employing recombinant activated C protein (Dotrecogin alpha), TLR4 receptor blockers (eritoren), lactoferrin, and anti-TNF antibodies were effective in the treatment of small subgroups, but have failed to show significant results in a large population and require improvement for expected positive outcome. One of the reasons that the diagnosis of sepsis and evaluation of its severity is complicated may be because of the highly variable and nonspecific nature of the signs and symptoms of sepsis. Sepsis response may vary even with the causative organisms, species, gender, age, etc. Hence, focusing on patient specific treatment, early diagnosis, and prediction of the severity of sepsis is very important, increasing the possibility of starting timely and specific treatment. Although these studies have helped in devising methods for treatment, yet the incidence of sepsis is rising. Sepsis often presents alongside other conditions; because of this complexity, a single "golden" biomarker may not exist, because of the heterogeneity in sample and methods employed. Thus, research should shift more focus on to assessing the combined diagnostic capabilities of multiple biomarkers.

In this review, we discuss the significance of approved biomarkers and give an overview on recently studied new biomarker and their possible role in sepsis diagnosis.

4.3.1 ROLE OF CYTOKINES AS MARKERS

Proinflammatory cytokines (IL1-beta, IL6, TNF) and anti-inflammatory cytokines (IL10, IL-1), receptor antagonists (interleukin 1 receptor antagonist (IL-1ra)), and soluable tumor necrosis factor receptors (sTNFRs) play an important role in mediating sepsis. Studies have reported that IL1 receptor antagonist correlates with Sequential Organ Failure Assessment (SOFA) score in sepsis, while high levels of IL1-beta help distinguish septic patients compared with non septic patients.[5] Processing of the cytokine precursors and regulation of cytokine production can be modulated by caspases (caspase-1, -4, -5, and -12), thus contributing to the pathogenesis of sepsis. Cleavage of IL-1b is mediated by caspase-1; IL-2 was reported to rise in parallel to the severity of sepsis, whereas IL-4 was associated with development of sepsis.[6] In a study involving 65 severe sepsis patients, the levels of both pro- and anti-inflammatory cytokines were found to be significantly elevated in patients with sepsis when compared with healthy subjects. It was reported that IL10 was high in non-survivors and a high IL-10 to TNF-α ratio was associated with death, indicating that anti-inflammatory cytokine IL-10 is the main predictor of severity and fatal outcome.[7] In another study, IL-6 and IL10 could distinguish survivors versus non-survivors of sepsis at 28 day mortality.[8] Development of multiorgan dysfunction and DIC (disseminated intravascular coagulation) were predicted by IL-8 and IL-12 helps predict mortality in the cases of postoperative sepsis. In severe sepsis, TNF levels are also known to elevate. Although Anti-TNF antibody therapy has been successful against other inflammatory diseases like rheumatoid arthritis, Crohn's disease, etc., it has failed to be effective in sepsis. Failure of the anti-TNF antibody strategy can be attributed to the reason that it is ineffective when administered after the acute expression of the cytokine. TNF production increases during sepsis and it correlates with acute development of septic shock. Hence, it is a good indicator of septic shock in critically ill patients. But, its levels do not correlate with characteristic slow progression of severe sepsis.

4.3.1.1   HIGH MOBILITY GROUP BOX PROTEIN-1 (HMGB1)

High Mobility Group Box Protein-1 (HMGB1) functions as an extracellular proinflammatory cytokine, produced by stimulated macrophages, monocytes, and dendritic cells. It is a nuclear DNA binding protein. It acts as a mediator of sepsis by binding to TLR4, which mediates HMGB1-dependent activation of macrophage cytokine release. Studies have reported significantly lower levels of HMGB1 in sepsis non-survivors compared with survivors. Unlike early cytokines IL1 and IL6, which rise in minutes after stimulation, HMGB1 is secreted by macrophages 20 h after the activation and serum HMGB1 is detected in 20–72 h after the onset of the disease symptoms[9] indicating HMGB1 as a late mediator of sepsis. Anti-HMGB1 antibodies provide a promising therapeutic approach. Studies involving animal models of anti-antibody treatment showed 55% survival. There are other studies which show that the inhibition of HMGB1 improves survival in experimental models of sepsis. *In vivo* experiments revealed acetylcholine as a physiological inhibitor of HMGB1 release by macrophages, and that nicotinic agonists may prove to provide therapeutic treatment of sepsis.[10]

4.3.1.2   MONOCYTE CHEMOATTRACTANT PROTEIN (MCP)-1

The *chemokine (C-C motif) ligand 2* (CCL2) or *monocyte chemotactic protein 1* (MCP1) is a small cytokine that belongs to the chemotactic cytokines (CC) chemokine family. Inflammation by recruiting monocytes, memory T cells, and dentritic cells to the sites of inflammation is mediated by CCL2. Associations between MCP-1 concentrations and disease status were evaluated statistically, where increased serum levels of MCP-1 were reported to be useful in assessing disease severity in critically ill dogs. Studies suggested a protective role of MCP-1/CCL2 in human sepsis, where it promotes the balance between anti-inflammatory and proinflammatory responses to infection as a positive regulator of IL-10 and a negative regulator of the proinflammatory cytokine macrophage migration inhibitory factor, indicating the immunomodulatory role for MCP-1/CCL2 in sepsis and that the absence of MCP-1/CCL2 reduces the survival rate in sepsis condition.[11]

4.3.2 ACUTE PHASE REACTANTS

Local or systemic disturbances in homeostasis of an organism caused by infection, tissue injury, surgery, trauma, or immunological disorders lead to the activation of a series of systemic reactions, which correspond to acute phase response (APR). Within a few hours after infection, the pattern of protein synthesis by the liver is drastically altered resulting in an increase or decrease in certain normal blood proteins. The APPs are mainly the proteins which are released by the hepatocytes in stimulation to cytokines, interleukins (IL), i.e., IL-1, IL-2, IL-6, tumor necrosis factor—alpha, and glucocorticoids. Documenting protein changes expressed in response to sepsis may prove to be a promising approach to elucidate pathophysiological, diagnostic, therapeutic, and prognostic aspects in this condition with a purpose of applying them to clinical practice.

4.3.2.1 C-REACTIVE PROTEIN

C-reactive protein (CRP) is an acute-phase protein released by the liver after the onset of inflammation or tissue damage. During infections, CRP is known produce both proinflammatory and anti-inflammatory effects. Studies report C-reactive protein as a clinical marker frequently used to assess the presence of infection and sepsis[12] and therefore, it is often used as a comparator in diagnostic studies. Pathogens may be recognized and adhered to by CRP; it may also adhere to damaged cells and mediate their elimination through interactions with inflammatory cells and mediators. C-reactive protein also prevents the adhesion of neutrophils to endothelial cells, inhibits superoxide production, and increases IL-1 receptor antagonist (IL-1ra) production.

C-reactive protein (CRP) is reported to rise up to 1000-fold in the blood in response to inflammation and infection.[12] Studies have shown CRP to be highly sensitive with values of 30 to 97.2%[13] or highly specific with values of 75–100% for infection. It was reported as a prognostic marker in a study relating to the severity of community acquired pneumonia, when using a CV (cut off value) of 110 mg/l. It is known to distinguish bacterial from fungal infections and is a relevant indicator of its own for viral infections, where levels above 100 mg/l are indicative of bacterial infection and elevated levels below 100 mg/l are indicative of fungal infection.[13,14] In spite of its applications as a marker for infection and other illnesses, it is still unclear if CRP can be used to distinguish Gram-positive from negative bacterial infections.

4.3.2.2   PROCALCITONIN

(PCT) is a precursor of the hormone calcitonin synthesized physiologically by thyroid C cells as 116 amino acid polypeptide. The PCT levels in the serum under normal conditions are low (0.033–0.1 ng/mL)[15] and rise above normal physiological conditions under infection where bacterial infections appear to cause the highest rises in PCT with lower or negligible rises in localized, viral, and intracellular bacterial (e.g., *Mycoplasma pneumoniae*) infections. Gram-negative bacterial infections are known to cause a higher rise in PCT levels than Gram-positive bacteria where, the serum PCT levels start to rise at 4 h after the onset of systemic infection, and peak between 8 and 24 h. Persistently high levels of PCT are indicative of continual presence of infection in sepsis subjects. Assicot et al. in 1993, first reported that higher levels of serum procalcitonin correlated with severity of sepsis, presenting PCT as a potential biomarker for sepsis.[16] It was found that PCT was produced mainly in response to gram negative bacterial LPS (endotoxin) or to mediators like IL-6, IL-1β, TNF released in response to bacterial infections. Reports indicated a strong correlation between PCT and severity of bacterial infection, suggesting it to be a biomarker for diagnosis and management of infection and sepsis as CRP.[17] Procalcitonin has been shown to have a high sensitivity of 74.8–100%, and specificity values of 70–100%.[18] Similar to the case of CRP, it is unclear if PCT can be used to distinguish between Gram-positive and Gram-negative bacterial infections, but is known to distinguish fungal and viral infections from bacterial infections, where the PCT levels are reported to remain at low levels during viral infection.[15] Fungal infections tend to cause mild elevations in PCT concentration compared with the levels seen in bacterial infections. Procalcitonin and CRP are the ideal biomarkers approved for clinical diagnosis and sepsis management. But, early diagnosis is still a focus of sepsis research; hence, there is a need to find a better biomarker or a combination of markers which can be used for early diagnosis and patient specific treatment in sepsis.

4.3.2.3   SERUM AMYLOID A (SAA)

Serum Amyloid A (SAA) is the precursor for Amyloid-A, which forms the bulk of fibrils in one form of secondary amyloidosis. Six different isoforms of SAA of pI 6.0, 6.4, 7.0, 7.4, 7.5, and 8.0 were identified. Three acute-phase SAA isoforms have been reported in mice called SAA1, SAA2, and SAA3. Among these, SAA1 and SAA2 are expressed and induced principally in the liver during inflammation and are induced up to a 1000-fold in mice under acute inflammatory conditions following exposure to bacterial LPS via the proinflammatory cytokines IL-1, IL-6, and TNF-α. Serum Amyloid A has shown to have immunomodulatory effect in the regulation of immune response to T-dependent antigens, production of prostaglandin-E2, and fever response. It has been reported that CRP and SAA have displayed a similar pattern in most inflammatory diseases, reaching a maximum serum concentration in about 24 hours after the inflammatory process sets in and slowly decreasing.[19] The liver produces CRP by in response to tissue injury or infection and is commonly used as a marker of an acute inflammatory state, where its plasma concentration has been reported to parallel the clinical course of infection and the fall of the protein level indicates the resolution of infection. The earliest and highest increase rate of all acute-phase proteins is shown by SAA, including CRP. Some authors have reported that SAA appears to be a clinically useful marker of inflammation in bacterial or viral infection, like CRP. It has been proved that SAA is a be potential marker in studies involving veterinary sepsis;[20] its relevance as an inflammatory marker superior to CRP in human sepsis can be further evaluated.

4.3.2.4   LIPOPOLYSACCHARIDE-BINDING PROTEIN (LBP)

Another acute phase protein that has been proved of its potential as a useful biomarker specific to bacterial infections is lipopolysaccharide binding protein. During acute phase response, LBP concentration increases because of transcriptional activation of the LBP gene mediated by IL-1 and IL-6. The LBP then binds to Gram-negative LPS to form LPS-LBP complex and activates cluster differentiation 14 (CD14) and TLR mediated activation of mitogen-activated protein kinase and nuclear factor κB pathways. Human lipopolysaccharide (LPS)-binding protein (LBP) is a serum glycoprotein synthesized by hepatocytes and intestinal epithelial cells. Its levels are known to increase in bacterial and fungal infections, whereas its levels showed no significant change in viral infections.[21] The LBP, like CRP did not distinguish between Gram-positive and Gram-negative infections. The normal physiological level of LBP in serum is 5 to 10 µg/ml; LBP levels increase during acute-phase reaction, reaching peak levels of up to 200 µg/ml. The LBP has been reported as a better biomarker than PCT, CRP, and soluble CD14 in determining the severity of infection in critically ill neonates and children.[21] Studies by Thomas Gutsmann et al. have shown concentration dependent dual role of LBP in the pathogenesis of Gram-negative sepsis, where low concentrations of LBP enhance the LPS-induced activation of mononuclear cells (MNC). On the other side, the acute-phase rise in LBP concentrations inhibits LPS-induced cellular stimulation.[22] Though LBP was considered as a potential biomarker for infection in neonates, certain studies showed that it could only moderately discriminate patients without infection from patients with severe sepsis in a surgical ICU and failed in patients with sepsis without organ dysfunction.[23] Hence, its role as a potential biomarker for early clinical diagnosis of sepsis in critally ill patients has to be evaluated further.

4.3.3   ACUTE PHASE PROTEINS AS MARKERS

APR leads to the rise in serum globular proteins, which are grouped into alpha 1 globulins, alpha 2 globulins, beta globulins secreted by the liver and gamma globulins. Here, we discuss the possible role and pattern of expression of alpha 1 globulins (alpha-1-antitrypsin, serum amyloid A, alpha-1-acid glycoprotein) and alpha 2 globulins (haptoglobin, transthyretin) which include some of the major acute phase proteins identified as potential markers of sepsis. Determination of APPs can help in monitoring the health of individual subjects especially when several acute phase variables are combined in an index. Well-chosen combinations of variables (which may differ for various species) result in a nutritional and acute phase indicator (NAPI), where both positive and negative acute phase proteins have to be evaluated. The acute phase reaction offers an effective mechanism that can be included in future systems for assessing health in human patients. Alpha 1 acid glycoprotein and transthyretin have been extensively studied and identified as markers for severity of sepsis in several studies.

4.3.3.1   ALPHA 1 ACID GLYCOPROTEIN

Identified originally as orosomucoid (abbr. ORM), it is one of the mucoproteins of human plasma and belongs to the immunoglobin superfamily. Alpha

1 acid glycoprotein (AGP) is one of the plasma positive acute phase proteins mainly secreted by the hepatocytes in the liver and its serum concentration was found to increase in response to systemic tissue injury, inflammation, infection, or cancer; t is shown to be a potential drug binding protein of plasma. In a study involving elderly subjects, both AGP and CRP showed an increased expression in response to sepsis, whereas only AGP independently correlated with other parameter for predicting mortality.[24] Recent study pointed AGP as a novel biomarker and independent predictor of 96 hour mortality in sepsis subjects, where non-survivors had lower AGP levels when compared with survivors at the time of admission.[25] These studies provide the hypothesis that AGP may be superior than C-reactive protein and PCT to study the progress of sepsis and mortality.

4.3.3.2   TRANSTHYRETIN

Known as Prealbumin (PAL), it is a thyroxine-binding protein which is a negative acute phase protein that functions also as a transporter of vitamin A by forming a complex with retinol-binding protein. It also acts as an indicator of nutritional status and responds quickly to low energy intake making it a sensitive indicator of protein deficiency, liver disease, and acute inflammation. Earlier studies indicated that transthyretin values remained lower during the recovery phase in patients who died than in patients who survived. In addition, persistently low transthyretin values were associated with sepsis in the survivors. Finally, transthyretin levels were only slightly dependent on the extent of the burn injury. These results emphasize the interest of transthyretin monitoring in patients with burn injuries.[26]

4.3.4   NEUROPEPTIDES AND THEIR ROLE AS SEPSIS MARKERS

Several neuropeptides have been found to be associated with inflammation, most of them being G-coupled receptors like vasoactive intestinal peptide (VIP), alpha-melanocyte-stimulating hormone (alpha-MSH), urocortin, adrenomedullin, cortistatin, and ghrelin which are produced by immune cells, whereas few are orphan receptors like noceceptin. Though the exact function of neuropeptides in pathophysiology and regulation of sepsis needs to be investigated further, their release by immune cells in response to stress provides evidence for the possible interlink between immune and neuro-endocrine systems. Anti-inflammatory activities of neuropeptides and their receptors have been reported earlier,[27] where they block the inflammatory mediators like cytokines and chemokines mostly via activation of cyclic andenosine monophosphate (cAMP)/protein kinase A (PKA) and down-regulation of transcription factors.

Noceceptin is a 17 amino acid neuropeptide, realeased by the immune cells in response to inflammation. Studies have investigated its potential role as a sepsis biomarker for early diagnosis. *In vivo* experiments involving human and rat models showed an increase in noceceptin levels in inflammation, and that targeted blocking of noceceptin receptor pathway may be a diagnostic tool to control inflammation as described by Thomas et al.[28] Studies have reported that Plasma Nociceptin/orphanin FQ (N/OFQ) and Serum SP concentrations were increased in critically ill patients with sepsis, where the results revealed significant high levels of N/OFQ and SP levels in non-survivors compared with the other group of survivors (N/OFQ $p < 0.031$) (SP $p < 0.001$). The levels were observed to be significantly high in the cases near death (non-survivors) compared with the survivors ($p < 0.012$) near recovery.[29]

4.3.5 RECENT BIOMARKERS

4.3.5.1 CD14 RECEPTOR MARKERS AND PRESEPSIN

CD14 is a 55-kDa glycosyl phosphatidyl inositol-anchored protein lacking a cytoplasmic domain, expressed on macrophages and monocytes, which serves as a receptor for lipopolysaccharide binding protein (LPS-LPB) of microorganisms and a potential sepsis biomarker. It exists either in an anchored membrane form (mCD14) or in a circulating soluble form (sCD14). During inflammation, plasma protease activity generates CD14 fragments; among the fragments, the soluble CD14 subtype presepsin (SCD14–ST) fragment is recognized as a sepsis marker called presepsin. It is a soluble CD14 subtype which is a 13 kDa truncated N terminal fragment of 64 amino acid residues. The body's response to bacterial infection, leads to an increased production of presepsin via proinflammatory cascade. The efficiency of presepsin as a marker for early diagnosis has been evaluated by several studies comparing presepsin and the available markers of sepsis like PCT, where presepsin was found to be a better predictor of survival, because of its higher concentration in non-survivors of sepsis. Presepsin was reported to distinguish between severe sepsis and SIRS, and was found to be superior than PCT in the identification of sepsis outcome.[30] Recently, in a study comparing the activities of PCT, presepsin, and interleukin-6 (IL-6), it was found that presepsin was found to be a useful predictor of bacterial and non-bacterial induced sepsis, where presepsin (91.9%) was reported to be more sensitive than blood culture (35.4%) in diagnosing sepsis[31] having a cut-off of 600 pg/mL when compared with other markers. Presepsin has a higher sensitivity (80.3%) and specificity (74.5%) with a cut-off of 399.8 pg/ml in the diagnosis of sepsis when compared with other biomarkers like PCT (89.9%), IL-6 (88.9%), and blood culture (35.4%), indicating its potential as a new biomarker, and predictor for the prognosis of sepsis. Though presepsin can be significantly considered as an early diagnostic marker of sepsis, further studies on the clinical values of presepsin are needed.

4.3.5.2   SOLUBLE UPAR

The urokinase-type plasminogen activator (uPAR) is a protease receptor expressed on various cell types including neutrophils, lymphocytes, monocytes/macrophages, endothelial, and tumor cells, where they mediate several immunological functions. The uPAR was first cloned in 1990 and in 1991, Ploug *et al.* identified its soluble form (soluable form of urokinase type plasminogen receptor (suPAR)).[32] Three forms (I-III, II-III, and I) of uPAR have been identified, where its soluable form suPAR can be found in the blood and other organic fluids in all individuals. In healthy adults, the median value of suPAR has been cited as 1.5 ng/ml (range: 1.2–1.9 ng/ml, N = 44), or 2.6 ng/ml (range: 1.5–4.0 ng/ml, N = 31), which is observed to be elevated in sepsis conditions. Higher serum suPAR concentrations were reported in critically ill patients at the time of admission than healthy controls. The area under receiver operating curve (AUROC) for suPAR when compared with CRP, PCT, and TNF in a study involving 273 critically ill patients, showed suPAR AUROC value for prediction of sepsis as 0.62, compared with 0.86 for CRP and 0.78 for PCT.[33] The optimum diagnostic cut-off value was 3.06 ng/ml, with AUROC of 0.94 in the study conducted on 100 patients with Crimean-Congo hemorrhagic fever (CCHF),[34] where serum suPAR levels were significantly higher in patients with infection than in the healthy controls. In another study involving critically ill sepsis patients with a cut-off value of 5.5 ng/ml, suPAR was reported to have a sensitivity of 75% and specificity of 72% for diagnosing sepsis.[35] High levels of suPAR have been widely demonstrated to correlate with the survival in critically ill subjects, supporting its value as a prognostic biomarker in various cohorts of infected patients; however, several studies have also reported that values greater than 10 ng/ml may be predictive of death.

4.3.5.3   SOLUBLE TREM-1

Another potential biomarker is Soluble Triggering Receptor Expressed on Myeloid cells-1 (sTREM-1). It may be released into the body fluids upon the upregulated expression of TREM-1. Significantly elevated plasma soluble triggering receptor levels have been reported in sepsis patients compared with systemic inflammatory response syndrome patients, as it is not upregulated during non infectious inflammatory diseases, a property not found in other biomarkers such as PCT. It has also been found that septic shock patients have high levels of serum sTREM-1 that correlate with the severity of infection, and sTREM-1 was reported to show positive correlation with the Sequential Organ Failure Assessment (SOFA) score.[36] This suggests that sTREM-1 may be a valuable diagnostic indicator for making distinctions between infectious and non-infectious diseases. In neonates, sTREM-1 was reported to have a potential to provide an excellent predictive value for septic shock/death. In a pilot study conducted on 73 septic shock subjects, the receiver operative characteristic (ROC) curve for a proposed sTREM-1 cut-off value of 300 pg/mL exhibited an area under the curve of 0.884 (95% CI = 0.73–1.0; p < 0.0001), with a sensitivity of 0.78 (95% CI = 0.46–0.94), and specificity of 0.97 (95% CI = 0.92–0.99). Urine sTREM-1 has been shown to be more sensitive than WBC counts, serum CRP, and serum PCT for the early diagnosis of sepsis, as well as for dynamic assessments of progress of severity and predicting mortality.[37]

4.3.5.4   NEUTROPHIL GELATINASE-ASSOCIATED LIPOCALIN (NGAL)

Neutrophil Gelatinase Associated Lipocalin (NGAL) is a 25KDa protein belonging to a unique family called lipocalin. Its secretory nature and small size makes it an ideal biomarker. The NGAL is up regulated in conditions like sepsis and is not categorized as an acute phase protein, which might have pro- or anti-inflammatory action whose function is a question mark. Moreover, NGAL emerged as an excellent biomarker for kidney injury (AKI). It is supposed that sepsis contributes about 30–50% of AKI in critically ill patients.[38] However, NGAL expression varies in patients with septic

AKI and non septic AKI which makes a controversial statement about NGAL being the marker of sepsis or AKI.[38] Recent study suggests a constant up regulation of NGAL for 24 hours in patients/subjects under observation and positive correlation of cytokines IL6, IL8, TNFα, and MCP1 explains higher levels of NGAL in septic AKI than in non septic AKI.[39] A clear case study should exert the actual molecular basis of NGAL in sepsis which will lay a proper foundation for NGAL to be an ideal biomarker for sepsis and to follow a profound analysis of patients entering into the critical phase.[39]

4.3.5.5   MICRO RNA (MI RNA)

The mi RNA as a biomarker of sepsis accounted several studies in the lime-light as they are small RNA molecules, 22 nucleotides in size and control high level cellular activities. Dissipating its use as a biomarker Jie Huang et al. identified 10 novel and reliable mi RNA biomarkers based on pathway analysis.[40] Wang HJ et al. compared few mi RNA among patients with mild sepsis, severe sepsis, septic shock, and normal controls. They found that miR-223 were higher in sepsis patients than in the normal controls and also studied that a few mi RNAs were lower in sepsis patients when compared with the normal controls.[41] A wide study of the genome revealed miR-297 and miR-574-5p as differently expressed miRNA between survival and non survival; miR-297 scored more for survival from sepsis and miR-574-5p expression lead to death from sepsis.[42] Further studies are being conducted on mi RNA as a signature of sepsis which will allow rapid diagnosis among patients in the ICU.

4.3.5.6   PRO ADM

Complete protein Adrenomedullin (ADM) is a hormone active in immune, metabolic, and vasodilation. Its pro hormone is active and more stable from ADM and is found in patients with sepsis. Pro ADM has a prognostic value in detecting patients with severe sepsis from pneumonia. Studies by Hagag et al. concluded that serum levels of pro ADM was higher in neonatal non survivals than sever sepsis, which in turn was higher in sepsis than the control neonatals. This was compared with the levels of antithrombin which were exactly the reverse from pro ADM.[43] Kang et al. studies also stated that pro ADM plasma concentration was significantly increasing among patients with septic shock to sepsis and its concentration was much higher when compared with other biomarkers standing as a good sepsis prognostic marker and risk stratificator.[44]

4.4   CONCLUSION

In this review, we have discussed the utility of a novel and emerging sepsis marker, which has a better diagnostic and prognostic value, and have been effective in predicting the survival of the sepsis patients. Clinical management of sepsis requires prompt laboratory diagnosis and devising effective patient management strategies, apart from antimicrobial therapy. Although many laboratory biomarkers are available for the diagnosis of sepsis, only a few markers have proven to be beneficial in differentiating sepsis due to infection and sepsis of non-infectious origin and in early identification of the degree of severity of disease to increase the chances of survival. Of the available markers, only a few have prognostic value. Further research on large scale multicentric and comparative studies in different geographical areas and patient groups is required to confirm the utility of these potential sepsis biomarkers.

KEYWORDS

- **Biomarkers**
- **CRP**
- **Lipopolysaccharide**
- **Sepsis**

REFERENCES

1. Todi, S. Sepsis: New horizons. *Indian J. Crit. Care Med.* **2010,** *14* (1), 1–2.
2. Martin, G. S.; Mannino, D. M.; Eaton, S.; and Moss, M. The epidemiology of sepsis in the United States from 1979 through 2000. *New Engl. J. Med.* **2003,** *348* (16), 1546–1554.
3. Wang, P.; Yang, Z.; He Y.; and Shu, C. Pitfalls in the rapid diagnosis of positive blood culture. *Rev. Med. Microbiol.* **2010,** *21* (3), 39–43.
4. Doherty, M.; Wallis, R. S.; Zumla, A.; and WHO Tropical Disease Research. Biomarkers for tuberculosis disease status and diagnosis. *Curr. Opin. Pulm. Med.* **2009,** *15* (3), 181–187.

5. Hynninen, M.; Valtonen, M.; Markkanen, H.; Vaara, M.; Kuusela, P.; Jousela, I.; Piilonen, A.; and Takkunen, O. Interleukin 1 receptor antagonist and E-selectin concentrations: A comparison in patients with severe acute pancreatitis and severe sepsis. *J. Crit. Care* **1999,** *14,* 63–68.

6. DiPiro, J. T.; Howdieshell, T. R.; Goddard, J. K.; Callaway, D. B.; Hamilton, R. G.; and Mansberger, A. R. Jr. Association of interleukin-4 plasma levels with traumatic injury and clinical course. *Arch. Surg.* **1995,** *130,* 1159–1162.

7. Gogos, C. A.; Drosou, E.; Bassaris, H. P.; and Skoutelis, A. Pro- versus anti-inflammatory cytokine profile in patients with severe sepsis: A marker for prognosis and future therapeutic options. *J. Infect. Dis.* **2000,** *181* (1), 176–180.

8. Wang, C. H.; Gee, M. J.; Yang, C.; and Su, Y. C. A new model for outcome prediction in intra-abdominal sepsis by the linear discriminant function analysis of IL-6 and IL-10 at different heart rates. *J. Surg. Res.* **2006,** *132,* 46–51.

9. Ombrellino, M.; Wang, H.; Ajemian, M. S.; Talhouk, A.; Scher, L. A.; Friedman, S. G.; and Tracey, K. J. Increased serum concentrations of high-mobility-group protein 1 in haemorrhagic shock. *Lancet* **1999,** *354,* 1446–1447.

10. Wang, H.; Yu, M.; Ochani, M.; Amella, C. A.; Tanovic, M.; Susarla, S.; Li, J. H.; Wang, H.; Yang, H.; Ulloa, L.; Al-Abed, Y.; Czura, C. J.; and Tracey, K. J. Nicotinic acetylcholine receptor alpha 7 subunit is an essential regulator of inflammation. *Nature* **2003,** *421,* 384–388.

11. Gomes, R. N.; Teixeira-Cunha, M. G. A.; Figueiredo, R. T.; Almeida, P. E.; Alves, S. C.; Bozza, P. T.; Bozza, F. A.; Bozza, M. T.; Zimmerman, G. A.; and Castro-Faria-Neto, H. C. Bacterial clearance in septic mice is modulated by MCP-1/CCL2 and nitric oxide. *Shock (Augusta, Ga)* **2013,** *39* (1), 63–69.

12. Barochia, A. V.; Cui, X.; Vitberg, D.; Suffredini, A. F.; O'Grady, N. P.; Banks, S. M.; Minneci, P.; Kern, S. J.; Danner, R. L.; Natanson, C.; and Eichacker, P. Q. Bundled care for septic shock: An analysis of clinical trials. *Crit. Care Med.* **2010,** *38,* 668–678.

13. Bernard, G. R.; Vincent, J. L.; Laterre, P. F.; LaRosa, S. P.; Dhainaut, J. F.; Lopez-Rodriguez, A.; Steingrub, J. S.; Garber, G. E.; Helterbrand, J. D.; Ely, E. W.; Fisher, C. J. Jr.; and PROWESS study group. Efficacy and safety of recombinant human activated protein C for severe sepsis. *N. Engl. J. Med.* **2001,** *344,* 699–709.

14. Brunkhorst, F. M.; Heinz, U.; and Forycki, Z. F. Kinetics of procalcitonin in iatrogenic sepsis. *Intensive Care Med.* **1998,** *24,* 888–889.

15. Becker, K. L.; Snider, R.; and Nylen, E. S. Procalcitonin assay in systemic inflammation, infection, and sepsis: Clinical utility and limitations. *Crit. Care Med.* **2008,** *36* (3), 941–952.

16. Assicot, M.; Gendrel, D.; Carsin, H.; Raymond, J.; Guilbaud, J.; and Bohuon, C. High serum procalcitonin concentrations in patients with sepsis and infection. *Lancet* **1993,** *341,* 515–518.

17. Gogos, C. A.; Drosou, E.; Bassaris, H. P.; and Skoutelis, A. Pro-versus anti-inflammatory cytokine profile in patients with severe sepsis: a marker for prognosis and future therapeutic options. *J. Infect. Dis.* **2000,** *181,* 176–180.

18. Enguix, A.; Rey, C.; Concha, A.; Medina, A.; Coto, D.; and Dieguez, M. A. Comparison of procalcitonin with C-reactive protein and serum amyloid for the early diagnosis of bacterial sepsis in critically ill neonates and children. *Intensive Care Med.* **2001,** *27* (1), 211–215.

19. Cicarelli, L. M.; Perroni, A. G.; Zugaib, M.; de Albuquerque, P. B.; and Campa, A. Maternal and cord blood levels of serum amyloid A, C-reactive protein, tumor necrosis

factor-alpha, interleukin-1beta, and interleukin-8 during and after delivery. *Mediators Inflamm.* **2005,** *2,* 96–100.

20. Christensen, M. B.; Langhorn, R.; Goddard, A.; Andreasen, E. B.; Moldal, E.; Tvarijonaviciute, A.; Kirpensteijn, J.; Jakobsen, S.; Persson, F.; and Kjelgaard-Hansen, M. Comparison of serum amyloid A and C-reactive protein as diagnostic markers of systemic inflammation in dogs. *Can. Vet. J.* **2014,** *55* (2), 161–168.

21. Pavcnik-Arnol, M.; Hojker, S.; and Derganc, M. Lipopolysaccharide-binding protein, lipopolysaccharide, and soluble CD14 in sepsis of critically ill neonates and children. *Intensive Care Med.* **2007,** *33* (6), 1025–1032.

22. Gutsmann, T.; Muller, M.; Carroll, S. F.; Mackenzie, R. C.; Wiese, A.; and Seydel, U. Dual role of lipopolysaccharide (LPS)-binding protein in neutralization of LPS and enhancement of LPS-induced activation of nononuclear cells. *Infect. Immun.* **2001,** 6942–6950.

23. Sakr, Y.; Burgett, U.; Nacul, F. E.; Reinhart, K.; and Brunkhorst, F. Lipopolysaccharide binding protein in a surgical intensive care unit: A marker of sepsis? *Crit. Care Med.* **2008,** *36,* 2014–2022.

24. Henry, O. F, Blacher, J.; Verdavaine, J.; Duviquet, M.; and Safar, M. E. Alpha 1-acid glycoprotein is an independent predictor of in-hospital death in the elderly. *Age Ageing* **2003,** *32* (1), 37–42.

25. Barroso-Sousa, R.; Lobo, R. R.; Mendonça, P. R.; Memória, R. R.; Spiller, F.; Cunha, F. Q.; and Pazin-Filho, A. Decreased levels of alpha-1-acid glycoprotein are related to the mortality of septic patients in the emergency department. *Clinics* **2013,** *68* (8), 1134–1139.

26. Cynober, L.; Prugnaud, O.; Lioret, N.; Duchemin, C.; Saizy, R.; and Giboudeau, J. Serum transthyretin levels in patients with burn injury. *Surgery* **1991,** *109* (5), 640–644.

27. Gonzalez-Rey, E.; and Delgado, M. Anti-inflammatory neuropeptide receptors: New therapeutic targets for immune disorders? *Trends Pharmacol. Sci.* **2007,** *28* (9), 482–491.

28. Thomas, R. C.; Batha, M. F.; Stoverb, C. M.; Lambert, D. G.; and Thompson, J. P. Exploring LPS-induced sepsis in rats and mice as a model to study potential protective effects of the nociceptin/orphanin FQ system. *Peptides* **2014,** *61,* 56–60.

29. Elkholy, M. T.; Hamid, H. S. A.; Ewees, I. E.; and Khan, T. A. Nociceptin/orphanin and substance P concentrations in critically Ill patients with sepsis. *Med. J. Cairo Univ.* **2009,** *77,* 473–477.

30. Spanuth, E.; Ebelt, H.; Ivandic, B.; and Werdan, K. Diagnostic and prognostic value of presepsin (soluble cd14 subtype) in emergency patients with early sepsis using the new assay PATHFAST presepsin [C]. In *21st International Congress of Clinical Chemistry and Laboratory Medicine*; IFCC-World Lab-Euro Med Lab: Berlin, 2011; 15–19.

31. Kibe, S.; Adams, K.; and Barlow, G. Diagnostic and prognostic biomarkers of sepsis in critical care. *J. Antimicrob. Chemother.* **2011,** *66,* 1133–1140.

32. Ploug, M.; Ronne, E.; Behrendt, N.; Jensen, A. L.; Blasi, F.; and Dano, K. Cellular receptor for urokinase plasminogen activator. Carboxyl-terminal processing and membrane anchoring by glycosyl-phosphatidylinositol. *J. Biol. Chem.* **1991,** *266,* 1926–1933.

33. Donadello, K.; Covajes, C.; Scolletta, S.; Taccone, F. S.; Santonocito, C.; Brimioulle, S.; Beumier, M.; Vannuffelen, M.; Gottin, L.; and Vincent, J. L. Clinical value of suPAR, a new biomarker. *Intensive Care Med.* **2011,** *37,* S199.

34. Yilmaz, G.; Mentese, A.; Kaya, S.; Uzun, A.; Karahan, S. C.; and Koksal, I. The diagnostic and prognostic significance of soluble urokinase plasminogen activator receptor in Crimean-Congo hemorrhagic fever. *J. Clin. Virol.* **2011**, *50*, 209–211.

35. Kofoed, K.; Andersen, O.; Kronborg, G.; Tvede, M.; Petersen, J.; Eugen-Olsen, J.; and Larsen, K. Use of plasma C-reactive protein, procalcitonin, neutrophils, macrophage migration inhibitory factor, soluble urokinase-type plasminogen activator receptor, and soluble triggering receptor expressed on myeloid cells-1 in combination to diagnose infections: a prospective study. *Crit. Care* **2007**, *11*, R38.

36. Dimopoulou, I.; Orfanos, S. E.; Pelekanou, A.; Kotanidou, A.; Livaditi, O.; Augustatou, C.; Zervou, M.; Douka, E.; Theodorakopoulou, M.; Karagianni, V.; Douzinas, E.; Armaganidis, A.; and Giamarellos-Bourboulis, E. J. Serum of patients with septic shock stimulates the expression of Trem-1 on U937 monocytes. *Inflamm. Res.* **2009**, *58*, 127–132.

37. Gibot, S.; Cravoisy, A.; Kolopp-Sarda, M. N.; Bene, M. C.; Faure, G.; Bollaert, P. E.; and Levy, B. Timecourse of sTREM (soluble triggering receptor expressed on myeloid cells)-1, procalcitonin, and C-reactive protein plasma concentrations during sepsis. *Crit. Care Med.* **2005**, *33*, 792–796.

38. Srinivasan, G.; Aitken, J. D.; Zhang, B.; Carvalho, F. A.; Chassaing, B.; Shashidharamurthy, R.; Borregaard, N.; Jones, D. P.; Gewirtz, A. T.; and Vijay-Kumar, M. Lipocalin 2 deficiency dysregulates iron homeostasis and exacerbates endotoxin-induced sepsis. *J. Immunol.* **2012**, *189* (4), 1911–1919.

39. Otto, G. P.; Busch, M.; Sossdorf, M.; and Claus, R. A. Impact of sepsis-associated cytokine storm on plasma NGAL during acute kidney injury in a model of polymicrobial sepsis. *Crit. Care* **2013**, *17* (2), 419.

40. Huang, J.; Sun, Z.; Yan, W.; Zhu, Y.; Lin, Y.; Chen, J.; Shen, B.; and Wang, J. Identification of microRNA as sepsis biomarker based on miRNAs regulatory network analysis. *BioMed. Res. Int.* **2014**, DOI: 10.1155/2014/594350.

41. Wang, H. J.; Zhang, P. J.; Chen, W. J.; Feng, D.; Jia, Y. H.; and Xie, L. X. Four serum microRNAs identified as diagnostic biomarkers of sepsis. *J. Trauma Acute Care Surg.* **2012**, *73* (4), 850–854.

42. Ma, Y.; Vilanova, D.; Atalar, K.; Delfour, O.; Edgeworth, J.; Ostermann, M.; Hernandez-Fuentes, M.; Razafimahatratra, S.; Michot, B.; Persing, D. H.; Ziegler, I.; Törös, B.; Mölling, P.; Olcén, P.; Beale, R.; and Lord, G. M. Genome-wide sequencing of cellular microRNAs identifies a combinatorial expression signature diagnostic of sepsis. *PLoS One* **2013**, *8* (10), e75918.

43. Hagag, A. A.; Elmahdy, H. S.; and Ezzat, A. A. Prognostic value of plasma pro-adrenomedullin and antithrombin levels in neonatal sepsis. *Indian Pediatr.* **2011**, *48* (6), 471–473.

44. Kang, F. X.; Wang, R. L.; Yu, K. L.; and Wei, Q. The study on pro-adrenomedullin as a new biomarker in sepsis prognosis and risk stratification. *Zhongguo Wei Zhong Bing Ji Jiu Yi Xue.* (*Chines Crit. Care Med.*) **2008**, *20* (8), 452–455.

CHAPTER 5

ROLE OF COPPER IN MODIFYING CISPLATIN INDUCED CYTOTOXICITY IN YEAST

B. SREEDHAR[1] and B. VIJAYA LAKSHMI[2*]

[1]*UGC-FRP Assistant Professor, Department of Biochemistry, Kakatiya University, Warangal, Telangana – 506009, India.*

[2]*Assistant Professor, Institute of Genetics and Hospital for Genetic Diseases, Osmania University, Hyderabad, Telangana – 500007, India.*

Corresponding author: B. Vijaya Lakshmi. E-mail: bodigavijayasri@ gmail.com

CONTENTS

Abstract 110
5.1 Introduction 111
5.2 Materials and Methods 113
5.3 Results 117
5.4 Discussion 125
5.5 Conclusion 127
Acknowledgments 127
Keywords 127
References 127

ABSTRACT

Cisplatin is an effective anticancer drug, whose efficacy is limited by the development of drug resistance in solid tumors. Although the cause for drug resistance is multifactorial, the most common underlying mechanisms include reduced drug uptake/accumulation, lowered bioavailability of the drug, increased efflux, and altered mitochondrial metabolism. These mechanisms can be manipulated by various biochemical/genetic means, to reinforce the therapeutic efficacy of cisplatin. However, this requires greater understanding of the mechanisms of drug resistance and new agents to overcome resistance. Considering the relation between cisplatin and copper in terms of overlapping transport pathways mediated by CTR1, an attempt was made to understand how the intracellular copper depletion affects the cisplatin uptake into the cells. A simple eukaryote *Saccharomyces cerevisiae*, has been established as a model system for cancer studies, because of the widely conserved family of genes involved in cell cycle progression, proliferation, and apoptosis. In the first investigation, we sought to determine whether copper deprivation affects sensitivity of the yeast to cisplatin. Yeast cultures grown in low copper medium and exposed to bathocuproiene disulfate (BCS) resulted in significant reduction in the intracellular copper, as assessed by atomic absorption spectrophotometry. The low copper medium rendered BY4741 strain hypersensitive to cisplatin (CDDP) as seen from the inhibited growth curves. Yeast grown in low copper medium exhibited ~2.0-fold enhanced cytotoxicity in survival and colony-forming ability compared with copper adequate control cells grown in yeast extract/peptone/dextrose (YPD) rich medium. The effect of copper restriction on CDDP sensitivity appeared to be dependent on the up-regulation of *CTR1*, facilitating enhanced uptake and accumulation of CDDP. In addition, CDDP further lowered copper deprivation-induced changes in *CUP1* metallothionein levels, superoxide dismutase (*SOD*) activity, and glutathione (GSH) levels, which are the general antioxidant molecules. These changes were associated with increased protein oxidation and lipid peroxidation, indicating that the copper-deficient cells are more prone to oxidative stress in the presence of cisplatin. These results, thus suggest that cisplatin cytotoxicity is potentiated under low copper conditions because of the enhanced uptake and accumulation of cisplatin and also in part because of the lowered antioxidant defense and increased oxidative stress imposed by copper deprivation.

5.1   INTRODUCTION

Cisplatin is one of the most successful anticancer agent used worldwide in almost 50% of solid tumor therapies, with annual sales of approximately $500 million (US).[1] Although the initial response rates can be high with cisplatin-based treatment regimen, the clinical effectiveness of the drug is often limited by the onset of intrinsic or acquired resistance. When cisplatin enters the cells, it is potentially vulnerable to cytoplasmic deactivation by many endogenous nucleophiles, such as glutathione (GSH), methionine, metallothionein (MT), protein, and other intracellular components, resulting in lowered bioavailability. The exact level of cisplatin resistance in patients is difficult to define, but at least a twofold resistance is inferred from clinical studies, primarily because responses have been observed when the standard clinical dose of cisplatin is doubled in drug-intensive therapy protocols.[2,3] In general, resistance to cisplatin may be substantially greater, as judged from studies with tumor cell lines established from clinically refractory tumors, which require cytotoxic concentrations as much as 50–100-fold in excess of those needed for sensitive tumor cells.[4–6] Thus, the problem posed by cisplatin resistance appears to be more severe than has been acknowledged in the past. Resistance mechanisms, therefore, arise as a consequence of intracellular changes that either prevent cisplatin from interacting with DNA, interfere with DNA damage signals from activating the apoptotic machinery, or both. Reducing the extent of DNA damage, therefore, increases resistance, and this can occur through changes in drug accumulation, intracellular thiol (-SH) levels, and/or DNA adduct repair.

Development of resistance is caused by reduced drug accumulation because of the decreased uptake/transport and/or increased efflux of the drug. Multidrug resistance-associated (MRP) has been found to be associated with cellular efflux of a variety of drugs' and appears to be important in cisplatin resistance. Cisplatin efflux has centered on *ATP7A* and *ATP7B*, two copper-transporting P-type ATPase genes that are overexpressed in cisplatin-resistant tumor cells. Interestingly, the current known mechanisms do not fully account for the observed resistance of tumors to cisplatin. Hence, additional mechanisms are believed to exist, that may aid in sophisticated understanding of the resistance phenomenon.

The budding yeast *Saccharomyces cerevisiae* is an effective model system to study modulation of drug sensitivity. The direct effects of standard cytotoxic and metabolic antineoplastic agents against fungi have been studied *in vitro* by several investigators.[8] By using transposon mutagenesis approach and screening for loss-of-function mutants for cisplatin resistance, mutation in the *MAC1* (multiple archesporial cells1) gene was identified to exhibit the highest level of cisplatin resistance.[9] Multiple archesporial cells1 (Mac1) consisting of a copper responsive trans-activation domain and DNA binding domain, is activated in response to copper deprivation, leading to the transcription of genes involved in high affinity copper uptake, such as *CTR1(Copper Transporter1)*, *CTR3(Copper Transporter3)*, and *FRE1 (Ferric Reductase 1)*.[10] In *Saccharomyces cerevisiae*, extracellular copper is reduced by cell surface reductases, Fre1 and Fre2, and is transported across the plasma membrane by the high affinity copper transporter Ctr1 or the functionally redundant Ctr3 protein.[11] Inside the cells, the Cox17 (Cytochrome c Oxidase 17) chaperone localized exclusively to the mitochondria facilitates the delivery of copper to the cytochrome c oxidase complex that is required for aerobic respiration.[12] Another copper chaperone Ccs1, provides copper for superoxide dismutase (SOD), an enzyme that protects cells against oxidative stress *via* the dismutation of superoxide to hydrogen peroxide.[13] Copper depletion results in increased DNA binding of Mac1 to copper-responsive element (CuRE) regions upstream of its target genes. Earlier studies have shown that Mac1 is a nuclear protein, suggesting that copper is either assembled co-translationally or is delivered to the nucleus to regulate Mac1.[14] Interestingly, there is virtually no free copper in the cell, as it is almost exclusively associated with either chaperones or copper-containing proteins.[15,16] Deletion of the yeast *CTR1* gene, which encodes a high-affinity copper transporter, resulted in an increased cisplatin resistance and reduced intracellular accumulation of cisplatin. Copper, which causes degradation of the Ctr1 protein, enhanced the survival of the wild-type yeast cells exposed to cisplatin and reduced cellular accumulation of the drug. Interestingly, cisplatin also caused degradation and delocalization of Ctr1 and interfered with copper uptake in yeast, thus suggesting that cisplatin uptake is mediated by the copper transporter Ctr1 in yeast. Subsequent studies confirmed that a functional human Ctr1 is indeed required for cisplatin uptake in mammalian cells and that Cu and Pt compete for transport by Ctr1.[9,17–25] Interactions between copper and platinum thus likely alter intracellular copper homeostasis and influence cisplatin cytotoxicity either way, through the generation of hydroxyl radicals by Fenton's reaction or by the induction of metallothionein, a potential scavenger of hydroxyl radicals.[26–28] Therefore, we hypothesized that copper deprivation may increase cisplatin cytotoxicity by enhanced accumulation of cisplatin and also by lowering metallothionein and SOD activity.

5.2 MATERIALS AND METHODS

5.2.1 MATERIALS

Yeast extract, peptone, and glucose were obtained from HiMedia (Mumbai, India), unless otherwise mentioned. Bacto agar was obtained from Difco, BD Biosciences, NJ. Bathocuproiene disulfonate (BCS) was obtained from Sigma Chemical Co. (St. Louis, MO). Cisplatin was obtained from Dabur Pharmaceuticals, India. All other chemicals were of analytical grade and were obtained from local sources.

5.2.2 GROWTH CONDITIONS OF BY4741

The *Saccharomyces cerevisiae* BY4741 (*MAT*a *his3Δ1 leu2Δ0 met15Δ0 ura3Δ0*) strain used in this study was a generous gift from Dr. Anand Bachchawat, Institute of Microbial Technology (IMTECH), Chandigarh, India. The BY4741 strain contains a wild-type copy of the *CTR1* gene. SD medium without copper (2% glucose, 0.17% yeast nitrogen base without Cu and Fe, 0.5% ammonium sulfate, amino acid, 1 μM $FeCl_3$), but containing 100 μM bathocuproine disulfonic acid disodium salt (BCS) was used as low copper medium (LCM). Yeast cultures were inoculated in standard YPD (1% yeast extract, 2% peptone, and 2% glucose) or LCM overnight at 30 °C and were seeded from single colonies grown on respective solid media. The effective copper concentration of LCM was <50 nM compared with 250 nM in YPD. The growth of experimental cultures was initiated at $A_{600 \, nm}$ = 0.05; the cultures were allowed to grow to $A_{600 \, nm}$ = 1.0 (log phase, 1.5×10^7 cells/mL) before use.

5.2.3 MEASUREMENT OF CELL GROWTH AND CISPLATIN CYTOTOXICITY

The yeast cells (5×10^4 cells/mL) were cultured in YPD or LCM containing various concentrations of CDDP for 2 h. The cell growth was monitored by measuring the absorbance of the culture at 600 nm. Sensitivity to the cytotoxic effect of CDDP was assessed using a colony formation assay. Cultures (1 ml) containing a total of 6×10^6 cells were exposed for 4 h to CDDP at concentrations of 0.1, 0.2, 0.4, 0.8, 1.6, and 3.2 mM, washed once in PBS, resuspended in YPD or LCM, diluted 1:4000–1:6000, and plated onto

100-mm agar plates. After 2 days of growth at 30 °C, the number of colonies was counted manually. The IC_{50} was defined as the drug concentration that reduced the number of colony-forming units to 50% of the value in a control culture not exposed to the drug. Each experiment was repeated three times with duplicate cultures for each drug concentration.

5.2.4   PREPARATION OF CELLS FOR PLATINUM AND COPPER MEASUREMENT

After cisplatin treatment, the cells were washed twice with PBS and lysed with 1% Triton X-100/0.1% sodium dodecyl sulphate (SDS). After a 3 min centrifugation at 15,000 rpm at 4 °C in a microcentrifuge, the supernatant was used to determine protein concentration by the Bio-Rad protein assay and copper/platinum content by atomic absorption spectrophotometry using a Perkin-Elmer Model 1100 atomic absorption spectrophotometer with a graphic furnace system. A volume of 20 μl was introduced into the graphite furnace and the peak area was read during a 5 s atomization step at 2500 °C. The amount of platinum and copper in the samples was determined from a calibration curve prepared by using standard solutions. The readings were normalized to protein concentration.

5.2.5   NORTHERN HYBRIDIZATION

Total RNA was extracted from mid-exponential phase cultures of *S. cerevisiae* strains (1×10^7 cells/ mL) by the SDS-hot phenol method described by Schmidt et al.[29] For Northern blot hybridization studies, 10 μg of *S. cerevisiae* total RNA were fractionated by electrophoresis in a formaldehyde-agarose gel (1.2% w/v) and blotted onto Hybond-membranes. The blots were prehybridised at 42 °C during 4 h in a high stringency solution containing 50% formamide, 1% SDS, 5X SSPE buffer - Sodium chloride, Sodium Phosphate and EDTA (Ambion Inc., Austin, TX), 5X Denhardt's solution, and 100 μg/ml denaturated sheared nonhomologous DNA. Northern hybridization was carried out at 42 °C for 12–14 h in the same solution containing 1×10^7 cpm/ml of [α-^{32}P] dCTP-labelled cDNA probe prepared with the direct 5'-AGTATGTCGATGGGAAGC-3' and reverse 5'-TCATAATGTTTTCCTT CG-3' for *CTR1*, direct (5'-TCAATCATCACATAAAATGTTC-3') and reverse (5'-CGTTTCATTCCCAGAGCAG-3') oligonucleotides for

CUP 1 and direct 5'-ACACGGTATTGTCACCAACTGGG-3' and reverse 5'-AGGACAAAACGGCTTGGATGG-3' for *ACT1*. After hybridization, the blots were washed in a 2X SSPE 1% SDS solution for 20 min at 42 °C and then washed again in a 0.1X SSPE 1% SDS solution for 10 min at 42 °C. The blots were then exposed on scientific autoradiographic imaging film (Kodak) at −80 °C for 24 h.

5.2.6 DETERMINATION OF GSH AND GSSG

For the determination of glutathione, the samples were suspended in 10% TCA and vortexed for 10 min to extract glutathione. After centrifugation at 10,000 g for 30 min, reduced glutathione (GSH) was estimated in a supernatant according to the method of Beutler et al.[30] This method is based on the reaction of the –SH groups of the reduced glutathione with 5,5'-dithiobis-2-nitrobenzoic acid. The total glutathione content was estimated by the method of Habeeb.[31] The levels of oxidized glutathione (GSSG) were calculated from the difference between the values of total glutathione and reduced glutathione (GSH), and the results were expressed as µg/mg protein.

5.2.7 EVALUATION OF LIPID PEROXIDATION

Lipid peroxidation was quantified by thiobarbituric acid (TBA)-reactive substances (TBARS). According to the method described by Buege and Aust,[32] TBARS was determined. At certain intervals, 0.5 ml of samples of the cell suspension were removed and added to 1 ml of TBA reagent (15% wt./vol. TCA and 0.375% wt./vol. TBA in 0.25 M HCl). Addition of the reagent terminated lipid peroxidation and initiated the assay. The samples were heated for 15 min in a boiling water bath and after cooling, were centrifuged at 10,000 g for 5 min to remove cell debris. Absorbance of the supernatant at 535 nm was measured by using a Shimadzu UV-1240 spectrophotometer, against a reference solution comprising 1 ml of TBA reagent with the sample replaced by an equal volume (0.5 ml) of distilled deionized water. The concentrations of TBARS in the samples were calculated using the molar extinction coefficient of malondialdehyde (MDA)–thiobarbituric chromophore 1.56×10^5 M^{-1} cm^{-1} and expressed as nmol of TBARS/mg protein.

5.2.8 QUANTIFICATION OF PROTEIN CARBONYLS

For the assessment of carbonyls, the reaction with dinitrophenylhydrazine was employed.[33] For each determination, samples containing 2–10 mg ml^{-1} of protein were treated with 4 ml of 10 mM dinitrophenylhydrazine in 2.5 M HCl for 1 h at room temperature. One tube, used as the blank, was incubated only with 2.5 M HCl. The reaction was stopped by the addition of 5 ml of 20% TCA. The pellets were washed twice with 3 ml of absolute ethanol/ ethyl acetate (1:1) solution. The protein pellets were finally dissolved in 6 M guanidine hydrochloride and the absorption at 375 nm (dinitrophenyl-hydrazine minus sample blank) was determined. The carbonyl content was calculated using the molar absorption coefficient of aliphatic hydrazones of 22,000 M^{-1} cm^{-1} and expressed as nmol carbonyl/mg of protein.

5.2.9 MEASUREMENT OF INTRACELLULAR OXIDATIVE STRESS

Intracellular ROS was detected by the oxidant-sensitive probe 2',7'-dichlo-rofluorescein diacetate (DCFH-DA, Sigma).[34] According to this procedure, 2',7'-dichlorofluorescein diacetate (DCFH-DA) was added from a fresh 5 mM stock (prepared in ethanol) to a final concentration of 10 mM in 1 ml of yeast cell culture (10^7 cells), then incubated at 28 °C for 20 min. Finally, the cells were cooled on ice, harvested by centrifugation and washed twice with distilled water. The pellet was resuspended in 500 µL of water and 1.5 g of glass beads were added. The cells were lysed by three cycles of 1 min agitation on a vortex mixer followed by 1 min on ice. The superna-tant was obtained after centrifugation at 25,000 rpm for 5 min. After appro-priate dilution with water, fluorescence was measured using a Shimadzu Spectrofluorophotometer (RF-5301PC) with an excitation at 502 nm and emission at 521 nm.

5.2.10 CU, ZN-SOD, AND MN-SOD ACTIVITY

The SOD activity gel assay was conducted as previously described.[35] The cells were homogenized by glass bead agitation briefly. The protein extracts (50 µg) were subjected to non denaturing polyacrylamide gel electropho-resis (PAGE) on 12% gels, and subsequently stained for SOD activity using Nitro Blue Tetrazolium (Sigma) as described previously.[36]

5.2.11   STATISTICAL ANALYSIS

Statistical analysis was performed using GraphPad Prism 4 (GraphPad Software, San Diego, CA, USA). The significance of the results was tested with the unpaired t-test or one-way Analysis of Variance (ANOVA) with Tukey test. For all graphs, bars with different symbols are significantly different from each other at $p < 0.05$.

5.3   RESULTS

5.3.1   EVALUATION OF LOW COPPER MEDIUM FOR INDUCTION OF COPPER-INSUFFICIENCY

The total average concentration of copper in standard YPD medium was 0.3 µM, while that in LCM was less than 0.02 µM. Exposure of cells to copper chelator, BCS in SD medium without copper (LCM) did not alter the growth curve (Figure 5-1A inset), but it showed a lower intracellular copper content (<50 nM) compared with those grown in YPD (>200 nM), confirming copper deprivation.

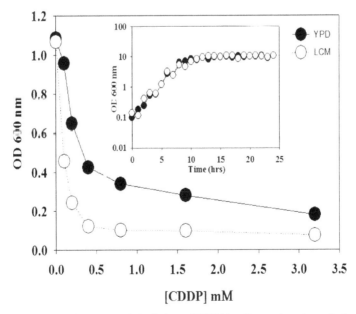

FIGURE 5-1   Effect of copper and cisplatin on BY4741 cell growth and survival.

(A) Determination of dose-ranging inhibition by CDDP at concentrations varying from 0.1 to 3.2 mM in YPD (●) or LCM (○). The cell growth was monitored by measuring the absorbance of the culture at 600 nm. The inset shows the growth curve of BY4741 cells as a linear plot of optical density at 600 nm (OD 600) as a function of time (in hours) in YPD (●) or LCM (○). (B) Survival of *BY4741* cells in YPD and LCM as a function of CDDP concentration as determined using colony-formation assay. Logarithmic-phase cells were exposed for 2 h to different concentrations of CDDP (0.1–3.2 mM) and incubated on YPD or LCM plates at 30 °C for 2 days to allow the formation of colonies. The data are expressed as percentages of colonies formed compared with BY4741 culture grown in YPD and not exposed to CDDP. Each data point represents the mean of three independent experiments with a variation of ±5%, each performed with duplicate cultures for each concentration.

5.3.2   EFFECT OF CDDP ON CELL GROWTH AND CYTOTOXICITY

Two independent methods, namely, growth inhibition and colony formation assays, were used to find out whether copper depletion causes any change in cisplatin cytotoxicity. The effect of varying concentrations of CDDP (0–3.2 mM) on the growth of BY4741 in YPD and LCM was examined over 24 h and is shown in Figure 5-1A. The dose-ranging inhibition curve confirmed that the growth of copper-sufficient and insufficient *S. cerevisiae* cells in cisplatin-free medium was comparable; however, the cells growing in LCM were distinctly more susceptible to cisplatin compared with those growing under copper-sufficient conditions. Cell growth in YPD decreased significantly in the presence of CDDP over 24 h in a dose-dependent manner. Similarly, cell growth decreased further in LCM in the presence of CDDP, suggesting that CDDP inhibits cell growth more in LCM than in YPD.

Colony-formation assay data in Figure 5-1B revealed that cells growing in LCM showed increased sensitivity to cisplatin compared with those growing under copper-sufficient conditions. The IC_{50} for the cells in YPD was 0.76 ± 0.06 µM, whereas for the cells in LCM was 0.32 ± 0.05 µM. Thus, the copper-deficient cells were more than twofold sensitive to CDDP. These results confirm further that copper-deprived cells are more sensitive to CDDP toxicity than copper-sufficient cells.

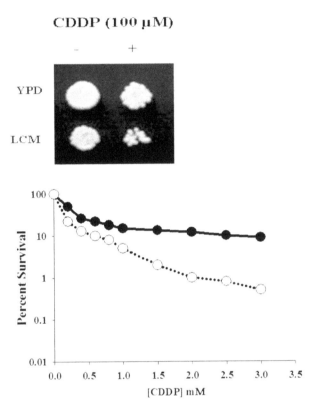

FIGURE 5-1B   Survival of *BY4741* cells in YPD and LCM.

Survival of *BY4741* cells in YPD and LCM as a function of CDDP concentration as determined using colony-formation assay. Logarithmic-phase cells were exposed for 2 h to different concentrations of CDDP (0.1–3.2 mM) and incubated on YPD or LCM plates at 30 °C for 2 days to allow the formation of colonies. Data are expressed as percentages of colonies formed compared with BY4741 culture grown in YPD and not exposed to CDDP. Each data point represents the mean of three independent experiments with a variation of ±5%, each performed with duplicate cultures for each concentration.

5.3.3   EFFECT OF COPPER STATUS ON CDDP UPTAKE AND ACCUMULATION

The cellular uptake and accumulation of platinum as a function of time in YPD and LCM is shown in Figure 5-2. Accumulation of platinum increased with time up to 1 h. The amount of cell-associated platinum concentration increased linearly as a function of time in both YPD and LCM grown cells. However, there was a substantial increase of platinum accumulation in LCM. Based on the slope of the plot, the accumulation of cisplatin in YPD was significantly less when compared with that in LCM. Thus, the uptake and accumulation of platinum increased significantly under low copper conditions.

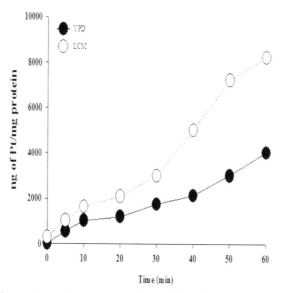

FIGURE 5-2   Copper deprivation results in intracellular CDDP accumulation.

Intracellular CDDP accumulation as a function of time after exposure to 100 µM CDDP concentration in YPD (●) or LCM (○). Each data point represents the mean of three experiments, each performed with duplicate cultures.

5.3.4   EFFECT OF COPPER AND CDDP ON CTR1 AND CUP1 mRNA LEVELS

To investigate the role of CTR1 in increased uptake and accumulation of CDDP, we determined the steady-state mRNA levels encoding *CTR1* by Northern blotting. Figure 5-3 shows that the levels of *CTR1* mRNA were elevated in LCM compared with YPD, indicating an up-regulation of *CTR1* mRNA in copper-deficient cells. Interestingly, the levels of *CTR1* mRNA increased with CDDP treatment in YPD and further increased in LCM. Expression levels of the internal control gene *ACT1*, which encodes actin, were equivalent across all the groups. To validate the copper-dependent changes in the expression, we also monitored the levels of metallothionein *CUP1* mRNA levels that are known to be directly influenced by intracellular copper status. The levels of *CUP1* mRNA decreased significantly in LCM compared with YPD. In the presence of CDDP, the levels of *CUP1* mRNA decreased dramatically in both YPD and LCM. These results demonstrate that up-regulated *CTR1* may be responsible for the increased transport of CDDP observed in LCM.

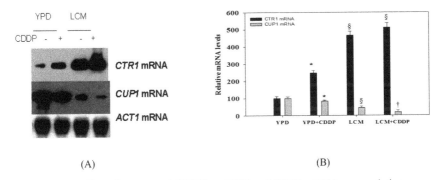

(A)   (B)

FIGURE 5-3   Effect of copper and CDDP on *CTR1* and *CUP1* mRNA accumulation.

BY4741 cultures were grown in YPD or LCM for 12 h. The aliquots were subsequently incubated in the presence of CDDP (100 μM) at 30 °C for 12 h. Fifteen μg of total RNA was prepared and analyzed by Northern hybridization using the radiolabeled *CTR1*, *CUP1*, and *ACT1* oligonucle-otide probes. (B) Densitometric analysis of the *CTR1* and *CUP1* mRNA levels are shown in (A). Relative values of *CTR1* and *CUP1* mRNA are expressed as the percentage of the *ACT1* mRNA value in each lane. The data represent the mean ± standard deviation of at least three independent experiments. Bars with a different symbol are significantly different from each other ($p < 0.05$).

5.3.5   EFFECT OF COPPER AND CDDP ON GSH LEVELS AND SOD ACTIVITY

As can be seen in Figure 5-4, copper deficiency did not alter GSH and GSSG levels. However, GSH levels decreased significantly with CDDP treatment in YPD and more so in LCM. The GSSG levels increased significantly with CDDP treatment in YPD and more so in LCM.

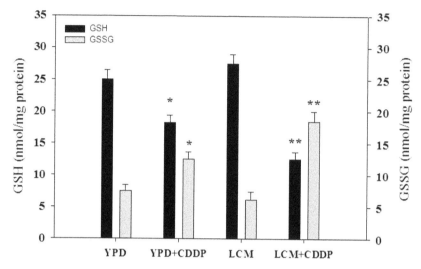

FIGURE 5-4   Levels of Glutathione.

GSH (reduced form), GSSG (oxidized form) of cells grown in YPD and LCM, in the presence or absence of 100 μM CDDP. The data represent the mean ± standard deviation of at least three independent experiments (* p < 0.05 vs YPD; ** p < 0.05 vs LCM).

We examined the enzymatic activity of the well characterized copper dependent enzyme, SOD1, which is reduced in copper deficiency. As shown in Figure 5-5, the activity of SOD1 was significantly reduced in LCM compared with that in YPD. The activity of Cu, Zn-SOD was not affected by CDDP in YPD, but was significantly lower in LCM. However, the analysis of Mn-SOD showed that its activity was similar under YPD and LCM, but reduced after CDDP exposure in LCM alone. Thus, it seems likely that copper directly affects Cu, Zn-SOD activity and Mn-SOD activity which is influenced only after a substantial accumulation of CDDP.

(A) Cells grown in YPD or LCM were treated with 100 μM CDDP for 2 h and total soluble extracts were electrophoresed for SOD gel activity. From each treatment, 50 μg of crude extract was analyzed on a 10% native acrylamide gel. The location of Mn SOD and Cu, Zn-SOD are indicated. (B) Densitometric quantitation of relative levels of SOD are shown in A. Bars of the same type with different symbols are significantly different (p < 0.05).

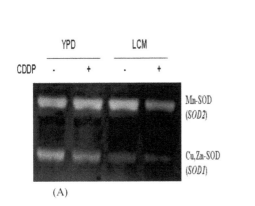

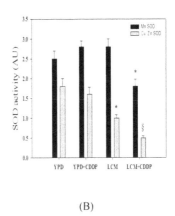

FIGURE 5-5   (A) SOD activity, (B) Densitometric quantitation.

5.3.6   EFFECT OF COPPER AND CDDP ON INTRACELLULAR OXIDATIVE STRESS

To understand if the altered copper status and *CUP1* mRNA levels affected intracellular oxidation differentially in YPD and LCM, we assessed DCF oxidation as a measure of intracellular oxidation. The cells grown in YPD showed no differences in DCF oxidation compared with those grown in LCM. Interestingly, 100 µM cisplatin challenge in LCM showed a dramatic increase in DCF oxidation compared with those grown in YPD, indicating that copper deprivation does indeed promote the intracellular oxidation levels when challenged with cisplatin (Figure 5-6).

Yeast extract/peptone/dextrose (YPD) or LCM grown cells were loaded with 10 µM H_2DCFDA were treated with 100 µM CDDP for 30 min, and intracellular peroxides were measured as an induction of fluorescent intensity with respect to control values in YPD. The results given are a mean of four independent experiments (* indicates $p < 0.05$ vs YPD; # indicates $p < 0.05$ vs YPD+CDDP).

Increased intracellular oxidation is associated with higher levels of oxidative stress markers, such as protein carbonylation and lipid peroxidation. The results show that protein carbonylation (Figure 5-7A) was not different between YPD and LCM. However, CDDP-induced protein carbonylation increased by 78% in YPD and 197% in LCM. Similarly, lipoperoxidation measured as TBARS increased by 75% in YPD and 280% in LCM, as shown in Figure 5-7B.

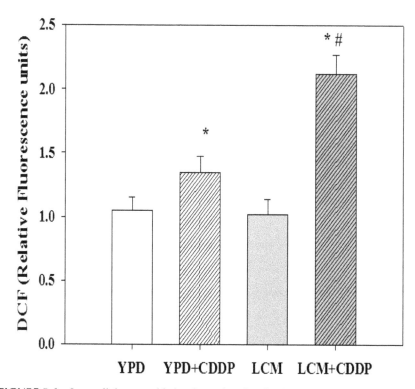

FIGURE 5-6   Intracellular peroxide levels produced under CDDP treatment.

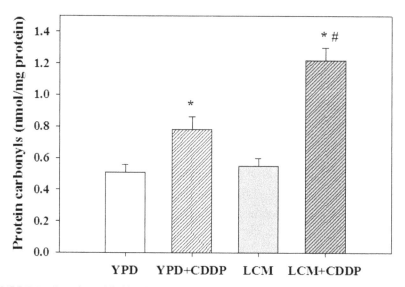

FIGURE 5-7   Protein and lipid oxidation.

(A) Protein carbonyl content, given in nmol of carbonyl/mg of protein, was determined after 4 h of treatment with CDDP. The values summarized here are mean values for three independent experiments. (B) Thiobarbituric acid-reactive substances (TBARS) values are given in nmols/mg protein; these were determined after 4 h of treatment with CDDP. The values summarized here are mean values for three independent experiments.

5.4   DISCUSSION

The aim of the present study was to determine the effect of Cu deficiency on CDDP sensitivity and cytotoxicity in yeast cells. The data show that a significant decrease in intracellular Cu levels can be observed by employing low copper medium, along with BCS, together with a parallel increase in CDDP levels resulting in cytotoxicity gain for CDDP. The decrease in intracellular copper was reflected in a decrement of copper-dependent metallothionein expression (*CUP1*) and Cu, Zn-SOD activity. Maximal *CTR1* expression was observed in the low copper medium compared with copper-sufficient YPD.

The severity of copper deprivation had no effect on the growth curve and cell density, without cisplatin challenge. Cell growth was not affected by CDDP at 0.1 mM in YPD, but decreased to 42% in LCM. Further, the cell growth decreased to 50% and 22% with 0.2 mM CDDP in YPD and LCM, respectively. Following exposure of BY4741 in YPD and LCM to 100 µM CDDP, platinum accumulation was linear over 60 min, but significantly higher in LCM compared with YPD at all time points. These data are consistent with the up regulation of *CTR1* mRNA under copper-deficient conditions and its role in increased uptake of CDDP. Our observations that the depleted pool of intracellular copper by BCS results in up-regulation of CTR1 expression is consistent with the well documented phenomenon that intracellular copper content plays an important role in regulating CTR1 expression. The available information regarding the mechanisms that regulate CTR1 expression in response to copper concentrations suggests that the regulation is mainly controlled by the transcription factor, MAC1. Increased sensitivity to CDDP correlated with an increased expression of CTR1 mRNA. These data extend previous observations that both Cu and CDDP share a common transporter, Ctr1. It is known that the levels of Ctr1 fall in the presence of copper[37] and thus, copper has an inhibitory effect on the uptake of cisplatin.[18,24] Conversely, addition of copper to the medium decreased cellular accumulation of CDDP twofold and increased survival twofold.[9] Thus, it can be understood that the underlying mechanism for such sensitization is the up-regulation and prevention of degradation of Ctr1 in LCM. On the contrary, CDDP treatment in the presence of copper ions results in reduced killing because of the competition for uptake, as seen in our study as well as reported by Ishida et al.[9] Earlier studies using yeast and mouse embryonic fibroblasts deleted for *CTR1* gene demonstrated a decreased sensitivity to CDDP, suggesting the correlation between copper status, Ctr1 levels, and cisplatin sensitivity.

Depletion of copper in our cell model did not alter growth curve, intracellular DCF oxidation, protein carbonylation, and lipid peroxidation, unless challenged with cisplatin. However, it did affect the *CUP1* mRNA levels and Cu, Zn-SOD activity significantly. A significant decrease in both these antioxidant factors can be attributed to enhanced cytotoxicity gain for CDDP in copper depleted cells, as observed from the reduced cytotoxicity in copper-sufficient cells with higher metallothionein levels and Cu, Zn-SOD activity. Yeast cells lacking SOD1 are highly sensitive to oxygen and to agents that lead to oxidative stress, such as paraquat and menadione.[13,28] Our observation that lowered SOD1 activity resulted in higher sensitivity to CDDP which suggests and confirms that CDDP leads to oxidative stress and that copper-sufficient cells are relatively protected against this stress.

Progressive accumulation of oxidatively damaged molecules may result from a decline in the levels of antioxidant defenses. The results clearly suggest that copper deficiency exacerbated the intracellular oxidation upon CDDP challenge. There is now abundant evidence for a link between copper status and metallothionein levels. Metallothioneins are a group of cysteine-rich proteins with antioxidant properties and have the capacity to buffer intracellular Cu. Yeast metallothioneins are encoded by *CUP1* and *CRS5* genes and have been shown to play a role in protecting yeast cells against oxidants, as the oxidant-sensitive phenotype of strains lacking Cu, Zn-SOD can be complemented by the over-expression of metallothionein.[38,24] These data were consistent with unaltered levels of GSH and GSSG in YPD and LCM, but a moderate to dramatic decrease in GSH levels with cisplatin in YPD and LCM, respectively. These data point out that a mere decrease in copper pool inside the cells is not sufficient to pose cytotoxicity, but can predispose cells to be more susceptible to CDDP because of the impaired antioxidant defense. While a few studies have shown that cisplatin can lead to the production of hydroxyl radicals *in vitro*, it is still not clear as to the precise mode of generation of these ROS. The increase in GSSG and decrease in GSH levels induced by CDDP suggests that oxidative stress is involved in cytotoxicity. It is understandable that the cisplatin challenge in copper-deprived cells causes an increase in steady state levels of lipoperoxides and carbonyls that exceeds the metabolic capacity of endogenous antioxidant systems, as they are compromised by copper-deficiency, and lack of metallothionein and SOD activity. In addition, cellular GSH levels have long been known to affect cellular sensitivity to cisplatin. Cisplatin is known to be detoxified by conjugation to GSH and is one of the mechanisms responsible for resistance.

5.5 CONCLUSION

The overall results suggest that copper-deficient cells exhibit a lower antioxidant defense and increased susceptibility to CDDP associated with increased uptake and accumulation of CDDP, that also resulted in an accumulation of oxidized proteins and lipids, exceeding the levels associated in YPD, because of a higher intracellular oxidation.

ACKNOWLEDGMENTS

Authors thank University Grants Commission for the financial assistance given to Dr. B. Vijaya Lakshmi in the form of start-up grant. UGC letter No.F.30-61/2014.

KEYWORDS

- **Apoptosis**
- **Cisplatin**
- **Copper**
- **Copper transporter**
- **SOD**

REFERENCES

1. Wong, E.; and Giandomenico, C. M. Current status of platinum-based antitumor drugs. *Chem. Rev.* **1999,** *99* (9), 2451–2466.

2. Ozols, R. F.; Corden, B. J.; Jacob, J.; Wesley, M. N.; Ostchega, Y.; and Young, R. C. High-dose cisplatin in hypertonic saline. *Ann. Intern. Med.* **1984,** *100* (1), 19–24.

3. Schilder, R. J. and Ozols, R. F. New therapies for ovarian cancer. *Cancer Invest.* **1992,** *10* (4), 307–315.

4. Hagopian, G. S.; Mills, G. B.; Khokhar, A. R.; Bast, R. C.; and Siddik, Z. H. Expression of p53 in cisplatin-resistant ovarian cancer cell lines: Modulation with the novel platinum analogue (1R, 2R-diaminocyclohexane)(trans-diacetato)(dichloro)-platinum(IV). *Clin. Cancer Res.* **1999,** *5* (3), 655–663.

5. Hills, C. A.; Kelland, L. R.; Abel, G.; Siracky, J.; Wilson, A. P.; and Harrap, K. R. Biological properties of ten human ovarian carcinoma cell lines: Calibration in vitro against four platinum complexes. *Br. J. Cancer* **1989,** *59* (4), 527–534.

6. Kelland, L. R.; Barnard, C. F.; Evans, I. G.; Murrer, B. A.; Theobald, B. R.; and Wyer, S. B. Synthesis and in vitro and in vivo antitumor activity of a series of trans platinum antitumor complexes. *J. Med. Chem.* **1995,** *38* (16), 3016–3024.

7. Borst, P.; Evers, R.; Kool, M.; and Wijnholds, J. A family of drug transporters: The multidrug resistance-associated proteins. *J. Natl. Cancer Inst.* **2000,** *92* (16), 1295–1302.

8. Ghannoum, M. A.; Abu-Elteen, K. H.; Motawy, M. S.; Abu-Hatab, M. A.; Ibrahim, A. S.; and Criddle, R. S. Combinations of antifungal and antineoplastic drugs with interactive effects on inhibition of yeast growth. *Chemotherapy* **1990,** *36* (4), 308–320.

9. Ishida, S.; Lee, J.; Thiele, D. J.; and Herskowitz, I. Uptake of the anticancer drug cisplatin mediated by the copper transporter Ctr1 in yeast and mammals. *Proc. Natl. Acad. Sci. U S A* **2002,** *99* (22), 14298–14302. DOI: 10.1073/pnas.162491399

10. Jamison McDaniels, C. P.; Jensen, L. T.; Srinivasan, C.; Winge, D. R.; and Tullius, T. D. The yeast transcription factor Mac1 binds to DNA in a modular fashion. *J. Biol. Chem.* **1999,** *274* (38), 26962–26967.

11. Dancis, A.; Haile, D.; Yuan, D. S.; and Klausner, R. D. The *Saccharomyces cerevisiae* copper transport protein (Ctr1p). Biochemical characterization, regulation by copper, and physiologic role in copper uptake. *J. Biol. Chem.* **1994,** *269* (41), 25660–25667.

12. Beers, J.; Glerum, D. M.; and Tzagoloff, A. Purification, characterization, and localization of yeast Cox17p, a mitochondrial copper shuttle. *J. Biol. Chem.* **1997,** *272* (52), 33191–33196.

13. Bermingham-McDonogh, O.; Gralla, E. B.; and Valentine, J. S. The copper, zinc-superoxide dismutase gene of *Saccharomyces cerevisiae*: Cloning, sequencing, and biological activity. *Proc. Natl. Acad. Sci. U S A* **1988,** *85* (13), 4789–4793.

14. Jensen, L. T.; Posewitz, M. C.; Srinivasan, C.; and Winge, D. R. Mapping of the DNA binding domain of the copper-responsive transcription factor Mac1 from *Saccharomyces cerevisiae*. *J. Biol. Chem.* **1998,** *273* (37), 23805–23811.

15. Jungmann, J.; Reins, H. A.; Lee, J.; Romeo, A.; Hassett, R.; Kosman, D.; and Jentsch, S. MAC1, a nuclear regulatory protein related to Cu-dependent transcription factors is involved in Cu/Fe utilization and stress resistance in yeast. *EMBO J.* **1993,** *12* (13), 5051–5056.

16. Rae, T. D.; Schmidt, P. J.; Pufahl, R. A.; Culotta, V. C.; and O'Halloran, T. V. Undetectable intracellular free copper: The requirement of a copper chaperone for superoxide dismutase. *Science* **1999,** *284* (5415), 805–808.

17. Blair, B. G.; Larson, C. A.; Adams, P. L.; Abada, P. B.; Safaei, R.; and Howell, S. B. Regulation of copper transporter 2 expression by copper and cisplatin in human ovarian carcinoma cells. *Mol. Pharmacol.* **2010,** *77* (6), 912–921.

18. Holzer, A. K.; Katano, K.; Klomp, L. W.; and Howell, S. B. Cisplatin rapidly down-regulates its own influx transporter hCTR1 in cultured human ovarian carcinoma cells. *Clin. Cancer Res.* **2004,** *10* (19), 6744–6749.

19. Holzer, A. K.; Samimi, G.; Katano, K.; Naerdemann, W.; Lin, X.; Safaei, R.; and Howell, S. B. The copper influx transporter human copper transport protein 1 regulates the uptake of cisplatin in human ovarian carcinoma cells. *Mol. Pharmacol.* **2004,** *66* (4), 817–823.

20. Howell, S. B.; Safaei, R.; Larson, C. A.; and Sailor, M. J. Copper transporters and the cellular pharmacology of the platinum-containing cancer drugs. *Mol. Pharmacol.* **2010,** *77* (6), 887–894.

21. Larson, C. A.; Adams, P. L.; Blair, B. G.; Safaei, R.; and Howell, S. B. The role of the methionines and histidines in the transmembrane domain of mammalian copper transporter 1 in the cellular accumulation of cisplatin. *Mol. Pharmacol.* **2010,** *78* (3), 333–339.

22. Larson, C. A.; Adams, P. L.; Jandial, D. D.; Blair, B. G.; Safaei, R.; and Howell, S. B. The role of the N-terminus of mammalian copper transporter 1 in the cellular accumulation of cisplatin. *Biochem. Pharmacol.* **2010,** *80* (4), 448–454.

23. Larson, C. A.; Blair, B. G.; Safaei, R.; and Howell, S. B. The role of the mammalian copper transporter 1 in the cellular accumulation of platinum-based drugs. *Mol. Pharmacol.* **2009,** *75* (2), 324–330.

24. Lin, X.; Okuda, T.; Holzer, A.; and Howell, S. B. The copper transporter CTR1 regulates cisplatin uptake in *Saccharomyces cerevisiae*. *Mol. Pharmacol.* **2002,** *62* (5), 1154–1159.

25. Safaei, R. Role of copper transporters in the uptake and efflux of platinum containing drugs. *Cancer Lett.* **2006,** *234* (1), 34–39.

26. Butt, T. R.; Sternberg, E. J.; Gorman, J. A.; Clark, P.; Hamer, D.; Rosenberg, M.; and Crooke, S. T. Copper metallothionein of yeast, structure of the gene, and regulation of expression. *Proc. Natl. Acad. Sci. U S A* **1984,** *81* (11), 3332–3336.

27. Carri, M. T.; Galiazzo, F.; Ciriolo, M. R.; and Rotilio, G. Evidence for co-regulation of Cu, Zn superoxide dismutase and metallothionein gene expression in yeast through transcriptional control by copper via the ACE 1 factor. *FEBS Lett.* **1991,** *278* (2), 263–266.

28. Tamai, K. T.; Gralla, E. B.; Ellerby, L. M.; Valentine, J. S.; and Thiele, D. J. Yeast and mammalian metallothioneins functionally substitute for yeast copper-zinc superoxide dismutase. *Proc. Natl. Acad. Sci. U S A* **1993,** *90* (17), 8013–8017.

29 Schmitt, M. E.; Brown, T. A.; and Trumpower, B. L. A rapid and simple method for preparation of RNA from *Saccharomyces cerevisiae*. *Nucleic Acids Res.* **1990,** *18* (10), 3091–3092.

30. Beutler, E.; Duron, O.; and Kelly, B. M. Improved method for the determination of blood glutathione. *J. Lab. Clin. Med.* **1963,** *61*, 882–888.

31. Habeeb, A. F. Reaction of protein sulfhydryl groups with Ellman's reagent. *Methods Enzymol.* **1972,** *25*, 457–464.

32. Buege, J. A.; and Aust, S. D. Microsomal lipid peroxidation. *Methods Enzymol.* **1978,** *52*, 302–310.

33. Reznick, A. Z.; and Packer, L. Oxidative damage to proteins: Spectrophotometric method for carbonyl assay. *Methods Enzymol.* **1994,** *233*, 357–363.

34. Davidson, J. F.; Whyte, B.; Bissinger, P. H.; and Schiestl, R. H. Oxidative stress is involved in heat-induced cell death in *Saccharomyces cerevisiae*. *Proc. Natl. Acad. Sci. U S A* **1996,** *93* (10), 5116–5121.

35. Luk, E. E.; and Culotta, V. C. Manganese superoxide dismutase in *Saccharomyces cerevisiae* acquires its metal co-factor through a pathway involving the Nramp metal transporter, Smf2p. *J. Biol. Chem.* **2001,** *276* (50), 47556–47562.
36. Flohe, L.; and Otting, F. Superoxide dismutase assays. *Methods Enzymol.* **1984,** *105,* 93–104.
37. Dancis, A., Haile, D., Yuan, D. S., and Klausner, R. D. The *Saccharomyces cerevisiae* copper transport protein (Ctr1p). Biochemical characterization, regulation by copper, and physiologic role in copper uptake. *J. Biol. Chem.* **1994,** *269* (41), 25660–25667.
38. Jensen, L. T.; Howard, W. R.; Strain, J. J.; Winge, D. R.; and Culotta, V. C. Enhanced effectiveness of copper ion buffering by CUP1 metallothionein compared with CRS5 metallothionein in *Saccharomyces cerevisiae*. *J. Biol. Chem.* **1996,** *271* (31), 18514–18519.

PART II

Role of Microorganisms in Agriculture and Plant Biotechnology

CHAPTER 6

ROLE OF MICROBIOME: INSIGHTS INTO THE KIN RECOGNITION PROCESS IN *ORYZA SATIVA* BY PROTEOMIC AND METABOLOMIC STUDIES

ANJANA DEVI TANGUTUR[1*], KOMMALAPATI VAMSI KRISHNA[1], AMRITA DUTTA CHOWDHURY[2], and NEELAMRAJU SARLA[2*]

[1]*Center for Chemical Biology, CSIR-Indian Institute of Chemical Technology, Hyderabad, Telangana – 500607, India.*

[2]*Biotechnology Laboratory, Directorate of Rice Research, Rajendranagar, Hyderabad, Telangana – 500030, India.*

*Corresponding authors: T. Anjana Devi. E-mail: anjana@iict.res.in; Neelamraju Sarla. E-mail: sarla_neelamraju@yahoo.com

CONTENTS

Abstract 134
6.1   Introduction 134
6.2   Materials and Methods 139
6.3   Results and Discussion 142
6.4   Conclusion 150
Acknowledgments 151
Keywords 151
References 151

ABSTRACT

Plants are affected by various biotic and abiotic factors. Within the biotic factors, plants are strongly influenced by their neighbors such as the associated microbes and the neighboring plants. This process is well-known to control/manipulate plant growth, development, and productivity. So, in the present study we made an effort to explore the role of associated microbes in kin recognition and growth within and among two selected varieties of *Oryza sativa* by *in vitro* and in field experiments. The initial study involved morphological observation of germinating seedlings of the two selected cultivars in petriplates (grown till 7 days) and also in pots (grown till 14 days) where parameters such as rate of germination, length of coleoptile, and radicle were recorded by applying a treatment of self/non self. The root associated microbes were also isolated in each treatment and further characterized by Gram staining, zone of inhibition studies, etc. The study of antimicrobial activity and metabolites produced by these organisms is in progress. Further, by proteomics it is possible to understand the effect of microbiome in each of the treatment. Therefore, proteome analysis of the two cultivars was carried out by 1D and 2D gel electrophoresis followed by matrix-assisted laser desorption ionization-time of flight mass spectrometry/mass spectrometry (MALDI MS/MS) analysis. Some of the protein spots showed a reproducible significant change between the control and experimental samples on 2D gels. These proteins identified belong to various functional categories such as carbohydrate and amino acid metabolism, cell stress, proteins of the electron transport, enzyme regulators, transcription, etc. The results provide some insights which aid to a better understanding of the role of associated microbes on the growth of plants to improve agricultural production. Also metabolomic studies of the root extracts are in progress to understand the role of microbiome in kin recognition in plants.

6.1   INTRODUCTION

Allelopathy phenomenon was defined for the first time in the late 1930s by Hans Molisch as the influence of one plant on another through the release of chemicals into the environment.[1] Although the definition of allelopathy includes both positive and negative aspect of allelochemical action, it is generally considered as a form of negative chemical communication between the organisms, whereby one participant, the donor, in an interaction produces a compound that is released in the environment in ecologically relevant quantities that negatively impacts the fitness of the other participants, the receivers; the effect presumably benefits the fitness of the donor. In 1998, the International Allelopathy Society confirmed the working definition of allelopathy. It is any process involved in the production of secondary metabolites by bacteria, algae, plants, and viruses that influences the growth and development of agricultural and biological systems; a study of the function of secondary metabolites and their importance in biological organization, their evolutionary origin and explication of the mechanisms involving plant-microorganisms, plant-virus, plant-insect, plant-plant, plant-soil-plant interactions.[2]

Theophrastus, who is considered as the Father of Botany, reported an example of the inhibitory effect of pigweed on alfalfa in 300 BC. In 1832, De Candolle, a Swiss botanist, suggested that the soil sickness problem in agriculture might be because of the exudes of crop plants.[3] Later, Hoy and Stickney in 1881 reported a deleterious effect by black walnut on the growth of plants nearby. Schreiner and Reed (1907, 1908) isolated soil organic acids released by plant roots that suppressed the growth of other crops. Molisch (1937) studied the effect of ethylene on plant growth and coined the word, "Allelopathy", from two Greek words "allelo" and "pathy" meaning "mutual harm", but, he at that time described allelopathy as both beneficial and harmful biochemical interaction between the organisms. The definition was later supported by Rice (1984). Muller (1966) used "interference" to explain the phenomenon of plant to plant interaction which involves both competition and allelopathy. Muller defined competition as a process in which one plant takes up a necessary factor from a habitat and thus, has a deleterious effect on other plants that require the same factor, whereas allelopathy is the process in which the plant releases toxic compounds into the environment, resulting in a negative effect on the neighboring plants par taking the same habitat. Allelochemicals with negative allelopathic effects are a central part of the plant defense against herbivory.[4] In addition, "Autointoxication", another phase of allelopathy, is defined as a process that chemicals produced by a plant or its decomposing residues in soil suppress the growth of its own, resulting in the decline of plant productivity in a natural vegetation or an agro-ecosystem.[5]

Allelopathy is considered as a multi-dimensional phenomenon occurring constantly in natural and anthropogenic ecosystems.[6] It is defined as the interaction between plants and microorganisms by a variety of compounds usually referred to as allelochemicals, allelopathic compounds, or allelopathins. Determination of direct or indirect effects of the quality and quantity of allelopathins on plant or microorganism communities in the natural environment is very difficult owing to the multidimensional nature of these interactions. Allelopathins are products of the secondary metabolism[7] and are non-nutritional primary metabolites.[8,9] These compounds belong to numerous chemical groups such as flavonoids, strigolactones, terpenes, benzoquinones, coumarins, terpenoids, tannins lignin, phenolic acids, nonprotein aminoacids, fatty acids, and ketones; simple lactones; long-chain fatty acids and polyacetylenes; quinines (benzoquinone, anthraquinone, and complex quinines); phenolics; cinnamic acid and its derivatives; coumarins; flavonoids; tannins; steroids and terpenoids.[10–12] Many reports have described plant reactions to allelochemicals during the last few decades.[13–16] Inderjit and Duke 2003, reviewed ecophysiological aspects of allelopathy, while some of the biochemical and physiological mechanisms mediated by allelochemicals have been discussed by others.[17–20]

Allelochemicals are released by plants into the environment by roots, leaves, flowers, fruits, seeds, bark, and stems. Allelopathic compounds affect mainly the germination and the growth of the neighboring plants by the combination of various physiological processes such as photosynthesis, respiration, water, and hormonal balance. The fundamental cause of their action is mainly to inhibit enzyme activity. The ability of an allelochemical to inhibit plant growth or seed germination is defined as its allelopathic potential or phytotoxic potential. The presence of allelochemicals in plants does not determine its allelopathic activity unless it is released into the soil environment and comes into contact with the roots of other plants.[21] However, the release of allelochemical in the soil undergoes several processes like sorption to clay and organic matter, transport, transformation which finally determines persistence, fate and toxicity of allelochemicals.[22,23] The best example of allelopathic interaction is seen in soil exhaustion because of the accumulation of chemical allelopathins which can be prevented by using fertilizers and rotating crops. Plants that produce allelopathins are considered as "donor" organisms, while the plants on which allelopathins are targeted are referred to as "acceptors" or "target" plants. Further, the effects and strength of these interactions are diverse because of the modifications of the allelopathins taking place in the soil. The allelochemicals penetrate the soil mostly as plant-active compounds such as cyanamide, phenolic acids, heliannuols, momilactones, etc. Some are converted into active form such as juglone, benzoxazolin-2-one (BOA), 2-amino-3-H-phenoxazin-3-one (APO) by microorganisms, or by specific environmental conditions like pH, moisture, temperature, light, oxygen, etc.

The microbial community is diverse and abundant[24-26] in the soil, the composition of which varies with space and time and can be influenced by a wide range of biotic and abiotic factors.[27] Among the more important types of soil biota with relevance to allelopathy, are the many free-living, symbiotic bacteria and fungi which are found in the plant rhizo and mycorrhizosphere.[28] Soil microbial community can greatly modify allelopathic effects of some plants, and sometimes beneficial microbes themselves appear to be directly negatively affected by the allelopathic compounds.[29] Microbial species may enhance the effectiveness of allelochemicals by transforming them into more toxic by-products. Alternatively, microbial species may reduce the effectiveness of allelochemicals by breaking down the compounds into less toxic forms, or by consuming them altogether.[30] It is widely known that plant species will allow the growth of some specific microbial populations in their rhizospheres that have subsequent feedbacks on nonspecific and heterospecific individuals grown in the same soil. This effect may be because of such factors as the amount and form of carbon and other nutrients that the plants provide to the soil, but is also because of the allelochemicals with direct positive and negative effects on microbes. Microbes are directly affected by allelopathic plants leading to indirect effects on competing plants, where microbes either protect plants from allelopathic competitors or enhance allelopathic effects, where soil microbial communities have changed through time in response to allelopathic plants with potential effects on plant communities. Microbes also play an important role in limiting allelopathic effects in natural environments, as has been demonstrated by comparing effects in sterile and nonsterile environments. The first demonstration of this phenomenon used an indirect approach in which the leaves of *Gmelina arborea* were incubated for varying times before adding corn seeds to perform a germination bioassay.[31]

When microbial degradation of leaf material was allowed to occur for 14 days before bioassays took place, the germination rates were significantly higher, suggesting that microbes degraded the allelopathic compounds. More direct demonstrations followed with Heisey[32] showed that ailanthone from *Ailanthus altissima* inhibited cress radicle growth more in sterile soil than in nonsterile soil. More recent work has demonstrated this phenomenon for other trees, as well as herbaceous plants and grasses producing a range of allelopathic compounds. The mechanisms by which microbes reduce allelopathic effects include degradation of allelochemicals, mounting the tolerance of the target plants to allelopathic effects, and altering phytochemical profiles of allelopathic plants to reduce the production of allelochemicals.

Microbes can mediate interactions in a number of ways with both positive and negative outcomes[33,34] for the surrounding plants and plant communities; they can serve as targets and mediators of allelopathic effects in plants. Allelopathic inhibition of beneficial microbes such as rhizobia, mycorrhizal fungi, and plant growth promoting rhizobacteria also indirectly limits the growth of the plants hosting those microbes. Microbes whose growth is not directly affected by the allelochemicals still often mediate effects of those compounds on other plants, both reducing and increasing the allelopathic effects. Reductions in allelopathic effects occur through several mechanisms that include microbial degradation of allelochemicals, an increasing tolerance of colonized plants to the stress of exposure to allelochemicals, and the alteration of the phytochemical profiles of allelopathic plants that reduces allelochemical production. Increases in allelopathic effects also can be driven by microbial degradation of natural products when the products of degradation are more toxic than the parent compounds, through modifications of plant microbe interactions, and by microbial induction of allelochemical production by plants. Furthermore, bioactive zones of allelochemicals are increased in soils with intact arbuscular mycorrhizal fungal networks, which seem to serve as highways for allelochemical movement directly from the donor to the target plants.

Plant roots must compete with the invading root systems of the neighboring plants for water, space, and mineral nutrients, with other soil borne organisms in the rhizosphere including bacteria and fungi. Root to root and root to microbe communications are continuous occurrences in this biologically active soil zone. Studies have shown that root exudates may initiate and manipulate biological and physical interactions between the roots and the soil organisms, and play an active role in root to root and root to microbe communication. Plants know more about their environment, that they might sense the presence of neighbouring plants through changes in water or nutrients available to them or through chemical indications in the soil, and can accordingly adjust their growth. The exact mechanism of the ability to recognize kin has not been demonstrated.

Therefore, the present study involved initial morphological observation of parameters such as germination rate, coleoptile and radicle length by applying a self, non self treatment of the two rice varieties *in vitro* (when grown in petriplates or pots) in soil and in field experiments for a period of 7 and/or 14 days. Also, the associated microbes in each treatment were isolated, identified, and further characterized. Simultaneously, the induced changes in the root samples were analyzed by one-dimensional and two-dimensional gel electrophoresis in conjunction with matrix-assisted laser desorption ionization-time of flight mass spectrometry (MALDI). The reasonable implications of the differentially expressed proteins in kin recognition will be found out. Further, the experimental data obtained by the proteomic analysis can be explored by bioinformatic studies. These bioinformatic studies are aimed at developing efficient multivariate statistical methodologies for extracting the relevant information from the large amount of analytical data resulted by proteomic studies. We are also carrying out metabolite analysis and/or the analysis of allelochemicals of root extracts/ root associated microbes of the varieties in each treatment (self, non self). The results will provide new insights that can lead to a better understanding of the molecular basis of sibling effect and competition and the role of associated microbiome in plants (in the cereal crop *Oryza sativa*) to improve agricultural production.

6.2 MATERIALS AND METHODS

6.2.1 CHOICE OF THE RICE CULTIVARS

The two rice cultivars Indian long grain indica rice variety (IR-64) and Nagina 22, Indian *aus* rice variety (N-22) were chosen for the present study on the basis of their genetic differences. These rice varieties were obtained from the Indian Institute of Rice Research (earlier, the Directorate of Rice Research), Hyderabad, India.

6.2.2 PLANT MATERIAL

Mature rice (*Oryza sativa* L.) seeds were surface-sterilized with 5% (v/v) sodium hypochlorite (NaOCl) for 15 min and thoroughly washed in distilled water. The seeds were randomly placed on the soil either in 90 mm petri dishes or disposable glasses and were germinated in the dark at 25°C with 50% air humidity. These were considered to have germinated when the shoots were longer than 2 mm. Once the seeds started germinating, 16 h light and 8 h dark conditions were maintained. The germination rate, radicle length, shoot length, and fresh weight were recorded at definite time intervals. Embryos including newly formed shoots and the radicle were isolated from the seeds after 7 days for proteome and metabolite analyses. Each experiment was performed in triplicates.

6.2.3 ISOLATION OF BACTERIA AND FUNGI FROM SOIL ADHERING TO THE ROOTS

Bacteria and fungi were isolated from the soil adhered to the plant root by serial dilution and agar plating method. For this, 1 ml of sterile double distilled water was aliquoted into 1.5 ml Eppendorf tubes and the root portion of the germinated seedlings (with adhered soil) was dipped into water thoroughly. This sample was further diluted (10^{-1} to 10^{-6} dilutions) and the dilutions were spread on sterile nutrient agar plates and potato dextrose agar (PDA) plates by spread plate method for the isolation of bacteria and fungi. The nutrient agar plates were then incubated at 37 °C overnight and the PDA plates were incubated at room temperature for 3–4 days or until the colonies appeared. The number of colonies obtained were counted in each case.

6.2.4 PROTEIN EXTRACTION AND QUANTITATION

Briefly, 0.1 g of the sample (embryo, shoot, or radical) was ground in 1.0 ml of pre cooled protein extraction buffer (50 mM Tris-HCl, pH 7.4, 1 mM ethylene diaminetetra acetic acid (EDTA), 2 mM $MgCl_2$, 2 mM dithio-threitol (DTT), 2.5 mM Phenylmethylsulfonyl fluoride (PMSF), 0.1% Triton-X-100). The homogenate was centrifuged at 15,000 rpm for 10 min at 4 °C. The supernatant was collected and mixed with 100% TCA to a final concentration of 10% and stored overnight at −70 °C. Then the sample was centrifuged at 15,000 rpm for 15 min at 4 °C, the supernatant was discarded. The pellet was washed thrice with cold acetone. After centrifugation, the pellet was vacuum dried and solubilized in a sample buffer (8.0 M urea, 2% (3-(3-cholamidopropyl)dimethylammonio)-1-propanesulfonate) CHAPS, 50 mM DTT). The protein concentration of these protein extracts was determined using the Bradford protein assay kit (Biorad, Hercules, CA, USA) according to the manufacturer's instructions and using BSA as a standard. This protein sample could be used for two-dimensional gel electrophoresis (2-DE) or stored in −70 °C until used. Biolyte 3/10 ampholytes (0.2%) and bromophenol blue (trace) was added to the protein sample before 2-DE (rehydration).

6.2.5 SDS-PAGE FOR PRELIMINARY PROTEOMIC PROFILING OF GERMINATING SEED, RADICLE AND SHOOT SAMPLES

The protein samples were mixed with SDS gel loading buffer to 1X concentration, heated to 100 °C for 3 min, loaded on 10% SDS-PAGE and electrophoresis was carried out. The apparent molecular weights of the polypeptides on the SDS gel were determined using prestained protein markers (Bangalore Genei, Bangalore, India) in the electrophoresis run.

6.2.6 TWO DIMENSIONAL GEL ELECTROPHORESIS (2-DE)

Protein samples (200 µg) were applied to an immobilized pH gradient (IPG) strip (7 cm, pH 5–8, Linear, BioRad, Hercules, CA, USA) using a passive rehydration method (16 h of rehydration at 20 °C) according to the manufacturer's instructions. The strips were then transferred to an isoelectric focusing (IEF) cell (BioRad, Hercules, CA, USA) and IEF was performed by applying a voltage of 200 V for 20 min, ramping to 4000 V over 2 h, and holding to 4000 V until 10 KVh was reached. Before the second dimension, the gel strips were equilibrated for 15 min twice in equilibration buffers as previously described.

The second dimension was performed using 10% SDS-PAGE at a constant current of 25 mA. The gels were stained by the fast-coomassie staining method and scanned with a BioRad GS-800 scanner. At least two independent runs were made for each sample to ensure the accuracy of analyses. The 2D gel maps were analyzed by the Progenesis same spots software (non linear dynamics, Durham, NC, USA). The quantity of each spot in a gel was normalized as a percentage of the total quantity of all spots in that gel and evaluated in terms of optical density (OD). Only the spots that showed consistent and significant differences (as determined using Progenesis same spots software) were selected for analysis with MS.

6.2.7 IN-GEL TRYPSIN DIGESTION AND MALDI-TOF ANALYSIS

In-gel digestion of proteins was carried out using MS-grade Trypsin Gold (Promega, Madison, WI, USA) according to the manufacturer's instructions. Briefly, the spots were cut out of the gel (1–2 mm diameter) using capillaries, and destained twice with 100 mM NH_4HCO_3/50% acetonitrile (ACN) at room temperature for 45 min in each treatment. After dehydration and drying, the gels were pre-incubated in 10–20 µl trypsin solution (20 ng/µl) for 1 h. Then the samples were added in adequate digestion buffer (25 mM NH$_4$HCO$_3$/50% ACN, 0.1% TFA) to cover the gels and incubated overnight at 37 °C. Tryptic digests (peptides) were extracted twice with 50% ACN/5% trifluoroacetic acid (TFA) for 30 min each time. The combined extracts were dried in a vacuum concentrator at room temperature. The extracted peptides were dissolved in 5 µl of 50% ACN/0.1% TFA, and then 0.8 µl of the digests was mixed with 0.8 µl of 5 mg/ml alpha-cyano-4-hydroxy-cinnamic acid (CHCA) in 50% ACN/0.1% TFA and spotted onto a MALDI target plate.

Mass spectra were acquired using a matrix assisted laser desorption ionization time-of-flight (MALDI-TOF) mass spectrometer (Applied Biosystems, 4800). The MS/MS was performed in a data-dependent mode in which the top ten most abundant ions for each MS scan were selected for MS/MS analysis. The MS/MS data acquired and processed using the global proteomics software and MASCOT search engine was used to search the database. Database searches were carried out using the following parameters: database, NCBI; taxonomy, *Oryza sativa*; enzyme, trypsin; and the allowance of one missed cleavage. Carbamido-methylation selected as a fixed modification and oxidation of methionine allowed being a variable. The peptide and fragment mass tolerance were set at 1 and 0.2 Da, respectively. Proteins with probability-based MOWSE scores (Molecular Weight Search score) exceeding their threshold ($P < 0.05$) were considered to be positively identified.

6.3   RESULTS AND DISCUSSION

6.3.1   *COLLECTION OF DATA ON GERMINATION RATE, SHOOT LENGTH, ROOT LENGTH, AND FRESH WEIGHT*

Five petriplates with respective seeds (surface sterilized) were kept for germination as follows—N22 (singles, 15 seeds), IR64 (singles, 15 seeds), N22 (doubles, 15 pairs), IR64 (doubles, 15 pairs), N22 and IR64 (as 15 pairs) (Figure 6-1). The seeds were allowed to germinate and grow upto 7 days in petriplates as described in materials and methods. The germination rate, radicle length, coleoptile length, and fresh weight were recorded upto and after 7 days. These experiments were performed in triplicates for accuracy of data. A seed was characterized as germinated after the protrusion of its radicle. The seeds were checked for germination daily and the following data was collected during and upto 7 days. Germination percentage (Table 6-1) represents the number of germinated seeds/total number of seeds taken, root length (Table 6-2) of each plant was measured in mm and their mean was calculated. Similarly, shoot length (Table 6-3) of each plant was measured in mm and their mean was calculated.

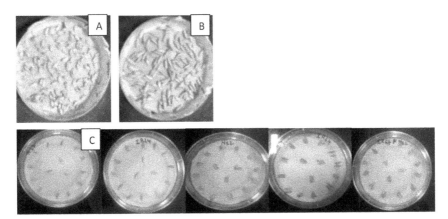

FIGURE 6-1   (A) N22 Seeds (B) IR64 seeds (C) *Oryza sativa* seeds were placed on sterilized and moistened filter paper in petriplates for germination—N22 singles, IR64 singles, N22 doubles, IR64 doubles, N22 and IR 64 grown together as pairs. In a similar fashion, the seeds were placed in petriplates with soil.

TABLE 6-1   Germination Percentage of the Two Varieties N22 and IR64

Germination Percentage			
Rice Variety	**Singles**	**Paired**	**Non -self**
IR64	96.6	100	96.6 (with N22)
N22	90	98.3	96.6 (with IR64)

TABLE 6-2   Root Length of the Two Varieties N22 and IR64

Root length (mm)			
Rice Variety	**Singles**	**Paired**	**Non -self**
IR64	56.1	61.5	68.1 (with N22)
N22	43.5	44.2	54.9 (with IR64)

TABLE 6-3   Shoot Length of the Two Varieties N22 and IR64

Rice Variety	Shoot length (mm)		
	Singles	Paired	Non-self
IR64	40.3	43.3	47.5 (with N22)
N22	42.6	47.2	53.9 (with IR64)

The germination percentage and the shoot and root length data clearly indicate that N22 and IR64 were affected by each other when grown together in pairs. They behave differently when grown singly, doubly, or in pairs as shown in the Figure 6-2. In addition, we observed that initially at and upto 3 days, the germination and growth of IR64 was better than N22, when grown singly, doubly, or in pairs, but at 7 days of germination, N22 competes and almost reaches the shoot length of IR64. Another important observation was that N22 grew better when paired with IR64 than by itself. This indicates that IR64 roots/root associated microbes may be releasing some growth promoting substances from the roots. However, the allelopathic and stimulatory effects of these interactions should be explored clearly by further experiments.

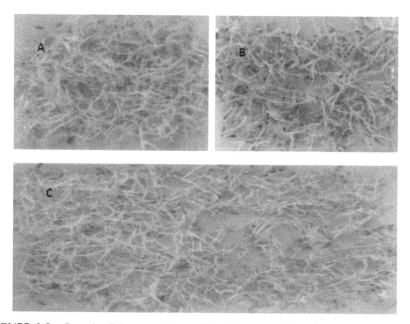

FIGURE 6-2   Growth of *Oryza sativa* seedlings in trays with moistened filter paper as observed after 7 days (A) N22, (B) IR64, and (C) N22 and IR64 grown together.

6.3.2 MICROBIAL ANALYSIS OF SOIL SAMPLES ADHERED TO THE ROOTS

To further investigate the kind of microbes associated with each variety in each of the treatments (single, double, or pairs), we isolated the bacterial and fungal colonies by serial dilution of the soil sample (adhered to the roots) and plated it on the respective nutrient agar (Figure 6-3) or potato dextrose agar plates (Figure 6-4). Table 6-4 gives details of the number of colonies obtained in each treatment. The bacterial sample from each of the colony (Figure 6-5) was stained by Gram staining method initially to identify the morphology and Gram staining nature (Gram-positive or negative) (Figure 6-6) of the bacteria. We also found the growth of actinomycetes in some plates based on the studies of colony morphology and staining.

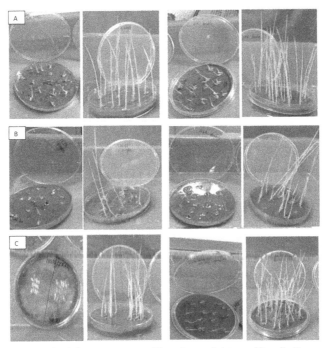

FIGURE 6-3   Growth of *Oryza sativa* seedlings in petriplates with soil. *Oryza sativa* seeds were placed on moist soil in petriplates for germination. (A) IR64 singles on day 3 after germination, on day 7 after germination, IR64 doubles on day 3 after germination, on day 7 after germination. (B) N22 singles on day 3 after germination, on day 7 after germination, N22 doubles on day 3 after germination, on day 7 after germination. (C) N22 and IR64 seeds were grown in the same plate, 15–20 seeds of each variety in each half of the plate on day 3 after germination, on day 7 after germination, N22 and IR 64 grown together as pairs 3 days after germination, 7 days after germination.

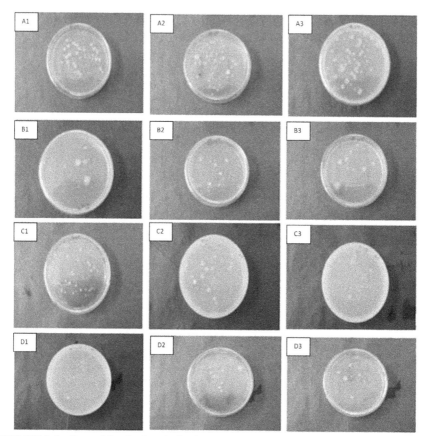

FIGURE 6-4   Bacterial colonies obtained on nutrient agar plates following dilution plating of soil samples from each treatment. Colonies obtained from IR64 singles are indicated by A1 and A2; A3 and B1 from IR64 doubles; B2 and B3 from N22 singles; C1 and C2 from N22 doubles; C3 and D1 from N22, IR64 pairs; D2 and D3 from N22 and IR64 grown in the same plates (half of the portion) (see Figure 6-3 for more details).

TABLE 6-4   Number of Colonies Obtained on Nutrient Agar Plates and Potato Dextrose Agar (PDA) Plates

S. No.	Sample Name	Dilution Factor	No. of Colonies on Nutrient Agar Plates	No. of Colonies on Potato Dextrose Agar Plates
1.	IR64-single-1	1×10^6	54×10^9	34×10^7
2.	IR64-single-2	1×10^6	9×10^8	23×10^8
3.	IR64-double-1	1×10^6	65×10^9	28×10^6
4.	IR64-double-2	1×10^6	10×10^7	25×10^6
5.	IR64-double-3	1×10^5	15×10^6	8×10^6
6.	IR64-double-4	1×10^6	18×10^9	8×10^7

TABLE 6-4 *(Continued)*

S. No.	Sample Name	Dilution Factor	No. of Colonies on Nutrient Agar Plates	No. of Colonies on Potato Dextrose Agar Plates
7.	N22-double-1	1×10^5	42×10^6	44×10^4
8.	N22-double-2	1×10^3	35×10^4	14×10^4
9.	N22-double-3	1×10^4	29×10^4	23×10^5
10.	N22-Single	1×10^6	15×10^7	16×10^5
11.	IRR64 + N22-1	1×10^5	10×10^6	16×10^5
12.	IR64 + N22-2	1×10^6	11×10^7	27×10^5

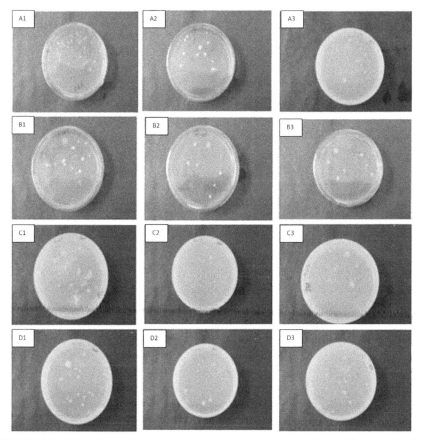

FIGURE 6-5 Fungal colonies obtained on potato dextrose agar plates following dilution plating of soil samples from each treatment. Colonies obtained from IR64 singles are indicated by A1 and A2; A3 and B1 from IR64 doubles; B2 and B3 from N22 singles; C1 and C2 from N22 doubles; C3 and D1 from N22, IR64 pairs; D2 and D3 from N22 and IR64 grown in same plates (half of the portion) (see Figure 6-3 for more details).

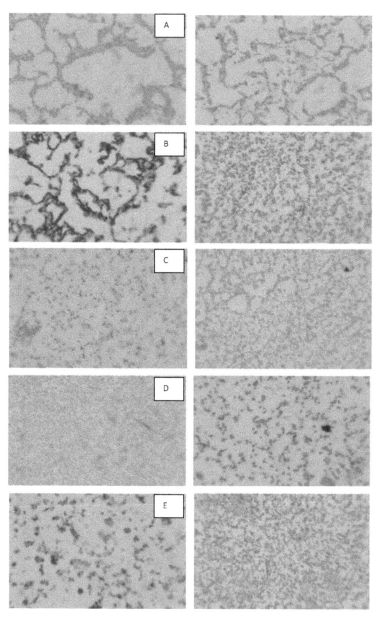

FIGURE 6-6   Gram staining of the bacterial colonies isolated from each treatment. (A) Gram-negative and Gram-positive rods from IR64 singles, (B) Gram-positive cocci in chains and Gram-negative rods from IR64 doubles, (C) Gram-positive cocci and Gram-negative rods from N22 singles, (D) Gram-negative cocci and Gram-positive monococcus from N22 doubles, (E) Gram-positive coccobacilli and Gram-positive rods from IR64 and N22 grown together as pairs.

Further, we also carried out studies on the inhibitory or stimulatory effects of these microorganisms by *in vitro* (such as zone of inhibition) studies.

6.3.3 CO-CULTURING OF ASSOCIATED MICROORGANISMS OF ONE VARIETY WITH MICROBES OF OTHER VARIETY

For understanding the microbial effect on one another and the total effect in kin recognition mechanisms, we also carryied out co-culturing studies, where the microbes associated with one rice variety i.e., N22 was grown with the microbes associated with IR64 (data not shown).

6.3.4 PROTEOMIC ANALYSIS OF PLANT SAMPLES

For preliminary analysis of the total proteins from these varieties, the protein extracts were prepared from germinating seeds, shoots, and the radicles of the rice varieties IR64, N22, IR64 and N22 grown together and subjected to 1D SDS-PAGE analysis. The apparent molecular weights of the polypeptides on SDS gel were determined using prestained protein markers in the electrophoresis run. These samples from the varieties N22 and IR64 grown together showed the presence of an extra protein band on SDS-PAGE. However, no such band was observed in the varieties grown in the presence of its own siblings.

The preliminary 2DE-experiments carried out using IPG strips with pI range 3–10 revealed that the vast majority of the protein spots focused between pI 5–8. Therefore, for better resolution of the protein spots, 5–8 pI IPG strips were used. Figure 6-7 shows preliminary one and two dimensional SDS-PAGE images of proteins extracted from these rice varieties.

Some of the protein spots on 2D gels showed a significant change between the control and other samples. These proteins identified may belong to various functional categories such as carbohydrate and amino acid metabolism, cell stress, electron transport proteins, enzyme regulator activity, transcription, etc.

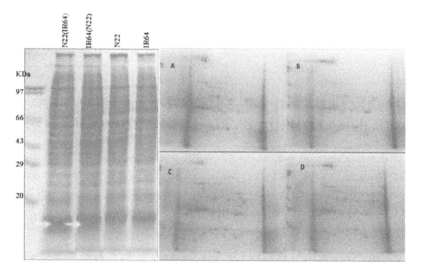

FIGURE 6-7   One-DE and 2 DE analysis of proteins from *Oryza sativa* samples. (A) N22, (B) Both N22 and IR64 grown together (Sample 1), (C) IR 64, and (D) Both N22 and IR64 grown together (Sample 2). The proteins were separated on a 7 cm immobilized pH gradient strip with a linear gradient of pH 5–8 followed by 10% SDS-PAGE in the first dimension and visualized by Coomasie R250.

6.4   CONCLUSION

Complex plant-plant and plant-microbe interactions in the ecosystems and the current studies on physiological, molecular, and cytological levels bring us to a better understanding of these processes. The traditional knowledge of the toxic properties of water extracts of a variety of allelopathic plants gives us a basis that could be used in the creation of a novel approach in weed control. Identification of allelochemicals in rice is also important to enable the toxicology studies needed before incorporating such traits into commercial germplasm and for improving competitive ability in rice.

Plant kin recognition will aid in better understanding of plant–plant communications and of plant evolution. Identification of the plant species with the ability to recognize kin, particularly amongst crop species and perennial species would be interesting. Investigation of the signaling mechanism involved in kin recognition, which includes root secretions and volatile emissions, and the role of microbiome in this will be crucial and may have implications on plant interactions on a larger multitrophic scale.

Allelopathy alone may not remove weed control practices because its effectiveness is influenced by many factors. But, it may slightly reduce the use of herbicide over time and will be a significant economical benefit to farmers with reduced ecological impact on the environment. If allelo-chemicals or genes responsible for the allelopathic effects are identified with proteomic and metabolomic profiling, these traits of allelopathy can be easily incorporated into plants and improve the cultivars through breeding techniques available presently.

ACKNOWLEDGMENTS

Anjana Devi Tangutur thanks Department of Biotechnology (DBT) (RGYI; Rapid Grant for Young Investigators) for funding the project entitled "Analysis and Identification of key proteins/factors involved in self/non self recognition in *Oryza sativa* (BT/PR6718/GBD/27/462/2012), to carry out this research work. Kommalapati Vamsi Krishna thanks University Grants Commission (UGC) for the fellowship.

KEYWORDS

- **Allelopathy**
- **Kin recognition**
- **MALDI-MS/MS analysis**
- **Microbiome**
- **Oryza sativa**

REFERENCES

1. Molisch, H. *The Influence of One Plant On Another: Allelopathy*; Narwal, S. S., Ed.; La Fleur, L. J., and Mallik, M. A. B., Translator; Scientific Publishers: Jodhpur, 2001; pp 1–132.
2. Inderjit, Mallik A. U. (Eds) Chemical ecology of plants: Allelopathy in aquatic and terrestrial ecosystems. *Ann. Bot.* **2003,** *92* (4), 625–626.
3. Rice, E. L. *Allelopathy*; Academic Press: New York, 1984; pp 1–422.
4. Fraenkel, G. S. The raison d'Etre of secondary plant substances. *Science* **1959,** *129* (3361), 1466–1470.

5. Borner, H. Liberation of organic substance from higher plants and their role in soil sickness problem. *Bot. Rev.* **1960**, *26*, 393–424.

6. Gniazdowska, A.; and Bogatek, R. Alleopathic interaction between plants. Multiside action of allelochemicals. *Acta Physiol. Plantar.* **2005**, *27*, 395–407.

7. Stamp, N. Out of the quagmire of plant defense hypotheses. *Q. Rev. Biol.* **2003**, *78* (1), 23–55.

8. Chou, C. H.; and Yao, C. Phytochemical adaptation of coastal vegetation in Taiwan: Isolation, identification and biological activities of compounds in *Vitex negundo* L. *Bot. Bull. Academia Sinica.* **1983**, *24* (2), 155–168.

9. Weir, T. L.; Park, S. W.; and Vivanco. J. M. Biochemical and physiological mechanisms mediated by allelochemicals. *Curr. Opin. Plant Biol.* **2004**, *7* (4), 472–479.

10. Macias, F, A.; Molinillo, J. M. G.; Galindo, J. C. G.; Varela, R. M.; Simonet, A. M.; and Castellano, D. The use of allelopathic studies in search for natural herbicides. *J. Crop. Prod.* **2001**, *4*, 237–256.

11. Oleszek, W.; and Stochmal, A. Triterpene saponins and flavonoids in the seeds of *Trifolium* species. *Phytochemistry* **2002**, *61*, 165–170.

12. Jose, S.; and Gillespie, A. R. Allelopathy in black walnut (*Juglans nigra* L.) alley cropping. II. Effects of juglone on hydroponically grown corn (Zea mays L.) and soybean (Glycine max L. Merr.) growth and physiology. *Plant Soil* **1998**, *203*, 199–205.

13. Czarnota, M. A.; Paul, R. N.; Dayan, F. E.; Nimbal, C. I.; and Weston, L. A. Mode of action, local isolation of production, chemical nature and activity of sorgoleone: A potent PSII inhibitor in *Sorghum* spp. root exudates. *Weed Tech.* **2001**, *15*, 813–825.

14. Bais, H. P.; Vepachedu, R.; Gilroy, S.; Callaway, R. M.; and Vivanco, J. M. Allelopathy and exotic plant invasion from molecules and genes to species interactions. *Science* **2003**, *301*, 1377–1380.

15. Abenavoli, M. R.; Sorgona, A.; Sidari, M.; Badiani, M.; and Fuggi, A. Coumarin inhibits the growth of carrot (*Daucus carota* L. cv. Saint Valery) cells in suspension culture. *J. Plant Physiol.* **2003**, *160*, 227–237.

16. Burgos, N. R.; and Talbert, R. E. Growth inhibition and root ultrastructure of cucumber (*Cucumis sativus*) seedlings exposed to allelochemicals from rye (Secale cereale). *J. Chem. Ecol.* **2004**, *30*, 671–689.

17. Inderjit.; and Duke, S. O. Ecophysiological aspects of allelopathy. *Planta* **2003**, *217*, 529–539.

18. Dayan, F. E.; Romagni, J. G.; and Duke, S. O. Investigating the mode of action of natural phytotoxins. *J. Chem. Ecol.* **2000**, *26*, 2079–2094.

19. Politycka, B. Peroxidase activity and lipid peroxidation in roots of cucumber seedlings influenced by derivatives of cinnamic and benzoic acids. *Acta Physiol. Plant* **1996**, *18*, 365–370.

20. Singh, H. P.; Daizy, D. R.; and Kohli, R. K. Allelopathic interactions and allelochemicals: New possibilities for sustainable weed management. *Cri. Rev. Plant Sci.* **2003**, *22*, 239–311.

21. Choesin, D. N.; and Boerner, R. E. J. Allyl isothiocyanate release and the allelopathic potential of *Brassica napus* (Brassicaceae). *Am. J. Botany* **1991**, *78*, 1083–1090.

22. Cheng, H, H. Characterization of the mechanisms of allelopathy modeling and experimental approaches. *ACS Symp. Ser.* **2009**, *582*, 132–141.

23. Inderjit.; and Weiner, J. Plant allelochemical interference or soil chemical ecology? *Persp. Plant Ecol. Evol. Syste.* **2001**, *4*, 3–12.

24. Torsvik, V.; Ovreas, L.; and Thingstad, T. F. Prokaryotic diversity: Magnitude, dynamics, and controlling factors. *Science* **2002,** *296,* 1064–1066.

25. Venter, J. C.; Remington, K.; Heidelberg, J. F.; Halpern, A. L.; Rusch, D.; Eisen, J. A.; *et al.* Environmental genome shotgun sequencing of the Sargasso Sea. *Science* **2004,** *304,* 66–74.

26. Whitman, W.; Coleman, D.; and Wiebe, W. Prokaryotes: The unseen majority. *Proc. Natl. Acad. Sci. USA* **1998,** *95,* 6578–6583.

27. Buckley, D.; and Schmidt, T. Exploring the biodiversity of soil—A microbial rain forest. In *Biodiversity of Microbial Life*; Staley, J., Reysenbach, A., Eds.; Wiley: New York, 2001; pp183–208.

28. Johansson, J.; Paul, L.; and Finlay, R. D. Microbial interactions in the mycorrhizosphere and their significance for sustainable agriculture. *FEMS Microbiol. Ecol.* **2004,** *48,* 1–12.

29. Ehlers, B. K. Soil microorganisms alleviate the allelochemical effects of a thyme monoterpene on the performance of an associated grass species. *PLoS One* **2011,** *6,* 1–5.

30. Inderjit. Soil microorganisms: An important determinant of allelopathic activity. *Plant Soil* **2005,** *274,* 227–236.

31. Hauser, S. Effect of Acioa barteri, Cassia siamea, Flemingia macrophylla and Gmelina arborea leaves on germination and early development of maize and cassava. *Agric. Ecosyst. Environ.* **1993,** *45,* 263–273.

32. Heisey, R. M. Identification of an allelopathic compound from *Ailanthus altissima* (Simaroubaceae) and characterization of its herbicidal activity. *Am. J. Botany* **1996,** *83,* 192–200.

33. Reinhardt, K. O.; and Callaway, R. M. Soil biota and plant invasions. *New Phyt.* **2006,** *170,* 445–457.

34. Emani, C.; Jiang, Y.; Miro, B.; Hall,T.; and Kohli, A. Rice. In *A Compendium of Transgenic Crop Plants, Vol. 1, Cereals and Forage Grasses*; Kole, C., Ed.; Wiley-Blackwell: New York, 2008; pp 1–47.

CHAPTER 7

ASSOCIATIVE EFFECT OF ARBUSCULAR MYCORRHIZAL FUNGI AND *RHIZOBIUM* ON PLANT GROWTH AND BIOLOGICAL CONTROL OF CHARCOAL ROT IN GREEN GRAM [*VIGNA RADIATA* L. (WILCZEK)]

A. HINDUMATHI[1], B. N. REDDY[1*], A. SABITHA RANI[2], and A. NARSIMHA REDDY[1]

[1]Mycology and Plant Pathology Laboratory, Department of Botany, Osmania University, Hyderabad, Telangana, – 500007, India.

[2]Mycology and Plant Pathology Laboratory, University College for Women, Koti, Hyderabad, Telangana, – 500095, India.

Corresponding author: B. N. Reddy. E-mail: reddybn1@yahoo.com

CONTENTS

Abstract 156
7.1 Introduction...156
7.2 Materials and Methods..158
7.3 Results and Discussion 161
7.4 Conclusion 168
Acknowledgments...168
Keywords 168
References 169

ABSTRACT

The present study was carried out in pot experiments to assess the response of green gram [*Vigna radiata* (L.) Wilczek] to the synergistic effects of treatments with predominantly occurring indigenous arbuscular mycorrhizal (AM) fungi (*Glomus constrictum* Trappe) and *Rhizobium* treated individually and in combination with or without the pathogen on the extent of nodulation, nodule dry weight, nodule nitrogen, mycorrhizal root colonization, plant growth, biomass production, nutrient uptake (NPK) into whole plant at 45 and 75 days old crop, grain yield, and suppression of the charcoal rot causing pathogen *Macrophomina phaseolina*. The experiment was repeated twice. The plants inoculated with *Glomus constrictum* significantly increased the shoot and root heights, and their fresh and dry weights. In dual inoculated plants, maximum growth parameters were recorded compared with plants that were treated with AMF or *Rhizobium* singly. The same trend was observed in the plants inoculated in combination with pathogen. The grain yield of the plants inoculated with AMF *Glomus constrictum* and *Rhizobium* alone or in combination was significantly greater than the uninoculated control plants. Mycorrhizal inoculated plants significantly reduced the disease incidence from 76.36 and 84.7% to 19.86 and 16.90% and increased the per cent disease control to 104.95 and 98.11% in experiment I and II, respectively. However, the plants co-inoculated with *Rhizobium* and *G. constrictum* were more tolerant to the fungal root rot pathogen as evidenced by a further reduction in the disease incidence by 13.10 and 12.60% and the per cent disease control increased to 113.80 and 103.19%, respectively. Hence, the synergistic effect of AM fungus *Glomus constrictum* and *Rhizobium* is beneficial to green gram for its increased nodulation, improved plant growth, biomass production, mycorrhizal root infection, and mobilized more N, P, and K compared with the control plants or those inoculated with the pathogen.

7.1   INTRODUCTION

The use of microbial inoculants plays an important role in sustainable agriculture[1] and is emphasized to play a major role in the current day research against the adverse effects of chemical inputs on soil health, fertility, and the environment. The role of Biological Nitrogen Fixing (BNF) is a nonpolluting and more cost-effective way to improve soil fertility. Most of the legumes possess two types of microbial symbionts, namely, mycorrhizal fungi and nitrogen fixing bacteria thereby establishing triple association capable of supplying N and P contents to the plants.[2,3]

Several legumes grow poorly and fail to nodulate even with Rhizobial inoculation in autoclaved soil unless they were inoculated with mycorrhizal fungi[4]. Because legumes can host both AM fungi and N_2-fixing bacteria, the tripartite symbiosis of AM fungi-rhizobia-legume assumes more significance in terms of improving soil fertility and crop productivity. A great deal of work has been carried out on the tripartite symbiosis legume-mycorrhiza-*Rhizobium*[5]. The AM fungi usually enhanced nodulation[6] and N_2-fixation in legumes. Khan et al.[7] reported that root infection by indigenous AM fungi resulted in the improvement of nodulation and attributed to the better uptake of phosphate.

It has been documented that plants with combined inoculation of arbuscular mycorrhizal fungi and *Rhizobium* have been found to have synergistic beneficial effects[8,9]. Furthermore, arbuscular mycorrhizae can influence the severity of disease in several crop plants[9–11]. Because these root symbionts, as well as soil-borne pathogens share a common niche, they influence the growth of the plant[12] by improved nutrient acquisition and inhibition of fungal plant pathogens.

Green gram is a protein rich pulse crop having a growing demand in Asia, especially in India because it provides a high quality of protein in the Indian diet. There is a need to enhance its productivity to meet the demand of the growing population. Researchers in the past few decades on various aspects of root symbionts have shown that dual interaction of AM fungi and *Rhizobium* have improved plant growth, nodulation, and yield[13,14] and also nutrient status.[14,15] In our earlier reports, the inoculation of mycorrhizal fungi alone and in combination with *Rhizobium* in the presence of charcoal rot pathogen *Macrophomina phaseolina* in soybean showed an increase in the yield, root colonization, nutrient status, and reduction in disease of up to 70%.[9] Charcoal rot of green gram caused by *Macrophomina phaseolina* (Tassi) Goid is a serious disease causing yield losses of up to 77%.

In light of these views, the present study was carried out as pot experiments under glasshouse conditions to assess the response of green gram [*V. radiata* (L.) Wilczek] to the synergistic effects of inoculation with predominantly occurring indigenous AM fungi (*Glomus constrictum* Trappe) and the introduced *Rhizobium* treated individually and in combination with or without the pathogen on the extent of nodulation, nodule dry weight, nodule nitrogen, mycorrhizal root colonization, biomass production, nutrient uptake (NPK) into whole plant, grain yield, and suppression of the charcoal rot causing pathogen *M. phaseolina*.

7.2   MATERIALS AND METHODS

7.2.1   PLANT MATERIAL

Green gram [*Vigna radiata* (L.) Wilczek] monolayer (ML) 267 seeds were obtained from the Department of Agriculture, Andhra Pradesh, India. The seeds used for experimental purpose were surface sterilized in 0.1% solution of mercuric chloride, then kept in 70% ethyl alcohol for 5 min and finally washed several times with sterile distilled water.

7.2.2   ISOLATION AND MULTIPLICATION OF THE TEST PATHOGEN MACROPHOMINA PHASEOLINA (TASSI) GOID

The test pathogen *Macrophomina phaseolina* (Tassi) Goid was isolated from the parts of diseased host plant on potato dextrose agar (PDA) medium. Pathogenicity test was conducted on the host plant, and the virulent strain thus proved positive which was multiplied by using sand and sorghum meal[16] and used for pot experimental trials.

7.2.3   ISOLATION OF MICROBIAL INOCULANTS

The indigenous arbuscular mycorrhizal fungal species was isolated from the rhizosphere soils of green gram cultivated in fields of Karimnagar, Nirmal, Adilabad, and Telangana by wet sieving and decanting technique.[17] The root nodule forming bacterial strain of *Rhizobium* was isolated from the mature and fresh root nodules of green gram host plant on Yeast Extract Mannitol Agar (YEMA) media (pH 7.2) using Congo red.[18] The bacterial strain was characterized by different biochemical tests and authenticated for its ability to form root nodules on green gram plants grown in sterilized sand and later screened for antifungal activity against the charcoal rot pathogen *M. phaseolina* by dual culture plate technique[19] under *in-vitro* conditions. The strains thus screened positive were maintained on yeast extract mannitol agar slants and stored at 4 °C for further use.

7.2.4 PREPARATION OF AM FUNGAL AND CARRIER BASED RHIZOBIUM INOCULUM

Predominantly occurring indigenous arbuscular mycorrhizal fungal spores of *G. constrictum* Trappe, were isolated, purified, and mass multiplied in sand and soil mixture (1:1 ratio), grown for four months using sorghum as the host plant.[20]

Carrier based inoculum of *Rhizobium* sp. ($2.5 - 3.5 \times 10^8$ cells g^{-1}) was prepared by using sterilized charcoal as the carrier material.[21] Seed bacterization was done (as per treatments), air-dried for 15 min under shade and was sown immediately.[22] The seeds for uninoculated control (no *Rhizobium* and no AM fungi) were treated with Chloramine-T (1%) for 10–12 min followed by rinsing 3–4 times in sterile distilled water, air drying and sowing.

7.2.5 SCREENING OF MICROBIAL INOCULANTS ON GREEN GRAM

Microbial inoculants were screened on green gram under glasshouse conditions in phosphorus deficient autoclave sterilized soil for mycorrhizosphere interaction studies to evaluate different effects of indigenous AM fungi *G. constrictum* individually and in combination with *Rhizobium* and the pathogen *M. phaseolina* on number, dry weight, and nitrogen content of root nodules, plant biomass, nutrient content (N, P, K) of whole plant (at 45 and 75 days), grain yield, and charcoal rot disease incidence (at harvest) on green gram.

The pot experiments were repeated twice. The soil used was of sandy loam, with pH 7.2. The treatments used were: (1) uninoculated control; (2) inoculated with AM fungi *Glomus constrictum* (Gc); (3) *Rhizobium* (Rhiz); (4) *Macrophomina phaseolina* (MP); (5) 2 + 3; (6) 2 + 4; (7) 3 + 4; (8) 2 + 3 + 4.

The inoculum of the test pathogen (10 g/kg soil consisting of 40 sclerotia /g inoculum) was mixed in the upper 2 inches layer of the pot and incubated for two days for stabilization of the pathogen. *Rhizobium* inoculation (@10^9 cfu/ml) was done by treating the seeds with charcoal based culture. Mycorrhizal inoculation was done by placing the inoculum (5 gm of 65 spores gm^{-1} soil) as a layer 2 cm below the seed hole and the seeds were placed 2 cm above the inoculum at the time of sowing so that the growing roots could pass through this inoculum layer. The mycorrhizal inoculum consisted of chlamydospores, mycelium, mycorrhizal infected root segments with vesicles and arbuscules, and soil from the pot culture of *Sorghum bicolor* infected with *G. constrictum* and grown for four months.

Surface sterilized seeds (5 seeds per pot) were sown in earthen pots filled with 4 kg soil. Thinning (3 plants per pot) was done on the 12th day after emergence. The plants were watered regularly as required by distilled water and Hoagland nutrient[23] solution was added once in every two weeks. The experiment was arranged in a complete randomized block design and three replications were maintained for each treatment. Watering was withdrawn from the post flowering to the grain filling stage. At this time, the stem reserves were depleted and translocated to the developing grains, weakening the stems and predisposing them to infection by the pathogen.

At two intervals i.e., 45 and 75 days after emergence, observations were recorded on plant height (shoot and root lengths in cm per plant), fresh and dry weights (dried at 70 °C for 72 h to a constant weight) grams per plant, number and dry weight of nodules per plant. The plant samples were determined for phosphorus calorimetrically by vanadomolybdate phosphoric yellow color method and triple acid digested sample was used to estimate potassium content by flame photometer[24], and nitrogen by Microkjeldahl method.[25] The roots were cleared and stained,[26] and per cent mycorrhizal infection (at 45 and 75 days crop growth period) of the root samples was determined by grid line intersect method.[27] The grain yield per plant was recorded at the time of harvest.

7.2.6 CHARCOAL ROT DISEASE INCIDENCE

The charcoal rot infected plants were scored and the percentage of infection was calculated[28] and expressed as the percentage of disease incidence (%).

Per cent disease incidence was calculated by using the following formula:

$$\text{Per cent disease incidence } (DI) = \frac{\text{Number of diseased plants}}{\text{Total number of plants}} \times 100$$

$$\text{Per cent disease control} = \frac{100 - DI \text{ in treatment}}{\text{Disease incidence in control}} \times 100$$

7.2.7 STATISTICAL ANALYSIS

Data analysis was performed using SPSS (Statistical Package for Social Sciences) software package, Version 17, an analysis of variances (ANOVA) was used to test whether treatment effects existed and the means were compared with the Least Significant Difference (LSD) at 1% and 5% level of significance.

7.3   RESULTS AND DISCUSSION

The synergistic effect of *Glomus constrictum* (Gc) and *Rhizobium* (Rhiz) was assessed on green gram (in Experiment I and Experiment II) under greenhouse conditions in relation to per cent AM fungal root colonization, plant growth (shoot and root height, fresh and dry weights), plant N, P, K, content, number, dry weight, and nitrogen content of root nodules at 45 and 75 days old crop. Grain yield and disease incidence were recorded at harvest. The Gc inoculation had a significant effect on all the plant growth parameters studied as summarized in Tables 7-1 and 7-2.

TABLE 7-1   Mycorrhizal Root Colonization (%), Plant Height (cm), Plant Fresh and Dry Weights (g/plant) in Green Gram as Influenced by *Glomus constrictum* and *Rhizobium* (Expt. I) Inoculation

Treatments	MC (%)		Plant height (cm)		Plant FW (g/plant)		Plant DW (g/plant)	
	45 d	75 d	45 d	75 d	45 d	75 d	45 d	75 d
Control	0	0	16.8	24.1	1.8	2.45	1.52	1.48
Gc	30.0	74.5	22.5	28.2	2.5	3.29	2.09	2.37
Rhiz	0	0	15.6	26.6	2.38	3.14	2.03	2.25
MP	0	0	13.0	18.0	0.63	1.18	0.41	0.83
Gc + Rhiz	36.7	86.2	23.3	34.6	2.88	3.39	2.37	2.97
Gc + MP	26.5	62.2	16.2	21.5	2.0	2.82	1.66	2.41
Rhiz + MP	0	0	11.0	21.9	1.98	2.88	1.68	2.1
Gc + Rhiz + MP	28.1	73.3	14.6	26.0	2.31	2.73	1.78	2.26
CD 5%	4.46	9.56	3.51	5.5	0.4	0.69	0.39	0.67
CD 1%	6.2	13.28	4.87	7.6	0.56	0.96	0.54	0.93

Gc - *Glomus constrictum*.

Rhiz - *Rhizobium*.

MP - *Macrophomina phaseolina*.

MC - mycorrhizal colonization (%).

FW - fresh weight.

DW - dry weight (g/plant).

CD - circular dichroism

TABLE 7-2 Mycorrhizal Root Colonization (%), Plant Height (cm), Plant Fresh and Dry Weights (g/plant) in Green Gram as Influenced by *Glomus constrictum* and *Rhizobium* (Expt. II) Inoculation

Treatments	MC (%)		Plant height (cm)		Plant FW (g/plant)		Plant DW (g/plant)	
	45 d	75 d	45 d	75 d	45 d	75 d	45 d	75 d
Control	0	0	15.2	23.7	1.65	2.36	1.25	1.96
Gc	30.7	73.4	22.5	33	2.81	3.52	2.42	2.9
Rhiz	0	0	20.2	29.4	2.01	3.06	1.86	2.64
MP	0	0	9.5	15.7	1.45	2.11	1.15	1.9
Gc + Rhiz	31.1	83.2	26.0	33.7	2.74	3.61	2.34	3.25
Gc + MP	21.9	68.0	20.0	29.7	2.49	3.15	2.14	2.78
Rhiz + MP	0	0	15.0	20.4	1.96	2.72	1.6	2.12
Gc + Rhiz + MP	22.6	68.9	19.2	26.96	2.42	2.86	2.02	2.26
CD 5%	4.56	4.51	4.46	4.45	0.67	0.66	0.47	0.57
CD 1%	6.33	6.26	6.19	6.17	0.94	0.91	0.65	0.79

Gc - *Glomus constrictum*.
Rhiz - *Rhizobium*.
MP - *Macrophomina phaseolina*.
MC - mycorrhizal colonization (%).
FW - fresh weight.
DW - dry weight (g/plant).

7.3.1   COLONIZATION BY AMF

The per cent mycorrhizal root colonization in plants inoculated with *G. constrictum* (Gc) showed significant increase in both the experiments conducted. The only pathogen inoculated and uninoculated control plants showed no root colonization as AM fungal inoculum was not added. The Gc inoculated plants showed significantly greater root colonization. However, combined inoculation of Gc with *Rhizobium* significantly enhanced per cent root colonization over plants inoculated with Gc alone (Tables 7-1 and 7-2). Co-inoculation of Gc plus *Rhizobium* in the presence of the pathogen showed greater root colonization over plants inoculated with Gc alone, and Gc plus pathogen and are statistically on par. *Rhizobium* co-inoculated with

Gc fungi significantly enhanced the root colonization ability of the mycor-rhizal fungi in both the experiments conducted. Further observations showed mycorrhizal inoculated plants alone and in combination showed heavy root infection quantitatively and qualitatively with abundant mycelium, vesicles, and arbuscules (Figure 7-1A–D).

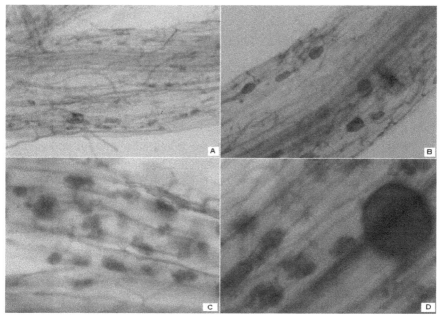

FIGURE 7-1   (A) showing mycelium, vesicle, and arbuscules, (B) mycelium and vesicles, (C) mycelium and arbuscules, (D) mycelium, vesicle, and arbuscules.

7.3.2   PLANT GROWTH

The ANOVA revealed highly significant differences among the treatments for plant height, fresh and dry weights at two different stages of plant growth period in both the experiments conducted. It is evident from the present study data that Gc inoculation improved the plant growth which was expressed as plant height compared with the uninoculated control plants (Tables 7-1 and 7-2). The plants inoculated with Gc significantly increased in plant fresh and dry weights. In dual inoculated plants, maximum growth parameters were recorded compared with plants treated with only Gc or *Rhizobium* alone. The same trend was observed in pathogen inoculated plants. However, it is evident from the data that plants on combined inoculation with Gc and *Rhizobium* enhanced plant height and promoted biomass production when compared with the control and inoculation with Gc or *Rhizobium* alone. Our results correlate with the findings [29–31] that combined application of rhizobia and AM fungi can act synergistically and enhance plant growth and biomass production to a greater extent than single inoculation as evidenced by maximum height attained in mycorrhizal and *Rhizobium* treatments in a combined form.

7.3.3 NODULE NUMBER, DRY WEIGHT, AND NODULE NITROGEN CONTENT

The number, dry weight, and nitrogen content of the root nodules in green gram as influenced by the inoculation of Gc and *Rhizobium* are presented in Tables 7-3 and 7-4 (Expt. I and II, respectively).

There was no nodulation at all in the treatments without *Rhizobium*. Plants treated with *Rhizobium* nodulated well. The number, size, dry weight, and nitrogen content of nodules inoculated with *Rhizobium* in combination with Gc were significantly greater than those of plants inoculated with *Rhizobium* alone. *Rhizobium* treated plants in combination with Gc significantly enhanced nodulation efficiency in green gram which may be attributed to the enhanced P supply because of AM fungal inoculation.

TABLE 7-3 Number, Dry Weight, and Nitrogen Content of Green Gram Root Nodules as Influenced by the Inoculation with *Glomus constrictum* and *Rhizobium* (Expt. I)

Treatment	Nodule number/plant		Nodule DW g/plant		Nodule N/plant	
	45 d	75 d	45 d	75 d	45 d	75 d
Control	0	0	0	0	0	0
Gc	0	0	0	0	0	0
Rhiz	14	20	0.14	0.28	1.37	5.65
MP	0	0	0	0	0	0
Gc + Rhiz	20.6	32.3	0.16	0.3	2.58	10.7
Gc + MP	0	0	0	0	0	0
Rhiz + MP	12	15	0.08	0.17	1.05	4.88
Gc + Rhiz + MP	15	28	0.13	0.23	1.96	8.21
CD 5%	4.89	4.34	0.04	0.04	0.3	0.95
CD 1%	6.79	6.02	0.06	0.04	0.42	1.32

Gc - *Glomus constrictum.*

Rhiz - *Rhizobium.*

MP - *Macrophomina phaseolina.*

TABLE 7-4 Number, Dry Weight, and Nitrogen Content of Green Gram Root Nodules as Influenced by the Inoculation with *Glomus constrictum* and *Rhizobium* (Expt. II)

Treatment	Nodule number/plant		Nodule DW g/plant		Nodule N/plant	
	45 d	75 d	45 d	75 d	45 d	75 d
Control	0	0	0	0	0	0
Gc	0	0	0	0	0	0
Rhiz	10	26.6	0.03	0.09	0.93	3.68
MP	0	0	0	0	0	0
Gc + Rhiz	19	45	0.11	0.35	1.31	7.98
Gc + MP	0	0	0	0	0	0
Rhiz + MP	10	15	0.08	0.1	0.99	4.44
Gc + Rhiz + MP	14	28	0.11	0.16	1.48	7.51
CD 5%	4.47	7.95	0.03	0.05	0.61	1.85
CD 1%	6.21	11.03	0.04	0.07	0.84	2.57

Gc - *Glomus constrictum.*

Rhiz - *Rhizobium.*

MP - *Macrophomina phaseolina.*

7.3.4 CHEMICAL ANALYSIS OF PLANT TISSUES

7.3.4.1 NUTRIENT UPTAKE

Nutrient content i.e., N, P, K in green gram inoculated with Gc and *Rhizobium* in Experiments I and II are presented in Tables 7-5 and 7-6. The data indicates that the plants inoculated with Gc had significantly higher N, P, and K content over uninoculated control plants. The co-inoculated plants with Gc and *Rhizobium* further enhanced plant nutrient content significantly. The P and K content showed a significant increase in the plants inoculated with Gc followed by Gc + *Rhizobium* and finally, Gc + *Rhizobium* + pathogen and were statistically on par. The *Rhizobium* inoculated plants showed a significant increase in N content, and combined inoculation with Gc and *Rhizobium* further enhanced N uptake. The plants inoculated with Gc + *Rhizobium* showed a significant increase in the N content.[15] The same trend was observed in pathogen treated plants with different combinations which were statistically on par with the other treatments. Mycorrhizal fungi are known to improve P nutrition, co-inoculation with *Rhizobium* further enhanced N uptake.

TABLE 7-5 Nitrogen, Phosphorus, Potassium, Disease Incidence, and Grain Yield of Green Gram as Influenced by the Inoculation with *Glomus constrictum* and *Rhizobium* (Expt. I)

Treatments	N (mg/plant)		P (mg/plant)		K (mg/plant)		DI (%)	DC (%)	SW
	45 d	75 d	45 d	75 d	45 d	75 d	At harvest		
Control	23.85	30.73	7.51	9.4	0.35	0.48	0	0	2.80
Gc	34.87	46.72	9.58	22.36	0.7	0.81	0	0	4.02
Rhiz	34.71	56.92	7.4	11.31	0.67	0.86	0	0	3.72
MP	15.18	18.8	5.14	7.42	0.39	0.47	76.36		2.10
Gc + Rhiz	48.45	97.23	10.17	15.64	0.87	0.93	0	0	5.64
Gc + MP	24.93	38.54	8.5	9.64	0.64	0.78	19.86	104.95	3.40
Rhiz + MP	33.14	45.06	6	8.85	0.58	0.8	26.2	96.65	3.18
Gc + Rhiz + MP	46	91	8.11	11.66	0.76	0.89	13.1	113.80	4.50
CD 5%	7.57	13.42	2.41	2.88	0.13	0.14	5.78		56.6
CD 1%	10.5	18.63	3.34	3.99	0.18	0.19	8.02		82.77

N - nitrogen.

P - phosphorus.

K - potassium.

DI - disease incidence.

DC - per cent disease control.

SW - seed weight in g/plant.

TABLE 7-6 Nitrogen, Phosphorus, Potassium, Disease Incidence, and Grain Yield of Green Gram as Influenced by the Inoculation with *Glomus constrictum* and *Rhizobium* (Expt. II)

Treatments	N (mg/plant)		P (mg/plant)		K (mg/plant)		DI (%)	DC (%)	SW
	45 d	75 d	45 d	75 d	45 d	75 d	At harvest		
Control	18.42	26.85	6.19	9.12	0.36	0.58	0	0	2.44
Gc	33.42	38.54	9.36	12.49	0.86	0.98	0	0	3.55
Rhiz	36.12	55.35	6.29	9.24	0.86	0.92	0	0	3.65
MP	12.62	19.62	4.84	7.24	0.23	0.38	84.7		1.92
Gc + Rhiz	46.11	59.72	12.92	18.28	1.04	1.49	0	0	5.65
Gc + MP	24.71	36.27	6.39	9.21	0.81	0.92	16.9	98.11	2.99

TABLE 7-6   *(Continued)*

Treatments	N (mg/plant)		P (mg/plant)		K (mg/plant)		DI (%)	DC (%)	SW
	45 d	75 d	45 d	75 d	45 d	75 d	At harvest		
Rhiz + MP	32.46	30.46	8.02	12.21	0.56	0.82	21.7	92.44	3.33
Gc + Rhiz + MP	40.26	52.07	7.89	11.89	0.84	1.01	12.6	103.19	5.17
CD 5%	7.74	8.97	1.98	2.63	0.21	0.27	5.43		76.22
CD 1%	10.74	12.45	2.76	3.65	0.29	0.37	7.54		105.8

N - nitrogen.

P - phosphorus.

K - potassium.

DI- disease incidence.

DC - per cent disease control.

SW - seed weight in g/plant.

7.3.5 GRAIN YIELD

The grain yield of the plants inoculated with Gc and *Rhizobium* singly or in combination was significantly greater (Tables 7-5 and 7-6) when compared with the uninoculated control plants. Data indicates that mycorrhizal inoculated plants recorded significantly higher grain yield when compared with the control plants. However, dual inoculated plants with Gc and *Rhizobium* (with and without pathogen) significantly enhanced grain yield when compared with plants treated with Gc or *Rhizobium* alone.

7.3.6 DISEASE INCIDENCE

The percentage disease incidence (Tables 7-5 and 7-6) in the pathogen inoculated plants was recorded as 76.36 and 84.7% (Expt. I and Expt. II, respectively). Mycorrhizal inoculated plants significantly reduced the disease incidence to 19.86 and 16.90%, and increased per cent disease control (DC) to 104.95 and 98.11% in Experiments. I and II, respectively. However, the plants co-inoculated with Gc and *Rhizobium* were more tolerant to the fungal root rot pathogen by further reducing the disease incidence (13.1 and 12.6%) and increased per cent disease control to 113.80 and 103.19% (Expt. I and Expt. II, respectively). Earlier reports reveal that mycorrhizal plants offer increased tolerance to fungal root pathogen.[32] Bacterization of legume seeds/ seedlings with *Rhizobium* significantly reduced some root diseases caused by soil borne fungal pathogens.[33–35] The results indicate that AMF have a vital role in inhibiting the fungal root rot pathogen from invasion, more so in the presence of *Rhizobium*.

7.4   CONCLUSION

In conclusion, the results of this study indicate that the synergistic effect of AM fungus, *Glomus constrictum* and *Rhizobium* is beneficial to green gram for its improved plant growth and development, mobilization of more N, P, and K as compared with control and pathogen inoculated plants. However, the mycorrhizal ability was significantly enhanced by co-inoculation with *Rhizobium*, in further increasing the nodulation ability and hence, a probable increase in the nodule N as also reported by different workers in other legumes. Therefore, the present study suggests that suitable combinations of arbuscular mycorrhizal fungus and *Rhizobium* inoculation may contribute efficiently to increase plant growth, nutrient uptake, and resistance to fungal root pathogens.

ACKNOWLEDGMENTS

Dr. A. Hindumathi is grateful to the Department of Science and Technology, New Delhi for providing fellowship under the Women Scientist Scheme-A (WOS-A) with a Grant No. SR/WOS-A/LS-498/2011(G).

KEYWORDS

- Biomass
- *Glomus constrictum*
- Green gram
- *Macrophomina phaseolina*
- *Rhizobium*

REFERENCES

1. Bagyaraj, D. J. Arbuscular mycorrhizal fungi in sustainable agriculture. In *Techniques in Mycorrhizae*; Bukari, M. J., and Rodrigues, B. P., Eds.; Department of Botany; Goa University: Goa, 2006.
2. Silveira, A. P. D.; and Cardoso, E. J. B. N. Arbuscular mycorrhiza and kinetic parameters of phosphorus absorption by bean plants. *Agricultural Sci.* **2004**, *61*, 203–209. DOI: 10.1590/S0103-90162004000200013.
3. Chalk, P. M.; de Souza, R. F.; Urquiaga, S.; Alves, B. J. R.; and Boddey, R. M. The role of arbuscular mycorrhiza in legume symbiotic performance. *Soil Biol. Bioch.* **2006**, *38*, 2944–2951.
4. Asai, T. Uber die Mykorrhizen-bildung der leguminosen Pflanzen. *Jpn. J. Bot.* **1944**, *13*, 463–485.
5. Barea, J. M.; Azcon, R.; and Azcon-Aguilar, C. Mycorrhizosphere interactions to improve plant fitness and soil quality. *Antonie Van Leeuwen.* **2002**, *81*, 343–351.
6. Olivera, M.; Tejera, N.; Iribarne, C.; Ocana, A.; and Lluch, C. Growth, nitrogen fixation and ammonium assimilation in common bean (*Phaseolus vulgaris*): Effect of phosphorus. *Physiologia Plantar.* **2004**, *121*, 498–505.
7. Khan, A. H.; Islam, A.; Islam, R.; Begum, S.; and Huq, S. M. I. Effect of indigenous VA mycorrhizal fungi on nodulation, growth and nutrition of lentil (*Lens culinaris* L.) and black gram (*Vigna mungo* L.). *J. Plt. Physiol.* **1988**, *133*, 84–88.
8. Bagyaraj, D. J.; Manjunath, A.; and Patil, R. B. Interaction between a vesicular-arbuscular mycorrhiza and *Rhizobium* and their effects on soybean in the field. *New Phytol.* **1979**, *82*, 141–145.
9. Hindumathi, A.; and Reddy, B. N. Synergistic effect of arbuscular mycorrhizal fungi and *Rhizobium* on the growth and charcoal rot of soybean [Glycine max (L.) Merr.]. *Wor. J. Sci. Technol.* **2012**, *2*, 63–70
10. Schenck, N. C.; and Kellam, M. K. The influence of vesicular-arbusclar mycorrhizae on disease development. *Flor. Agri. Exper. Stat. Bull.* **1978**, 798.
11. Reddy, B. N.; Raghavender, C. R.; and Sreevani, A. Approach for enhancing mycorrhiza-mediated disease resistance of tomato damping-off. *Ind. Phytopathol.* **2006**, *59*, 299–304.
12. Azcon-Aguilar, C.; and Barea, J. M. Arbuscular mycorrhizas and biological control of soil borne plant pathogens-an overview of the mechanisms involved. *Mycorr.* **1996**, *6*, 457–464.
13. Gill, T. S.; and Singh, R. S. Effects of *Glomus fasciculatum* and *Rhizobium* inoculation on VA. mycorrhizal colonization and plant growth of chickpea. *Ind. Phytopathol.* **2002**, *32*, 162–166.
14. Talaat, N.; and Abdallah, A. M. Response of faba bean (*Vicia fava* L.) to dual inoculation with *Rhizobium* and VA mycorrhiza under different levels of N and P fertilization. *J. App. Sci. Res.* **2008**, *4*, 1092–1102.
15. Chakrabarty, J.; Chatterjee, N. C.; and Dutta, S. Interactive effect of VAM and *Rhizobium* on nutrient uptake and growth of *Vigna mungo*. *J. Mycopathol. Res.* **2007**, *45*, 289–291.
16. Menge, J. A.; Nemec, S.; Davis, R. M.; and Minassian, V. Mycorrhizal fungi associated with citrus and their possible interactions with pathogens. *Proc. Int. Soc. Citricul.* **1977**, *3*, 872–876.

17. Gerdemann, J. W.; and Nicolson, T. H. Spores of mycorrhizal endogone species extracted from soil by wet sieving and decanting. *Trans. of the British Mycol. Soc.* **1963,** *46,* 235–244.

18. Vincent, J. M. A manual for the practical study of root nodule bacteria. In *International Biology Program, Handbook No. 15;* Blackwell Scientific Publications: Oxford; 1970.

19. Dennis, C.; and Webster, J. Antagonistic properties of species groups of *Trichoderma.* III. Hyphal interaction. *Trans. Brit. Mycol. Soc.* **1971,** *57,* 363–369.

20. Schenck, N. C. *Methods and Principles of Mycorrhizal Research;* The American Phytopathological Society: America, 1982.

21. Jauhri, K. S.; Bhatnagar, R. S.; and Ishwaran, V. Associative effect of inoculation of different strains of *Azotobacter* and homologous *Rhizobium* on the yield of mung (*V. radiata*), soybean (*Glycine max*) and pea (*Pisum sativum*). *Plant and Soil* **1979,** *53,* 105–108.

22. Tilak, K. V. B. R.; Pal, K. K; and Dey, R. *Microbes for Sustainable Agriculture*; IK International Pub Ltd.: New Delhi, 2010, p 200.

23. Hoagland, D. R.; and Arnon, D. I. *The Water-culture Method for Growing Plants Without Soil*; University of California, College of Agriculture, Agricultural Experiment Station: Berkeley, California, 1950; Vol. 347, pp 1–32.

24. Jackson, M. L. *Soil Chemical Analysis*; Prentice Hall Pvt. Ltd.: New Delhi, India, 1973.

25. Bremner, J. M. Determination of nitrogen in soil by Kjeldahl method. *J. Agric. Sci.* **1960,** *55,* 11–33.

26. Phillips, J. M.; Hayman, D. S. Improved procedures for clearing of roots and staining parasitic and vesicular-arbuscular mycorrhizal fungi for rapid assessment of infection. *Trans. British Mycol. Soc.* **1970,** *55,* 158–161.

27. Giovannetti, M.; Mosse, B. An evaluation of techniques for measuring vesicular-arbuscular mycorrhizal infection in roots. *New Phytol.* **1980,** *84,* 489–500.

28. Ledingham, R. J.; Atkinson, T. G.; Horrocks, J. S.; Piening, L. J.; Mills, J. T.; Tinline, R. D. Wheat losses due to common root rot in the prairie provinces of Canada. *Can. J. Pland. Dis. Sur.* **1973,** *53,* 113–122.

29. Mortimer, P. E.; Pérez-Fernández, M. A.; Valentine, A. J. The role of arbuscular mycorrhizal colonization in the carbon and nutrient economy of the tripartite symbiosis with nodulated *Phaseolus vulgaris. Soil Biol. Biochem.* **2008,** *40,* 1019–1027.

30. Rajasekaran, S.; Nagarajan, S. M. Effect of dual inoculation AM fungi and *Rhizobium* on chlorophyll content of *Vigna unguiculata* (L.) Walp. var. pusa 151. *Mycorrhiza* **2005,** *17,* 10–11.

31. Nazir, H.; Hassan, B.; Habib, R.; Chand, L.; Ali, A.; Hussain, A. Response of bio-fertilizers on growth and yield attributes of blackgram. *Inter. J. Current Res.* **2011,** *2,* 148–150.

32. Singh P. K. To study Microbial Interaction of Below ground organisms with special reference to Fusarium wilt of chickpea. Ph.D. Thesis, unpublished, Dr. H. S. Gour University, Sagar, MP, India, 2008.

33. Chakraborty, U.; Purkayatha, R. P. Role of rhizotoxin in protecting soybean roots from *Macrophomina phaseolina* infection. *Can. J. Microbiol.* **1984,** *30,* 285–289.

34. Chakraborty, U.; Chakraborty, B. N. Interaction of *Rhizobium leguminosarum* and *Fusarium solani* f.sp. *pisi* on pea affecting disease development and phytolaexin productions. *Can. J. Bot.* **1989,** *67,* 1698–1702.

35. Dar, H.; Zargar, G. M. Y.; Beigh, G. M. Biocontrol of Fusarium root rot in the common bean (*Phaseolus vulagaris* L.) by using symbiotic *Glomus mosseae* and *Rhizobium leguminosarum. Microb. Ecol.* **1997,** *34,* 74–80.

CHAPTER 8

MORPHOLOGICAL CHANGES IN *VIGNA RADIATA* ROOT UNDER CADMIUM INDUCED STRESS IN THE PRESENCE OF PLANT GROWTH PROMOTING *ENTEROBACTER* SP. C1D

RAKESH KUMAR SHARMA[1], KAVITA BAROT[1], and G. ARCHANA[1*]

[1]*Department of Microbiology and Biotechnology Center, The Maharaja Sayajirao University of Baroda, Vadodara, Gujarat – 390002, India.*

*Corresponding Author: G. Archana. E-mail: archanagayatri@yahoo.com

CONTENTS

Abstract 172
8.1   Introduction 172
8.2   Materials and Methods 173
8.3   Results and Discussion 176
8.4   Conclusion 181
Acknowledgments 181
Keywords 182
References 182

ABSTRACT

Enterobacter sp. C1D is a multi-metal tolerant plant growth promoting bacterium. The aim of the present work was to understand the effect of Cd induced morphological changes on the roots of *Vigna radiata* and the influence of *Enterobacter* sp. C1D on root development. The experiments were conducted under hydroponics in which germinated seedlings of *Vigna radiata* were exposed to a range of different Cd concentrations to determine 50% inhibitory concentration (IC_{50}), Cd accumulation, and cell death. Hematoxylin staining revealed that Cd accumulation was deceased by 90% using bacterial treatment. A novel dual agar plate method has been developed in which one side of the germinated seedling was exposed to Cd containing medium, while the other side was without Cd. This method allows the visualization of the effect of Cd on the overall growth of the plant as well as on the tissues directly exposed to Cd as compared with the unexposed tissues of the same plant. In this method, bacterial effect can be studied by incorporating the culture in the medium. Cadmium was found to inhibit the overall root elongation and it negatively influences root anatomy and lateral root development. Some clear morphological responses observed were: (i) the roots were quite sensitive with the root apex tending to turn toward the no Cd zone, (ii) higher numbers of lateral roots were seen on the side of the Cd free medium, (iii) microscopic images of the transverse section of the root showed Cd mediated damages on one side of the root, (iv) phloroglucinol staining of the section showed high lignin formation only one side of the root. On the other hand, coinoculation with bacterium *Enterobacter* sp. C1D alleviated Cd stress. The number and the length of the lateral roots were much higher as compared with the uninoculated control.

8.1   INTRODUCTION

Heavy metal contamination of agricultural lands around industrial areas is a serious problem. Metal contaminations arise because of the activities such as the use of phosphate fertilizers and pesticides containing heavy metals, burning of fossil fuels, discharge of industrial wastes and effluents, mining, etc.[1] Cadmium (Cd) is a carcinogenic metal and is toxic to almost all the living forms including microorganisms, plants, and humans in high concentrations. Cadmium is biologically nonessential but is readily absorbed by plants and affects almost all the stages of plant development from seed germination to fruit ripening.[2] It can be accumulated by many plant species including cereals, pulses, vegetables, and fruits. This accumulated Cd in the edible parts of the plants enters the human food chain, which may cause severe chronic diseases. Several physiological and morphological changes have been demonstrated in plants upon exposure to Cd.[3] Plant response to Cd is very specific as some plants have efficient mechanisms to cope up with Cd stress, while others are very sensitive. Cadmium affects pulses like mung bean in various ways: First, it negatively affects the growth of the plant and enters in the plant trough roots as there is no well defined mechanism to cope up with the Cd stress. Second, the entered Cd may be stored in the grains, which is the main route of entry into the human body.[4] On the other hand, some plants are used for soil remediation technologies based on Cd phytoextraction in the phytoremediation process. Techniques like scanning electron microscopy and energy-dispersive X-ray microanalysis revealed that poplar plants used for phytoremediation show Cd accumulation by storing Cd in minor toxic forms in cell compartments weakly sensitive to Cd.[5]

Plant-associated microorganisms play a vital role in promoting plant health by various mechanisms (production of plant hormones, stress relieving enzymes, organic acids, siderophores, phosphate solubilization, etc.) which are known as plant growth promoting rhizobacteria (PGPR). Some of these bacteria possesses Cd resistance and colonize metal contaminated rhizospheric soil. Metal resistance in bacteria is either because of metal efflux system or sequestration of metal ions.[6] Many Cd resistant PGPR show a number of responses to metal ions, e.g., metal biosorption, metal precipitation, enzymatic metal transformation, and rendering them unavailable; thereby reducing the metal ion toxicity toward the plant. Application of these metal resistant bacteria alleviates metal toxicity and enhances the growth and productivity of plants in metal polluted environments.[7,8] The present study demonstrates the morphological effect of Cd on mung bean and the effect of *Enterobacter* sp. C1D inoculation. A dual agar plate method was designed to study the effect of Cd simultaneously on the same plant in the presence and absence of *Enterobacter* sp. C1D.

8.2 MATERIALS AND METHODS

8.2.1 *SURFACE STERILIZATION AND GERMINATION OF MUNG BEAN SEEDS*

Vigna radiata (mung bean) seeds were procured from Pulse Research Station, Anand Agricultural University, Model Farm, Vadodara. The seeds were surface sterilized by soaking in 0.1% (w/v) $HgCl_2$ for 10 min. The seeds were then thoroughly washed with sterile water and place onto a pre-sterilized, moistened filter paper in a petri dish and germinated for 24–30 h at 30 °C ± 2 °C in dark.

8.2.2   BACTERIAL STRAIN

A heavy metal tolerant bacterial strain, *Enterobacter* sp. C1D, was earlier isolated from a heavy metal contaminated field. Identification was done by 16S rRNA gene sequence (NCBI Accession No. JN936958). The bacterial isolate was earlier studied for its plant growth promoting properties and its ability to alleviate Cr stress in plants.[9]

8.2.3   HYDROPONIC EXPERIMENTS

Germinated seedlings were transferred on floats in autoclaved plant tissue culture jars containing 0.05% $CaCl_2$ solution with varying concentrations of $CdCl_2$ and incubated at 30 °C ± 2 °C for 5 days under natural day light conditions.

8.2.4   DUAL AGAR PLATE METHOD FOR STUDYING THE EFFECT OF CD ON PLANT ROOTS

A dual agar plate method was designed to observe the effect of Cd and bacterial culture on plants. The plates were prepared as follows (Figure 8-1): A sterilized separator was fixed in a pre sterilized empty petri plate to divide the plate into two sections. Sterilized Murashige-Skoog (MS) agar medium was poured on one side of the plate, allowed to solidify and then MS medium containing Cd (10 or 50 µg/ml) was poured on the other side. The removal of separator after the solidification of the agar formed a gap of approximately 2 mm in width along the diameter of the plate. A germinated seedling was placed in such a manner that its root was aligned in the gap between the two agar halves. The plates were incubated at 30 °C ± 2 °C for 5 days. A similar set of plates was prepared by the addition (1% v/v) of 6 h grown culture of *Enterobacter* sp. C1D to see the effect of bacteria on seedling development in the presence and absence of Cd. After incubation, various parameters including total length, number of lateral roots, and weight were recorded.

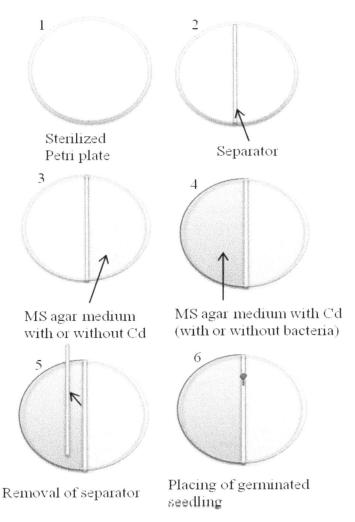

FIGURE 8-1   Schematic presentation of dual agar plate method to study the effect of metal and PGPR on the same plant.

8.2.5   *MICROSCOPIC AND STAINING STUDIES*

The roots were stained by soaking them in hematoxylin[10] or Evan's blue[11] solution for 15 min, washed with distilled water three to four times to remove excess stain. Transverse sections were cut (10 mm from for the root apex) and observed under a bright field microscope. The sections were also stained with phloroglucinol to detect lignification.

8.3   RESULTS AND DISCUSSION

8.3.1   EFFECT OF CD ON V. RADIATA

Cadmium negatively affected the seedling growth under hydroponic conditions from 1 µg/ml, the lowest concentration used. A 50% growth inhibitory concentration of Cd (Cd-IC$_{50}$) was 4 µg/ml (Figure 8-2A). Several morphological and physical changes were observed during seedling germination in the presence of Cd as compared with the control. Damage of seedlings was evident as browning of root tip, brown patches on the root surface, and severe cuts were seen under the microscope (Figure 8-2B). It is well known that Cd reduces plant growth and influences various visual growth abnormalities such as leaf chlorosis, root browning, etc.[12]

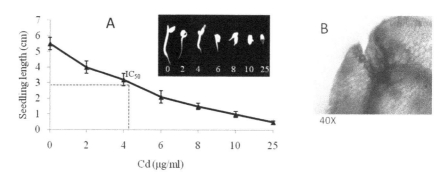

FIGURE 8-2   (A) Effect of Cd on *V. radiata* seedling germination (B) Cd mediated root tissue damage.

8.3.2   V. RADIATA ROOT DAMAGE AND CD ACCUMULATION

Evan's blue staining is widely used for the detection of dead cells in plants.[13] The cell membrane is impermeable to Evan's blue, but, when it undergoes an injury or loses its selective permeability, the dye can penetrate it. Thus, the amount of Evan's blue entering and binding dead cells reflects the degree of loss of cell membrane integrity and is an estimate of cell death.[14] Staining of Cd exposed roots of *V. radiata* with Evan's blue showed a dark blue appearance; particularly the fine root hair were evident. However, the control plants and Cd-treated plants inoculated with *Enterobacter* sp. C1D did not show blue color; the staining of the fine root hair, which were apparently present, also did not show a blue color as they remained unstained (Figure 8-3).

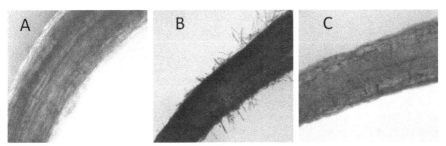

FIGURE 8-3   *Vigna radiata* root after Evan's blue staining showed (A) no color in the control untreated seedling (B) dark blue stained cells in the presence of Cd (C) very less color in the *Enterobacter* sp. C1D treated seedling.

Hematoxylin root staining was developed by Polle et al.[15] to evaluate the metal tolerance mechanism in cereal crop plants. Hematoxylin binds metal to produce a purple complex. The absence of color in root tips treated with hematoxylin indicates that these plants either exclude or bind metal in complexes that are unavailable to hematoxylin. The method has been used extensively for quick evaluation of and screening for metal tolerance including Cd tolerance. Cadmium treated plant roots were dark pink in color when stained with hematoxylin as compared with the control (Figure 8-4A). The plant may accumulate Cd in the shoot and roots. Lux et al.[16] reported that Cd concentrations were greater in the root apoplasm than in the root symplasm, and tissue Cd concentrations decreased from peripheral to the inner root tissues. Similarly, in the present study, hematoxylin staining revealed that the peripheral tissues were stained dark pink as compared with the inner tissues, which were lighter in color.

Thus, the Cd:hematoxylin complexes were more pronounced in outer tissues than the inner ones (Figure 8-4B). On the other hand, there was lesser deposition of Cd inside the plant cells as evident by lesser color formation in the presence of *Enterobacter* sp. C1D (Figure 8-4C). On the contrary, some reports suggest higher Cd accumulation inside the plant, while some demonstrated lower accumulation.[17,18] Thus, this Cd accumulation phenomenon depends upon the plant, associated microbe and their environment.

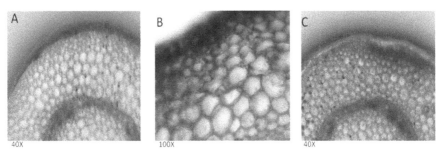

FIGURE 8-4   Hematoxylin staining of *V. radiata* root showed (A) no color in untreated seedling (B) intense pink color shows higher Cd content (C) very less color in the *Enterobacter* sp. C1D treated seedlings.

8.3.3   DUAL AGAR PLATE METHOD FOR THE EFFECT OF CD AND ENTEROBACTER ON MUNG BEAN

In the plates with no Cd, the root and overall seedling growth was normal, with the root growing along the gap in between the agar (Figure 8-5A). As seen in Figure 8-5B, the addition of *Enterobacter* sp. C1D into the agar resulted in a longer root as well as more pronounced lateral root development. It is interesting to note that the changes in lateral rooting occurred on both the sides of the root although, *Enterobacter* was added only on one side. In the presence of Cd in one half of the plate, the main root length was reduced and the root apex turned towards the agar with no Cd (Figure 8-5C). Lateral roots also predominantly emerged on the Cd-free side. In the presence of *Enterobacter* and Cd, the lateral rooting was enhanced; however, the roots developed only in the agar with no Cd. Seedling growth including total length, number of lateral roots, and weight was significantly reduced in the presence of Cd on one side of the dual agar (Figure 8-5). *Enterobacter* sp. C1D supported better growth along with higher lateral roots (50% enhancement under 10 µg/ml Cd); however, the number of lateral roots was higher in agar containing no Cd (Table 8-1).

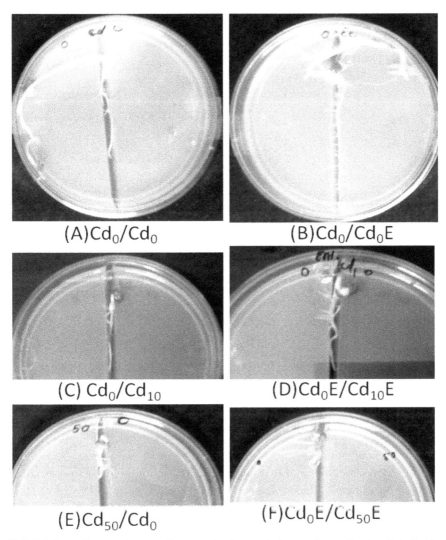

FIGURE 8-5   *Vigna radiata* seedling germination on dual agar plate; subscript digit (0, 10, 50) shows Cd concentration (µg/ml) in the medium, E: *Enterobacter* sp. C1D.

TABLE 8-1   Plant Development Parameters during Dual Agar Plate Method

S. No.	Cd concentration (μg/ml) (S1/S2)	Total length (cm)	No. of lateral roots		No. of leaves	Weight (mg)
			S1	S2		
1	Cd_0/Cd_0	12 ± 2	5 ± 0.7	7 ± 0.8	2 ± 0.0	0.31 ± 0.05
2	Cd_0/Cd_{10}	8 ± 1.2	5 ± 0.5	2 ± 0.1	1 ± 2	0.51 ± 0.02
3	Cd_0/Cd_{50}	2 ± 0.4	6 ± 1.5	0	0	0.20 ± 0.04
4	Cd_0/Cd_0E^*	12.5 ± 2.3	7 ± 1.2	11 ± 1.4	2 ± 0.0	0.32 ± 0.05
5	$Cd_0E/Cd_{10}E$	10 ± 1.8	9 ± 1.1	4 ± 0.5	2 ± 0.2	0.39 ± 0.07
6	$Cd_{10}/Cd_{10}E$	3 ± 0.2	3 ± 0.5	6 ± 0.8	0	0.22 ± 0.03
7	$Cd_0E/Cd_{50}E$	5 ± 0.25	10 ± 1.5	0	0	0.19 ± 0.02
8	$Cd_{50}/Cd_{50}E$	3.5 ± 0.2	1 ± 1	5 ± 1.2	0	0.17 ± 0.03

S1: side 1 of the dual agar plate.

S2: side 2 of the dual agar plate.

*E indicates the addition of *Enterobacter* sp. C1D to the agar.

8.3.4   MORPHOLOGICAL CHANGES IN ROOT STRUCTURE

Significant Cd mediated damages could be seen in half of the root exposed to Cd (Figure 8-6A). Sections were cut at a distance of 10 mm from the root apex and stained by phloroglucinol–HCl to detect lignin. Lignification was asymmetrical, occurring primarily in the half of the root exposed to Cd (Figure 8-6B). It was known that Cd stimulates lignification in a dose-independent manner by enhancing the activity of enzymes involved in this pathway. It is reported that the sequence of reactions leading to Cd-stimulation of lignification might not be direct, rather the metal toxicity in plants results in the accumulation of endogenous phenolic compounds. These phenolics (intermediates of lignin biosynthesis) can be oxidized to phenoxyl radicals by peroxidases to form polymers such as lignin in the apoplast.[19] The hampered root growth that was observed during Cd exposure can easily be correlated with lignification of the plant cell wall. The restricted growth of the roots was earlier explained as the effects caused by Cd which may be because of the excessive production of monolignols forming lignin, which solidifies the cell wall and restricts the root growth.[20] Lignification may also prevent lateral root emergence on the side exposed to Cd. *Enterobacter* sp. C1D reduced the lignification of the root as evident by phloroglucinol staining (Figure 8-6C). Thus, *Enterobacter* sp. C1D not only enhanced plant growth, but also protected the plant from Cd induced damages. Earlier, different strains of *Enterobacter* had been reported to reduce metal induced stress in plants, which enhanced plant growth as well as the productivity of the agriculturally important crops.[21] The present work addresses a detailed effect of *Enterobacter* on root development in the presence of Cd.

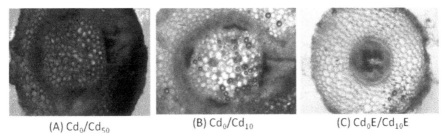

(A) Cd_0/Cd_{50} (B) Cd_0/Cd_{10} (C) $Cd_0E/Cd_{10}E$

FIGURE 8-6   (A) Significant Cd mediated damages could be seen in half of the root (B) phloroglucinol staining showing asymmetric lignin formation (C) no detectable lignification was seen on *Enterobacter* sp. C1D containing medium.

8.4   CONCLUSION

A novel method for studying the effect of heavy metal and metal tolerant PGPR on plant growth and root development in short term experiments has been demonstrated. In this study, *Enterobacter* sp. C1D, a PGPR isolate was shown to alleviate Cd stress in mung bean plants by the stimulation of root growth and lateral root development. It prevented Cd-mediated morphological deformities in root architecture by reducing lignification and cell death.

ACKNOWLEDGMENTS

Rakesh Kumar Sharma is thankful to University Grants Commission (UGC), India for awarding UGC-Dr. D. S. Kothari Postdoctoral Fellowship (F.4-2/2006(BSR)/13-728/2012(BSR).

KEYWORDS

- **Cadmium toxicity**
- **Dual agar plate method**
- *Enterobacter sp.*
- *Vigna radiata*

REFERENCES

1. Chopra, A. K.; Pathak, C.; and Prasad, G. Scenario of heavy metal contamination in agricultural soil and its management. *J. Appl. Natural Sci.* **2009,** *1,* 99–108.
2. Mondal, N. K.; Das. C.; Roy, S.; Datta, J. K; and Banerjee, A. Effect of varying cadmium stress on chickpea (*Cicer arietinum* L) seedlings: An ultrastructural study. *Annals Environ. Sci.* **2013,** *7,* 59–70.
3. Benavides, M. P.; Gallego, S. M.; and Tomaro, M. L. Cadmium toxicity in plants. *Brazalian J. Plant Physiol.* **2005,** *17,* 21–34.
4. Wahid, A.; Ghani, A.; and Javed, F. Effect of cadmium on photosynthesis, nutrition and growth of mungbean. *Agron. Sustain. Develop.* **2008,** *28,* 273–280.
5. Cocozza, C.; Minnocci, A.; Tognetti, R.; Iori, V.; Zacchini, M.; and Scarascia Mugnozza, G. Distribution and concentration of cadmium in root tissue of *Populus alba* determined by scanning electron microscopy and energy-dispersive X-ray microanalysis. *iForest Biogeosci. Forest.* **2008,** *1,* 96–103.
6. Nies, D. H. Microbial heavy-metal resistance. *Appl. Microbiol. Biotechnol.* **1999,** *51,* 730–750.
7. Prapagdee, B.; Chanprasert, M.; and Mongkolsuk, S. Bioaugmentation with cadmium-resistant plant growth-promoting rhizobacteria to assist cadmium phytoextraction by *Helianthus annuus. Chemosphere* **2013,** *92,* 659–666.
8. Ganesan, V. Rhizoremediation of cadmium soil using a cadmium-resistant plant growth-promoting rhizopseudomonad. *Curr. Microbiol.* **2008,** *56,* 403–407.
9. Subrahmanyam, G.; and Archana, G. Plant growth promoting activity of *Enterobacter* sp. C1D in heavy metal contaminated soils. Proceedings of the 2nd Asian PGPR Conference, Beijing, P.R. China, **2011,** 440–446.
10. Kavita, B.; Shukla, S.; Naresh Kumar, G.; and Archana, G. Amelioration of phyto-toxic effects of Cd on mung bean seedlings by gluconic acid secreting rhizobacterium *Enterobacter asburiae* PSI3 and implication of role of organic acid. *World J. Microbiol. Biotechnol.* **2008,** *24,* 2965–2972.
11. Yin, L.; Mano, J.; Wang, S.; Tsuji, W.; and Tanaka, K. The involvement of lipid peroxide-derived aldehydes in aluminum toxicity of tobacco roots. *Plant Physiol.* **2010,** *152,* 1406–1417.
12. Yang, X.; Baligar, V. C.; Martens, D. C.; and Clark, R. B. Cadmium effects on influx and transport of mineral nutrients in plant species. *J. Plant Nut.* **1996,** *19,* 643–656.

13. Steffens, B.; and Sauter, M. Epidermal cell death in rice is confined to cells with a distinct molecular identity and is mediated by ethylene and H_2O_2 through an autoamplified signal pathway. *Plant Cell* **2009**, *21*, 184–196.

14. Gonzalez-Mendozaa, D.; Quiroz-Moreno, A.; Medrano, R. E.; Grimaldo-Juarez, O; and Zapata-Perez, O. Cell viability and leakage of electrolytes in *Avicennia germinans* exposed to heavy metals. *Z. Naturforsch. C.* **2009**, *64*, 391–394.

15. Polle, E.; Konzak, C. F.; and Kittrick, J. A. Visual detection of aluminum tolerance levels in wheat by hematoxylin staining. *Crop Sci.* **1978**, *18*, 823–827.

16. Lux, A.; Martinka, M.; Vaculik, M; and White, P. J. Root responses to cadmium in the rhizosphere: A review. *J. Exp. Bot.* **2011**, *62*, 21–37.

17. Ahmad, I.; Akhtar, M. J.; Zahir, Z. A.; Naveed, M.; Mitter, B; and Sessitsch, A. Cadmium-tolerant bacteria induce metal stress tolerance in cereals. *Environ. Sci. Pollut. Res.* **2014**, *21*, 11054–11065.

18. Jeong, S.; Moon, H. S.; Nam, K.; Kim, J. Y.; and Kim, T. S. Application of phosphate-solubilizing bacteria for enhancing bioavailability and phytoextraction of cadmium (Cd) from polluted soil. *Chemosphere* **2012**, *88*, 204–210.

19. Chaoui, A.; El Ferjani, E. Effects of cadmium and copper on antioxidant capacities, lignification and auxin degradation in leaves of pea (*Pisum sativum* L.) seedlings. *C. R. Biologies.* **2005**, *328*, 23–31.

20. Finger-Teixeira, A.; Ferrarese Mde, L.; Soares, A. R.; da Silva, D.; and Ferrarese-Filho, O. Cadmium-induced lignification restricts soybean root growth. *Ecotox. Environ. Safe.* **2010**, *73*, 1959–1964.

21. Dimkpa, C.; Weinand, T.; and Asch, F. Plant–rhizobacteria interactions alleviate abiotic stress conditions. *Plant Cell Environ.* **2009**, *32*, 1682–1694.

CHAPTER 9

POTENTIAL USE OF *TRICHODERMA* SPECIES AS PROMISING PLANT GROWTH STIMULATOR IN TOMATO (*LYCOPERSICUM ESCULANTUM L.*)

B. N. REDDY[1*], K. SARITHA[1], and A. HINDUMATHI[2]

[1]Mycology and Plant Pathology Laboratory, Department of Botany, Osmania University College for Women, Koti, Hyderabad, Telangana – 500095, India.

[2]Mycology and Plant Pathology Laboratory, Department of Botany, Osmania University, Hyderabad, Telangana, – 500007 India.

*Corresponding author: B. N. Reddy. E-mail: reddybn1@yahoo.com

CONTENTS

Abstract 186
9.1   Introduction 186
9.2   Materials and Methods 187
9.3   Results and Discussion 189
9.4   Conclusion 196
Acknowledgments 196
Keywords 197
References 197

ABSTRACT

This paper deals with six *Trichoderma* species screened for plant growth response in tomato (*Lycopersicum esculantum L.*) under pot experimental conditions. The inoculation of different *Trichoderma* species alone and in combination showed a positive influence on plant growth and biomass production. *Trichoderma viride*(Tv), *T. harzianum* (Th), and dual inoculated plants with *Trichoderma viride* + *T. harzianum* showed a significant increase in plant growth measured in terms of shoot and root lengths, their fresh and dry weights as well as yield parameters. The dual inoculated plants with Tv + Th showed an enhanced shoot and root height by 24.46 and 27.23% and by 15.85 and 14.73%, respectively over *Trichoderma* species treated singly either with *T. viride* or *T. harzianum*. The growth response was observed to increase proportionately with an increase in the age of the crop. The plants treated with Tv, Th, and dual inoculated plants showed statistically significant increase in shoot and root fresh and dry weights over control and other treatments of *Trichoderma* species. Fruit yield showed statistically significant increase in the plants treated with five *Trichoderma* species. However, plants treated with *T. pseudokoningi* showed a marked increase in fruit yield over the control, but were not significantly increased.

9.1   INTRODUCTION

Trichoderma species are free-living fungi that are common in the soil and root ecosystems. They have been widely studied for their capacity to produce antibiotics, parasitize other fungi, and compete with deleterious plant microorganisms.[1] In addition to their biocontrol activities, *Trichoderma* spp. has been reported to have shown beneficial effects on plants, increasing their growth potential. Possible explanations of this phenomenon include: control of minor pathogens leading to stronger root growth and nutrient uptake,[2] secretion of plant growth regulatory factors such as phytohormones[3–5] and the release of soil nutrients and minerals by increased saprophytic activity of *Trichoderma* in the soil.[6]

However, it is becoming increasingly clear that certain *Trichoderma* spp. have substantial direct influence on plant growth, development, and crop productivity.[7] The influence of *Trichoderma* on plant growth enhancement has been known for many years and can occur in axenic systems or in soils.[4,8,9] The fungus, *Trichoderma harzianum* is an efficient biological control organism against a wide range of soil-borne pathogens and has plant growth promoting capacity. It has been reported that *T. harzianum* stimulated the growth of tomato plants.[10–12]

Earlier results indicate that *Trichoderma* promotes growth responses in radish, pepper, cucumber, and tomato.[9,13] Further studies demonstrated that *Trichoderma* increases root development and crop yield, proliferation of secondary roots, seedling fresh weight, and foliar area[14], enhances root biomass production, and increases root hair development.[15,16]

Trichoderma is widely used as a biocontrol agent against phytopathogenic fungi, and also as a biofertilizer because of its ability to establish mycorrhiza-like association with plants. Different species of *Trichoderma* exhibit plant-growth promoting activity.[17] Today, it is considered that the plants benefit from this relationship through increased plant growth and development. Hence, direct effects of these fungi are crucially important for agricultural uses. Therefore, *Trichoderma* sp. may benefit as a growth promoter (biofertilizer) and also as a pathogen control agent (mycofungicide/biocontrol). Their application may lower the production costs and also environmental impact. Tomato (*Lycopersicon esculentum*), the "Love Apple" is the most popular vegetable in the world. It is one of the most important protective food crops of India and is an important commercial vegetable crop for farmers because of its multipurpose uses. Tomato contains amazing amounts of lycopene which is said to have antioxidant and defense properties. The nutritional benefits are many, so there is no reason to avoid this simple, cost effective, and widely available fruit/vegetable. To remain competitive with the leading countries in tomato production, growers must increase yields and offset production costs. Hence, in the present study, six native *Trichoderma* species were evaluated for their efficacy in stimulating plant growth and the yield of tomato under pot experimental conditions.

9.2 MATERIALS AND METHODS

9.2.1 MICROBIAL ISOLATES

Seven different species of *Trichoderma* were isolated from rhizosphere soils of tomato plants obtained from vegetable garden, Women's College, Koti, Hyderabad using Trichoderma Specific Agar (TSA) and Potato Dextrose Agar (PDA) media. *Trichoderma* spp. were identified based on the distinct morphological and cultural characters using standard manuals.[18–21] The cultures of *Trichoderma* spp. were purified by single spore isolation technique, maintained on PDA slants and stored in the refrigerator at 4 °C for further use.

9.2.2   PREPARATION OF ANTAGONIST INOCULUMS

From among the seven *Trichoderma* species isolated, six species were selected to study their efficacy on plant growth promotion and fruit yield. Inoculum of *Trichoderma* spp. was prepared by liquid fermentation method using Czapek Dox broth medium.[22] The fungal biomass and broth were mixed with talc powder in the ratio of 1 : 2. The mixture was air dried and mixed with carboxy methyl cellulose (CMC) @ 5 g/kg of the product. It was packed in sterilized polythene covers and used within four months.

9.2.3   SEED TREATMENT WITH TRICHODERMA FORMULATION

Tomato seeds were surface disinfected with 2% sodium hypochloride solution followed by serial washings with sterile distilled water. The surface disinfected seeds were inoculated first with six different bioagents (*Trichoderma* spp.). The tomato seeds were mixed thoroughly with 2 g of the formulation (2×10^3 cfu/seed). The seeds were air dried for 30 min and planted directly. Surface disinfected seeds sown in sterilized soil served as the control and were applied as 2 g/kg of soil.

9.2.4   POT EXPERIMENT

A pot experiment was designed under greenhouse conditions using earthen pots (15 cm, diameter). The soil used for the pot experiment was sterilized at 15 lb pressure, 121 °C for 20 min for three consecutive days. The soil was analyzed for physico-chemical characters. The sterilized soil was filled in earthen pots of 5 kg capacity. The pots which were filled with soil, were watered daily for one week before sowing the seeds. The seeds treated with six different species of *Trichoderma* were sown. Seeds without any *Trichoderma* spp. were sown in the control pots. In each pot, six seeds were sown and three replications were maintained for each treatment. The pots were arranged in completely randomized block design. Thinning was done to four plants per pot four weeks after germination.

The experiment included the following treatments: (1) non infested soil (control), (2) *T. atroviride* (Tatr), (3) *T. koningii* (Tk), (4) *T. reesei* (Tr), (5) *T. viride* (Tv), (6) *T. pseudokoningi* (Tps), (7) *T. harzianum* (Th), (8) Tv + Th. The Pots were kept under greenhouse conditions till the end of the experiment. The observations were recorded on 30, 60, and 90 days after sowing (DAS). On each sampling date, the following parameters were recorded: (a) shoot and root length (b) shoot and root fresh and dry weights (g/plant) (c) fruit yield (g/plant).

9.2.5   STATISTICAL ANALYSIS

The parameters recorded to study plant growth enhancement included shoot and root length, their fresh and dry weights, and yield. These were statistically analyzed for significant differences by Analysis Of Variance (ANOVA) with mean separation using the least significant difference (LSD) at $p \leq 0.05$.

9.3   RESULTS AND DISCUSSION

9.3.1   ISOLATION AND IDENTIFICATION OF TRICHODERMA SPP.

Green colonies of *Trichoderma* species which appeared on TSA and PDA media were mounted on lactophenol with trypan blue and observed under the microscope to confirm *Trichoderma* colonies. Based on the distinct morphological and cultural characters, seven *Trichoderma* spp. were identified, that is, *T. atroviride*, *T. koningii*, *T. reesei*, *T. viride*, *T. pseudokoningi*, *T. harzianum*, and *T. longibraculum*. These *Trichoderma* cultures were transferred into PDA slants, incubated for a week at room temperature and preserved in a refrigerator for further use.

9.3.2   PHYSICO-CHEMICAL FACTORS OF THE SOIL USED FOR POT EXPERIMENTS

The physico-chemical factors of the soil used for the pot experiments were analyzed and the results are presented in Table 9-1. The soil was of red sandy loam type, with pH 7.4 (slightly alkaline), the amount of available P and available K were 22 and 38 kg/Acre, respectively.

TABLE 9-1   Physico-chemical Factors of the Soil used for Pot Experiments

Color	Texture	Ca status	pH	pH status	EC	EC status	OC (%)	Avail. P	Avail. K
Red	Sandy loam	Low	7.4	Slightly alkaline	260	Normal	Low	22	38

EC - Electron Conductivity (mS/cm).

OC - Organic carbon.

Avail. - Available P and K- kg/Ac.

The inoculation of six different *Trichoderma* species alone and in combination showed significant ($p < 0.05$) the influence on plant growth promotion of tomato measured in terms of shoot and root lengths, their fresh and dry weights as well as yield parameters. The growth parameters were observed to increase proportionately along with an increase in the age of the crop.

9.3.3   EFFECT OF TRICHODERMA SPECIES ON SHOOT AND ROOT LENGTH

In the present study, the analysis of variance (ANOVA) data indicated significant ($p < 0.05$) differences among different treatments for shoot and root height at 30, 60, and 90 days plant growth period (Table 9-3). The maximum increase in shoot and root height was observed when the soil was infested with *T. viride* (65.00 and 22.4 cm, respectively), followed by *T. harzianum* (60.50 and 20.20 cm, respectively) compared with the uninoculated control (33.20 and 12.70 cm, respectively) plants (Table 9-2). The plants treated in combination (Tv plus Th) further enhanced shoot and root height (75.30 and 25.70 cm, respectively). The response of shoot and root length in the plants treated with Tv, Th and Tv + Th showed statistically significant ($p < 0.05$) increase over the other treatments of *Trichoderma* spp. This indicates that the interaction between different treatments and different plant growth period are highly significant (Tables 9-2 and 9-3). The plants inoculated with *T. atroviride*, *T.koningi*, *T. reseeii*, and *T. pseudokoningi* also showed a marked increase in shoot and root heights, but the increases were not statistically significant ($p \geq 0.05$) (Table 9-2).

TABLE 9-2 Effect of Six *Trichoderma* Species on the Shoot and Root Length (in cm/plant) of Tomato

Interval	30 DAS		60 DAS		90 DAS		MD	Sig.	MD	Sig.
Treatment	S	R	S	R	S	R	SL		RL	
Control	9.4	5.0	19.6	6.5	33.2	12.7				
T.atroviride	10.8	6.2	20.9	7.5	39.0	16.2	−2.400	0.83	−1.90	0.67
T.koningii	11.6	6.46	22.2	8.1	40.1	17.2	−2.187	0.91	−2.52	0.21
T.reesei	10.8	5.8	20.0	7.6	38.5	15.3	−.700	1.00	−1.50	0.92
T.viride	20.1	9.0	40.2	10.7	65.0	22.4	−17.33*	0.00	−3.60*	0.01
T.ps	10.1	5.5	20.0	6.9	33.5	14.8	1.067	1.00	−1.00	1.00
T.har	15.4	8.0	35.0	9.80	60.5	20.2	−13.77*	0.00	−4.60*	0.00
T.v + T.h	23.3	9.5	43.0	13.7	75.3	21.2	−21.87*	0.00	−6.73*	0.00

MD - Mean difference.

S - Shoot.

R - Root.

Sig. - Significant at $p \leq 0.05$.

DAS - Days after sowing.

SL - Shoot Length

RL - Root Length

T.ps - T.pseudokoningii, *T.har- T.harzianum*.

* - The mean difference is significant at the 0.05 level.

TABLE 9-3 Multiple Comparisons Between Days (shoot and root length, Turkey HSD)

Day	Day	MD	SE	Sig.	MD	SE	Sig.
30d	60	−14.816*	0.5539	0.00	−1.48*	0.388	0.00
	90	−33.203*	0.5539	0.00	−8.24*	0.388	0.00
60d	30	14.816*	0.5539	0.00	1.48*	0.388	0.00
	90	−18.387*	0.5539	0.00	−6.75*	0.388	0.00
90d	30	33.203*	0.5539	0.00	8.24*	0.388	0.00
	60	18.387*	0.5539	0.00	6.75*	0.388	0.00

HSD – honest significant difference

SE - Std. Error.

MD - mean difference.

Sig. - Significant at $p \leq 0.05$.

* - The mean difference is significant at the 0.05 level.

Further observations showed that dual inoculation with Tv + Th enhanced shoot and root length over inoculation with *Trichoderma* sp. alone by 15.85% and 14.73%, and 24.46% and 27.23% compared with Tv and Th, respectively. Similar results were reported on increased plant growth because of the application of *T. gamsii* in cereals and legume crops.[23] The increase in plant growth might be associated with the secretion of auxins, gibberellins, and cytokinins.

The shoot and root fresh and dry weights showed statistically significant ($p < 0.05$) differences (Tables 9-4–9-7) at 30, 60, and 90 days plant growth period (Tables 9-5 and 9-7) as compared with the uninoculated control. The statistically significant increase in shoot fresh weight was observed when the plants were treated with *T. koningi* (Tk), Tv, Th, and Tv + Th (18.00, 26.20, 23.13, and 28.30 g, respectively) over uninoculated control (10.00 gm) treatment (Table 9-4). In a similar way, significant ($p \leq 0.05$) differences were observed in root fresh weight in different treatments. The plants treated with Tv, Th, and Tv + Th showed statistically significant increase ($p < 0.05$) in root fresh weights (9.50, 6.75, and 11.38 g, respectively) compared with the control (1.80 g) treatment. However, the plants inoculated with Tatr, Tk, Tr, and Tps showed prominent increases in root fresh weights (3.82, 3.71. 2.88, and 2.16 g, respectively) compared with the control (1.80 gm), but the increases were not statistically significant ($p \geq 0.05$) (Table 9-4). This indicates that Tk, Tv, Th, Tv + Th application stimulated plant growth. The effect of different *Trichoderma* spp. on plant growth was observed as early as 30 days growth period which increased with an increase in crop growth period and are statistically significant (p < 0.05) (Table 9-5).

The shoot dry weights were statistically significant ($p < 0.05$) among the tested six *Trichoderma* spp. except for Tps (Table 9-6). A maximum increase in shoot dry weight was observed in dual inoculated (Tv + Th) treatments (10.42 g), followed by *Trichoderma* sp. inoculated singly with Tv, Th, Tk, and Tatr. (9.05, 6.59, 3.80, and 3.32 g, respectively) (Table 9-6). However, plants inoculated with Tps showed a marked increase in shoot dry weight, but were not statistically significant ($p > 0.05$). The root dry weights were significantly ($p < 0.05$) increased in Tv and Tv + Th applied treatments compared with the control (Table 9-6) showing a maximum increase (2.18 g) in dual inoculated (Tv + Th) plants. The root dry weights in plants treated with other *Trichoderma* spp. were not statistically significant (p < 0.05) though prominent increases (Table 9-6) were observed. The statistically significant ($p \leq 0.05$) differences were recorded at different stages of the plant growth period (Table 9-7) among all the treatments studied. The data showing statistically significant ($p < 0.05$) increases in shoot and root fresh weights on application with Tv, Th, Tv + Th among the different treatments evaluated indicates that the significant effect of *Trichoderma* on plant growth and development consistently depends on the *Trichoderma* species applied. This finding is consistent with the results of other authors.[8,24 25]

TABLE 9-4   Effect of Six *Trichoderma* Species on Shoots and Root Fresh Weights (g/plant) in Tomato

Treatments	30 DAS		60 DAS		90 DAS		MD	Sig.	MD	Sig.
	S	R	S	R	S	R	S		R	
Control	0.19	0.08	2.8	0.12	10.0	1.80				
T.atroviride	0.35	0.08	5.10	0.15	15.46	3.82	−2.862	0.13	−.6800	0.62
T.koningii	0.64	0.10	5.8	0.15	18.0	3.71	−4.014*	0.00	−.6533	0.68
T.reesei	0.28	0.08	4.7	0.15	14.1	2.88	−1.991	0.68	−.3678	1.00
T.viride	3.78	0.30	9.7	0.57	26.2	9.50	−8.853*	0.00	−2.7811*	0.00
T.ps	0.196	0.06	3.5	0.13	11.73	2.16	−.7687	1.00	−.1033	1.00
T.har	1.95	0.22	7.6	0.34	23.13	6.75	−6.563*	0.00	−1.7667*	0.00
Tv + Th	4.08	0.56	10.7	1.80	28.3	11.38	−9.999*	0.00	−3.9133*	0.00

MD - Mean difference.

S - Shoot.

R - Root.

Sig. - Significant.

DAS - Days after sowing.

T.ps - T.pseudokoningii.

T.har - T.harzianum.

*- The mean difference is significant at the 0.05 level.

TABLE 9-5   Multiple Comparisons Between Days (shoot and root fresh weights, Turkey HSD)

Day	Day	MD	SE	Sig.	MD	SE	Sig.
30d	60	−4.4066*	0.40774	0.00	−0.1698	0.13449	0.42
	90	−16.4534*	0.40774	0.00	−4.3702*	0.13449	0.00
60d	30	4.4066*	0.40774	0.00	0.1698	0.13449	0.42
	90	−12.0469*	0.40774	0.00	−4.2004*	0.13449	0.00
90d	30	16.4534*	0.40774	0.00	4.3702*	0.13449	0.00
	60	12.0469*	0.40774	0.00	4.2004*	0.13449	0.00

SE - Standard Error.

MD - Mean difference.

* - The mean difference is significant at the 0.05 level.

TABLE 9-6   Effect of Six *Trichoderma* Species on Shoot and Root Dry Weights (g/plant) in Tomato

Treatment↓	30 DAS		60 DAS		90 DAS		MD	Sig.	MD	Sig.
	S	R	S	R	S	R	S		R	
Control	0.24	0.010	0.49	0.023	1.15	0.053				
T.atroviride	1.35	0.017	1.73	0.040	3.32	0.103	−1.190*	0.00	−.0257	1.00
T.koningii	1.26	0.023	2.09	0.053	3.80	0.123	−1.567*	0.00	−.0436	1.00
T.reesei	0.26	0.013	1.43	0.050	2.64	0.087	−.840*	0.00	−.0223	1.00
T.viride	2.2	0.137	4.34	0.327	9.05	1.773	−4.587*	0.00	−.7180*	0.00
T.ps	0.08	0.010	1.12	0.030	1.75	0.083	−.373	0.45	−.0133	1.00
T.har	2.09	0.077	3.41	0.143	6.59	0.290	−3.384*	0.00	−.1428	0.18
Tv + Th	2.08	0.533	5.38	1.310	10.42	2.177	−5.350*	0.00	−1.3133*	0.00

The values are a mean of four replicates.

T.ps - T.pseudokoningii.

T.har - T.harzianum.

S - shoot.

R - root.

Sig. - Significant.

DAS - Days after sowing.

TABLE 9-7   Multiple Comparisons Between Days (shoot and root dry weights, Turkey HSD)

Day	Day	MD	SE	Sig.	MD	SE	Sig.
30d	60	−1.3618*	0.06684	0.00	−0.09351*	0.02138	0.00
	90	−3.4793*	0.06684	0.00	−0.31242*	0.02138	0.00
60d	30	1.3618*	0.06684	0.00	0.09351*	0.02138	0.00
	90	−2.1176*	0.06684	0.00	−0.21891*	0.02138	0.00
90d	30	3.4793*	0.06684	0.00	0.31242*	0.02138	0.00
	60	2.1176*	0.06684	0.00	0.21891*	0.02138	0.00

SE - Standard Error.

MD - Mean difference.

Sig. - Significant.

* - The mean difference is significant at the 0.05 level.

On the contrary, in the plants treated with Tatr, Tk, Tr, Tps, though marked increases were recorded, the effect was not significant. These results clearly indicate that the increased growth response of plants, caused by *Trichoderma* sp. depended mainly on the ability of *Trichoderma* to survive and develop in the rhizosphere.[1,7] The work of other researchers shows that the rhizosphere competent isolates produced diffusible metabolites in the rhizosphere which actively influence the growth of *Trichoderma* colonized plant because of their action as plant growth regulators (auxin and/or auxin-like compounds).[26 27]

In the present study, the inoculation of six different *Trichoderma* species showed a positive influence on plant growth and biomass production of the tomato plant. However, there were differences in the growth promoting efficiency of different *Trichoderma* spp. *Trichoderma viride* was observed to be most effective, followed by Th over the other inoculated *Trichoderma* species in significantly increasing the plant growth and development. The increase in biomass production might be because of the production of plant growth promoters or through indirect stimulation of nutrient uptake and by the production of siderophores or antibiotics to protect plants from deleterious effects of the rhizosphere organisms.

In the present study, the results indicated that *T. viride* and *T. harzianum* isolates had a positive effect on plant growth in tomato singly and in combination (Tv + Th). Numerous other species such as *T. harzianum*,[4,11,12,28] *T. viride*,[2,29] and *T. koningii*[3,30] have also been reported to promote plant growth.

9.3.4   EFFECT OF TRICHODERMA SPECIES ON YIELD PARAMETERS

The six *Trichoderma* spp. inoculated plants resulted in greater yield compared with the control. The application of different *Trichoderma* species was observed to be effective and showed statistically increased yield parameters in tomato. The analysis of variance of the data showed significant differences ($p < 0.05$) between the treatments (Table 9-8). Maximum increase in fruit yield was observed when the plants were inoculated with Tv (1016.66 g/plant), followed by Th, Tk, Tatr, Tr, and the least by Tps (820.365, 771.665, 633.335, 580.00, and 468.333 g/plant, respectively). Dual inoculation with Tv + Th further enhanced plant yield which was statistically significant. The yield per plant was statistically significant ($p < 0.05$) in all the treatments except Tps over the control treatment. The plants treated with Tps recorded a marked increase in yield parameters over the control, but were not statistically significant ($p < 0.05$). Similar results were reported earlier.[27]

TABLE 9-8   Effect of *Trichoderma* Species on the Yield of Tomato

Treatment ↓	Yield g/plant	MD	SE	Sig.
Control	388.3300			
T.atroviride	633.3300	−245.00000*	37.19066	0.00
T.koningii	771.6600	−383.33000*	37.19066	0.00
T.reesei	580.0000	−191.67000*	37.19066	0.00
T.viride	1016.6600	−628.33000*	37.19066	0.00
T.pseudokoningii	468.3300	−80.00000	37.19066	0.69
T.harzianum	820.3600	−432.03000*	37.19066	0.00
Tv + Th	1200.0300	−811.70000*	37.19066	0.00

SE - Standard Error.

MD - Mean difference.

Sig. – Significant.

* - The mean difference is significant at the 0.05 level.

9.4 CONCLUSION

The growing interest in the beneficial microorganisms for their ability to enhance plant growth is increasing, particularly with respect to their eco-friendly nature, and the constraint with regard to the use of high price fertilizers. Therefore, *Trichoderma viride* is recommended to be used commercially because of its significant effect on plant growth and fruit yield which eventually improves agricultural production and better yield.

ACKNOWLEDGMENTS

Dr. A. Hindumathi is very grateful to the Department of Science and Technology, New Delhi for providing fellowship under the Women Scientist Scheme-A (WOS-A) with a grant No. SR/WOS-A/LS-498/2011(G).

KEYWORDS

- **Biomass**
- **Plant growth**
- **Tomato**
- *Trichoderma*

REFERENCES

1. Harman, G. E.; Howell, C. R.; Viterbo, C. A.; Chet, I.; and Lorito, M. *Trichoderma* species—opportunistic, avirulent plant symbionts. *Nat. Rev. Microbiol.* **2004a,** *2*, 43–56.
2. Ousley, M. A.; Lynch, J. M.; and Whips, J. M. Effect of *Trichoderma* on plant growth: A balance between inhibition and growth promotion. *Microb. Ecol.* **1993,** *26*, 277–285.
3. Windham, M. T.; Elad, Y.; and Barker, R. A mechanism for increased plant growth induced by *Trichoderma spp. Phytopathol.* **1986,** *76*, 518–521.
4. Chang, Y. C.; Chang, Y. C.; Baker, R.; Kleifeld, O.; and Chet, I. Increased growth of plants in the presence of the biological control agent *Trichoderma harzianum. Plant Dis.* **1986,** *70*, 145–148.
5. Baker, R. *Trichoderma* spp. as plant stimulants. *Crit. Rev. Biotech.* **1988,** *7*, 97–106.
6. Ousley, M. A.; Lynch, J. M.; and Whips, J. M. Potential of *Trichoderma* spp. as consistent plant growth stimulators. *Biol. Fertil. Soils* **1994a,** *17*, 85–90.
7. Harman, G. E. Overview of mechanisms and uses of *Trichoderma* spp. *Phytopathol.* **2006,** *96*, 190–194.
8. Yedidia, I.; Srivastva, A.; Kapulnik, Y.; and Chet, I. Effect of *Trichoderma harzianum* on microelement concentrations and increased growth of cucumber plants. *Plant Soil* **2001,** *235*, 235–242.
9. Adams, P.; De-Leij, F. A.; and Lynch, J. M. *Trichoderma harzianum* Rifai 1295-22 mediates growth promotion of Crack willow (*Salix fragilis*) saplings in both clean and metal-contaminated soil. *Microb. Ecol.* **2007,** *54*, 306–313.
10. Chet, I. Biological control of soil-borne plant pathogens with fungal antagonists in combination with soil treatments. In *Biological Control of Soil-borne Plant Pathogens*; Hornby, D., Ed.; C.A.B. International: Wallingford, UK, 1990; pp 15–25.
11. McGovern, R. J.; Datnoff, L. E.; and Tripp, L. Effect of mixed infection and irrigation method on colonization of tomato roots by *Trichoderma harzianum* and *Glomus intraradix. Proc. Fla. State Hort. Soc.* **1992,** *105*, 361–363.
12. Datnoff, L. E.; and Pernezny, K. L. Effect of bacterial and fungal microorganisms to colonize tomato roots, improve transplant growth and control of Fusarium Crown and Root Rot. *Proc. Florida Tomato Inst.* **1998,** *111*, 26–33.
13. Baker, R.; Elad, Y.; and Chet I. The controlled experiment in the scientific method with special emphasis on biological control. *Phytopathol.* **1984,** *74*, 1019–1021.
14. Harman, G. E. Myths and dogmas of biocontrol Changes in perceptions derived from research on *Trichoderma harzianum T*-22. *Plant Dis.* **2000,** *84*, 377–393.

15. Bjorkman, T.; Blanchard, L. M.; and Harman, G. E. Growth enhancement of shrunken-2 (sh2) sweet corn when colonized with *Trichoderma harzianum* 1295-22: Effect of environmental stress. *J. Am. Soc. Hort. Sci.* **1998,** *123,* 35–40.

16. Harman, G. E.; Petzoldt, R.; Comis, A.; and Chen, J. Interactions between *Trichoderma harzianum* T22 and maize inbred line Mo17 and effects of this interaction on diseases caused by *Pythium ultimtum* and *Colletotrichum graminicola. Phytopathol.* **2004b,** *94,* 147–53.

17. Harman, G. E.; and Bjorkman, T. Potential and existing uses of *Trichoderma* and *Gliocladium* for plant disease control and plant growth enhancement. In *Trichoderma and Gliocladium*; Kubicek, C. P., and Harman, G. E., Eds.; Taylor and Francis: London, 1998; Vol. 2, pp 229–265.

18. Rifai, M. A. *Revision of the Genus Trichoderma*; Commonwealth Mycological Institute: England, 1969; Vol. 116, p 56.

19. Domsch, K. H.; Gams, W.; and Anderson T. H. *Compendium of Soil Fungi*; Academic Press: London, 1980, 809.

20. Martha, P. K. Influence of some physico-chemical factors on the germination and growth of biotype of *Trichoderma harzianum* and *Gliocladium virens*. M.Sc. dissertation, B.C.K.V., Mohanpur, 1992.

21. Majumdar, D. Hyperparasitic potential of few biotypes of *T. harzianum* and *G. virens* against two major pathogens of bêtel vine (Piper bêtel L.). M.Sc. dissertation, B.C.K.Y. Mohanpur, 1993.

22. Harman, G. E.; Jin, X.; Stasz, T. E.; Peruzzotti, G.; Leopold, A. C.; and Taylor, A. G. Production of conidial biomass of *Trichoderma harzianum* for biological control. *Biolog. Con.* **1991,** *1,* 23–28.

23. Rinu, K.; Sati, P.; and Pandey, A. *Trichoderma gamsii* (NFCCI 2177): A newly isolated endophytic, psychrotolerant, plant growth promoting, and antagonistic fungal strain. *JBM.* **2013,** DOI: 10.1002/jobm.

24. Hajieghrari B.; Torbi-Gigou, M.; Mohammadi, M. R.; and Davari, M. Biological potential of some Iranian *Trichoderma* isolates in the control of soil borne plant pathogenic fungi. *Fr. J. Biotechnol.* **2010,** *7,* 967–972.

25. Ozbay, N.; and Newman, S. E. Effect of *Trichoderma harzianum* strain to colonize tomato roots and improve transplant growth. *Plant Dis.* **2004,** *81,* 492–496.

26. Vinale, F.; Sivasithamparam, K.; Ghisalberti, E. L.; Marra, R.; Barbetti, M. J.; Li, H.; Woo, S. L.; and Lorito, M. A novel role for *Trichoderma* secondary metabolites in the interactions with plants. *Physiol. Mol. Plant Pathol.* **2008a,** *72,* 80–86.

27. Vinale, F.; Sivasithamparam, K.; Ghisalberti, E. L.; Marra, R.; Barbetti, M .J.; Li, H.; Woo, S. L.; and Lorito, M. *Trichoderma*-plant-pathogen interactions. *Soil Biol. Biochem.* **2008b,** *40,* 1–10.

28. Inbar, J.; Abramsky, M.; Cohen, D.; and Chet, I. Plant growth enhancement and disease control by *Trichoderma harzianum* in vegetable seedlings grown under commercial conditions. *Eur. J. Plant Pathol.* **1994,** *100,* 337–346.

29. Ousley, M. A.; Lynch, J. M.; and Whipps, J. M. The effects of addition of *Trichoderma* inocula on flowering and shoot growth of budding plants. *Sci. Hortic.* **1994b,** *59,* 147–155.

30. Liu, S. Growth promotional effect of *Trichoderma* species. Unpublished work, 1990.

PART III

Microbial Enzymes and Their Potential Industrial Applications

CHAPTER 10

STRAIN IMPROVEMENT OF *ASPERGILLUS NIGER* FOR THE ENHANCED PRODUCTION OF CELLULASE IN SOLID STATE FERMENTATION

G. PRAVEEN KUMAR REDDY[1], A. SRIDEVI[2],
KANDERI DILEEP KUMAR[1], G. RAMANJANEYULU[1], A. RAMYA[1],
B. S. SHANTHI KUMARI[1], and B. RAJASEKHAR REDDY[1*]

[1]*Department of Microbiology, Sri Krishnadevaraya University, Anantapur, Andhra Pradesh – 515591, India.*

[2]*Department of Applied Microbiology, Padmavati Mahila University, Tirupathi, Andhra Pradesh – 517502, India.*

*Corresponding Author: B. Rajasekhar Reddy. E-mail: rajasekharb64@ gmail.com

CONTENTS

Abstract 202
10.1 Introduction 202
10.2 Materials and Methods 203
10.3 Results and Discussion 209
10.4 Conclusion 215
Keywords 216
References 216

ABSTRACT

Cellulases are a group of hydrolytic enzymes that are capable of releasing soluble sugars from lignocellulosic materials which provide a platform for a wide range of applications. The sole objective of the present investigation was strain improvement of *Aspergillus niger* for the enhancement of production of cellulases. The spores of *A. niger* were subjected to mutation by exposure to ultraviolet (UV) radiation and chemical treatment with Ethyl Methane Sulfonate (EMS). The percent survival of spores was measured after the exposure of spores to mutagenic agents. The survival percentage of spores to physical mutation and chemical mutation was between 10–90% and 10–20%, respectively. The mutants generated were screened on carboxymethyl cellulose—congo red plates on the basis of clear zone formation in comparison with wild type *A. niger*. The potential mutant—EMS5 enhanced cellulase production over the parent/wild type strain on a combination of lignocelluloses (rice bran + wheat bran) at equal proportion in Solid State Fermentation (SSF) by 30–50% and was further optimized for different factors—carbon source, nitrogen source, temperature, moisture content, and inoculum density. The mutant—EMS5 yielded 50.16, 24.20, and 35.60 U/g of Filter Paper Activity (FPase), carboxymethylcellulase (CMCase), and β-glucosidase under optimized conditions in SSF.

10.1   INTRODUCTION

Cellulose is the most abundant renewable organic molecule in the biosphere with an estimated annual production of 4.0×10^7 tons[1]. Cellulose is a highly hydrogen bonded biopolymer consisting of thousands of D-glucose units linked with β-1,4 glycosidic bonds. Large quantities of lignocellulosic wastes are generated through agricultural practices, forestry, and industrial processes, particularly from agro-allied industries such as breweries, paper-pulp, textile, and timber industries. However, the plant lignocellulosic biomass regarded as "waste" is biodegradable and can be converted into valuable products such as biofuels, chemicals, and cheap energy sources for fermentation, improved animal feeds, and human nutrients.[2] However, the exploitation of lignocellulosic biomass for conversion to value-added products rests on its depolymerization to fermentable sugars through microbial cellulases.[3–10] The conversion of cellulosic mass to fermentable sugars through the biocatalyst cellulase derived from cellulolytic organisms has been suggested as a feasible process and offers a potential to reduce the use of fossil fuels thereby reducing environmental pollution.[11] Complete enzymatic hydrolysis of cellulose requires the synergistic action of three types of enzymes, namely, cellobiohydrolase, endoglucanase, or carboxymethylcellulase (CMCase), and β-glucosidases.[12] Though many organisms (pro and eukaryotes) have the property to produce cellulase, fungi are well known for the secretion of cellulase enzymes in large quantities. Technology for the conversion of lignocelluloses to value added products through the use of cellulase is not economical at present because of the high cost production of cellulase.[13] To date, the production of cellulase has been extensively studied in submerged fermentation with different microorganisms in comparison with the solid state fermentation.[13] Solid state fermentation is an attractive process for the economical production of cellulase because of its low capital investment and facile operating conditions.[14] One of the approaches to cut down cost for the production of cellulase enzyme is strain improvement of cellulolytic organisms for enhanced production of cellulase by a site—directed mutagenesis or random mutagenesis.[15–17] Most of the *Trichoderma reesei* mutants such as Rut C-30, CL-847, VTT-D, and KY 746 useful for the commercial production of cellulases were derived from the wild strain *T. reesei* 6a/wild type QM6a that was originally isolated from the US army laboratories.[18–23] Improvement of strains was extended to few other cellulolytic organisms — *Cellulomonas biazotea*,[24] *Bacillus pumiluss*,[25] *Trichoderma aureoviridae* 7-121,[26] and *Bacillus cereus*.[27] Systematic improvement of the production strains by mutagenesis and screening resulted in the selection of the mutant strains producing high levels of extracellular protein of over 40 g/liter with approximately half of the protein being the main cellulase, and cellobiohydrolase I.[22]

The aim of the present study was to improve the strain of *A. niger* through random mutagenesis by UV irradiation and EMS treatment for the enhancement of the production of cellulase and to optimize cellulase production under laboratory conditions with an improved strain of *A. niger* on a combination of rice bran and wheat bran in equal proportions.

10.2   MATERIALS AND METHODS

10.2.1   MICROORGANISM

A local isolate of *A. niger* isolated from the soil contaminated with the effluents of cotton ginning industry in Nandyal, India[28] was used in the present study. The fungal culture was maintained on Czapek-Dox agar slants.

Composition of the medium was in g/L sucrose - 30, $NaNO_3$ - 1.0, K_2HPO_4 - 1.0, $MgSO_4.7H_2O$ - 0.5, KCl - 0.5, $FeSO_4.7H_2O$ - 0.01, cellulose - 0.5%, distilled water - 1000 ml, and pH 5.

10.2.2   LIGNOCELLULOSIC SUBSTRATES

The lignocellulosic substrates used in this study were rice bran and wheat bran. These lignocellulosic substrates were washed with running tap water to remove all residual sugars. The samples were dried in an oven at 70 °C for 4 hours and then used in this study.

10.2.3   PRODUCTION OF CELLULASE IN SOLID STATE FERMENTATION BY WILD TYPE AND MUTANTS OF A. NIGER

Solid state fermentation was carried out in 250 ml Erlenmeyer flasks. Ten grams of the sample of rice bran and wheat bran in equal proportion was taken in flasks and was moistened with basal mineral salt solution to attain 70% initial moisture content. The basal mineral salt solution used for the experiment had the following composition (g/L): K_2HPO_4 - 6.3, KH_2PO_4 - 1.8, NH_4NO_3 - 1.0, $MgSO_4.7H_2O$ - 1.0, $CaCl_2$ - 0.1, $FeSO_4.7H_2O$ - 0.1, $MnSO_4.7H_2O$ - 0.1, $NaMo_7O_4$ - 0.006. The initial pH of the mineral salt solution was adjusted to 5.0. After sterilization, all the flasks were inoculated with the spore suspension of the wild type or the mutant of *A. niger*. Spore suspension was prepared by flooding a seven day old slant with 2 ml of sterile distilled water containing 0.01% Triton × 100 and the resulting spore suspension was used to inoculate the flasks. All the inoculated flasks were incubated at 30 °C for seven days.

10.2.4   ENZYME EXTRACTION

After seven days of incubation period, the enzyme extraction was carried out by mixing the fermented mass with sodium acetate buffer (0.2 M, pH - 4.8, 1:10 w/v) for 1 hour in a rotary shaker at 180 rpm. The contents of the flask were filtered through muslin cloth and the filtrate was centrifuged at 9,000 rpm for 10 min. The supernatant obtained after centrifugation served as the enzyme source.

10.2.5   MUTAGENESIS

10.2.5.1   UV IRRADIATION

Spores were scrapped off from a completely sporulated slant of *A. niger* (7-day old) into 5 ml of sterile distilled water containing Triton X-100 to obtain a uniform suspension. The suspension was transferred into a sterile conical flask and thoroughly shaken for 30 min on a rotary shaker to break the spore chains. The spore suspension was then filtered through a thin sterile cotton wad into a sterile tube to remove the vegetative mycelium from the suspension. The density of spore suspension was measured with the help of a haemocytometer. Three milliliters of the spore suspension was aseptically pipetted into sterile petri dishes (*100 mm × 15 mm*). Exposure of the spores to UV light (2 μJ mm^{-1}) was carried out at distance of 15 cm away from the center of the UV light source with an occasional shaking for 10–60 min. The same number of spores in the suspension without exposure to UV radiation served as the control. Each UV exposed spore suspension was stored in the dark overnight to avoid photoreactivation. Aliquots from the spore suspension with/without UV exposure was serially diluted in phosphate buffer (molarity - 0.1 M, pH - 7.6) and plated on Czapek-Dox agar. The colonies which appeared on the plates after incubation at 30 °C were counted. The percentage survival of the spores was expressed on the basis of the results of the control.

10.2.5.2   CHEMICAL MUTAGENESIS

Spore suspension of *A. niger* with the desired density, prepared in the same way as described earlier, was treated with ethyl methane sulphonate (EMS) at different final concentrations within a range of 1–10 ug/ml for two periods (5 and 10 min). The same number of spores in suspension without EMS served as the control. At the end of 5 and 10 min incubation, the spores in the control and EMS treatment were recovered by decanting the supernatant after centrifugation and finally suspended in saline phosphate buffer after washing. Spores from the control and EMS treatments after serial dilution were plated on Czapek Dox agar plates. The plates were incubated for seven days at 30 °C. The colonies that appeared on the plates were counted and percentage survival of the spores was expressed on the basis of the results of the control.

10.2.6   SCREENING FOR CELLULASE HYPER-PRODUCING MUTANTS

Mutants, generated by the two methods, were screened on Czapek-Dox agar supplemented with 2% (w/v) carboxymethyl cellulose (CMC). The CMC agar plates were inoculated with mutants of *A. niger* and incubated at 30 °C for seven days after which they were stained with an aqueous solution of congo red (1% w/v) for 15 min. The excess dye was removed by washing with 1.0 M NaCl[29] and the production of extracellular cellulase by the wild type or mutant of *A. niger* was indicated by a zone of clearance around the colony. The ratio of the clear zone diameter to the colony diameter was measured to select the highest cellulase activity producer.

10.2.7   OPTIMIZATION FOR CELLULASE ENZYME PRODUCTION IN SSF BY EMS5 MUTANT STRAIN

Optimization studies were carried out with only a potential mutant strain (EMS5) of *Aspergillus niger* because of its high cellulase productivity. The potential mutant strain was grown on a 10 g mixture of rice bran and wheat bran in equal proportion moistened with 20 ml of mineral salt solution (equivalent to 75% moisture content).

10.2.7.1   EFFECT OF CARBON SOURCE ON CELLULASE PRODUCTION

To select the best carbon for cellulase production, the nutrients (cellobiose, carboxymethyl cellulose, and glucose as a carbon source) were included in a mineral salt solution so as to add to the solid substrate medium (rice bran + wheat bran) at a final concentration of 1% (w/w) in the flasks. Flasks with solid medium devoid of carbon source served as the control. After sterilization at 121 °C for 20 min, the flasks were inoculated with 2 ml of EMS5 strain spore suspension and incubated at 30 °C for seven days. After incubation, the enzyme activity was determined in the culture extract of the fermented bran.

10.2.7.2 EFFECT OF NITROGEN SOURCE ON CELLULASE PRODUCTION

For assessing the influence of different nitrogen sources on cellulase production in solid state fermentation by the potential EMS5 mutant strain, the ingredient ammonium nitrate in mineral salt solution was replaced with nitrogen sources such as $NaNO_3$, KNO_3, or Urea at 1% w/w to the solid substrate fermentation medium (rice bran + wheat bran) for not only moistening, but also adding nitrogen at a final concentration of 1% (w/w) to the medium. After sterilization, the flasks with the medium were inoculated with 2 ml of EMS5 strain spore suspension and incubated at 30 °C for seven days. Cellulolytic enzymes from the fermented bran were extracted in acetate buffer for the determination of the enzyme activity as specified elsewhere.

10.2.7.3 EFFECT OF INCUBATION TEMPERATURE ON CELLULASE PRODUCTION

To determine the influence of incubation temperature on cellulase production by the strain EMS5 in solid state fermentation, 10 g mixture of rice bran + wheat bran with 75% moisture content in 250 ml Erlenmayer conical flasks was inoculated with the spore suspension of EMS5 mutant. The flasks were then incubated at different temperatures — 25, 30, and 40 °C for seven days. Activities of the cellulolytic enzymes were determined after extraction from the fermented bran as mentioned earlier.

10.2.7.4 EFFECT OF MOISTURE CONTENT OF THE MEDIUM ON CELLULASE PRODUCTION

To determine the optimum initial moisture content for cellulase production, 10 g sample of rice bran + wheat bran was distributed into 250 ml conical flasks. Twenty milliliters of the mineral salt solution (equivalent to 75% moisture content) was added to all the flasks. To raise the initial moisture content from 75 to 80% and 75 to 85%, additional requirement of 1.3 and 2.7 ml distilled water was provided to only the flasks which were earmarked for 80 and 85% moisture content, respectively. All the flasks were incubated at 30 °C for 7 days after inoculation with the spore suspension of EMS5 as described earlier. Activities of the cellulolytic enzymes in the fermented bran were determined as described elsewhere.

10.2.7.5 EFFECT OF SPORE DENSITY IN THE INOCULUM ON CELLULASE PRODUCTION

A suitable volume of sterile distilled water was aseptically poured on the surface of potato dextrose agar slants containing seven days old culture of EMS5 and a suspension of the spores was prepared. The spore density of the suspension was measured with the help of haemocytometer. Conical flasks of the capacity 250 ml were maintained with 10 g of the solid substrate (rice bran + wheat bran) at 75% initial moisture content and were inoculated with the density of spores (1×10^7, 1×10^8, 2×10^7, 2×10^8) and incubated for growth in the manner as described elsewhere. After seven days of incubation, activities of the cellulolytic enzymes were determined from the culture extracts of the fermented bran.

10.2.8 ANALYTICAL METHODS

10.2.8.1 ENZYMES ASSAYS

Activities of the individual enzyme components of the cellulase system secreted into the culture medium of *Aspergillus niger*, were estimated in accordance with the methods listed by Wood and Bhat[30]. Filtrate recovered after filtration of the culture medium was used as a source of enzyme.

10.2.8.2 FILTER PAPER ASSAY

Filter paper activity is a measure of the total cellulolytic activity resulting from the combined action of the different enzyme components present in the culture filtrate. Filter paper activity of the culture filtrates of *Aspergillus niger* was determined according to the method of Mandels and Weber.[31] Whatman filter paper strip (1×6 cm) containing 50 mg weight was suspended in 1 ml of 0.05 M sodium citrate buffer (pH 4.8) at 50 °C in a water bath. Suitable aliquots of the enzyme source with/without dilution was added to the above mixture and incubated for 60 min at 50 °C. Enzyme blanks (without the enzyme) were run simultaneously in the same manner as specified earlier. After incubation, addition of 3,5-dinitrosalicylic acid (DNS) was made and the contents were mixed. All the samples, enzyme blanks, and glucose standards were vigorously boiled for exactly 5 min in a boiling water bath. After cooling, the color developed in the tubes was read at 540 nm in a spectrophotometer (ELICO-SL164). The activity of cellulase was expressed in filter paper units. One unit of filter paper unit (FPU) was defined as the amount of enzyme releasing μmole of reducing sugar from the filter paper per minute.

10.2.8.3 CARBOXYMETHYL CELLULASE METHOD

The reaction mixture contained 1.0 ml of 1% carboxymethyl cellulose in 0.2 M acetate buffer (pH 5.0). This reaction mixture was preincubated at 50 °C in a water bath for 20 min. An aliquot of 0.2 ml of the culture filtrate with appropriate dilution was added to the reaction mixture and incubated at 50 °C in a water bath for one hour. Appropriate control devoid of the substrate or enzyme was simultaneously run. The reducing sugar produced in the reaction mixture was determined by dinitrosalicylic acid method.[32]. The reagent 3, 5-dinitrosalicylic acid was added to the aliquots of their action mixture and the color developed was read at 540 nm in a spectrophotometer (ELICO-SL164). One unit of endoglucanase activity was defined as the amount of enzyme releasing one μmole of reducing sugar per minute.

10.2.8.4 β-D-GLUCOSIDASE ASSAY

The activity of β-D-glucosidase in the culture filtrate was determined according to the method of Herr.[33] The activity of β-D-glucosidase (EC 3.2.1.21) was measured in an assay mixture containing 0.2 ml of 5 mM *p*-nitrophenyl-β-D-glucopyranoside (PNPG) (Merck) dissolved in 0.05 M citrate buffer with a pH 4.8 and 0.2 ml of diluted enzyme solution with appropriate controls. After incubation for 30 min at 50 °C, the reaction was stopped by adding 4 ml of 0.05 M NaOH-glycine buffer (pH-10.6). The yellow colored *p*-nitrophenol which was liberated, was determined by spectrophotometer (ELICO-SL164) at 405 nm. The activity of β-glucosidase was defined as the amount of enzyme liberating one μmole of *p*-nitrophenol per minute under standard assay conditions.

10.3 RESULTS AND DISCUSSION

The present study was aimed at improving the local isolate of *Aspergillus niger* through random mutagenesis for the enhancement in cellulase production. Initially, spores of the wild type of *A. niger* were exposed to UV radiation and the spores which survived after the exposure gave rise to colonies on a suitable medium like Czapek Dox agar medium. The survival data obtained by the counting of colonies on plates from the control (without exposure to UV radiation) and UV-exposed spores are presented in Figure 10-1.

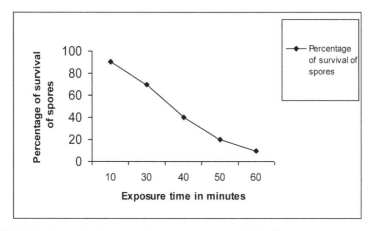

FIGURE 10-1   Survival of spores of *A. niger* exposed to UV radiation.

The increase in exposure time of UV radiation resulted in a decrease in per cent survival of the spores. About ninety percent of the spores exposed to UV radiation for 60 min were killed with only 10% survival of the spores. A similar observation of the increase in per cent killing of the spores of *Penicillium chrysogenum* with an increase in the exposure time of UV radiation was made.[34] The number of survivors of conidial spores of *T. viride* kept on decreasing with an increase in the exposure time of the UV radiation.[34] Figure 10-2 represents data on the viability of spores upon treatment with EMS at different concentrations for two periods (5 and 10 min). The viability of the spores was dependent on the dosage of EMS used in the present study. About 2 and 10% viability of spores were recorded when treated with EMS at concentrations of 1 and 10 mg/ml, respectively. Similarly, killing of about 93% of spores of *Penicillium chrysogenum* at 1 mg/ml concentration of EMS was reported.[35] The number of surviving spores increased with a decreasing concentration of EMS from 6 to 1 µg/ml.[34]

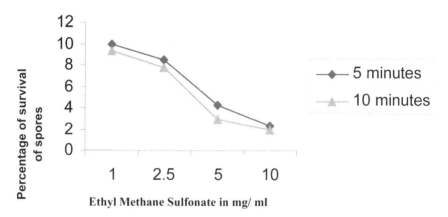

FIGURE 10-2   Survival of spores of *A. niger* treated with Ethyl Methane Sulfonate (EMS).

Isolated colonies from the plates in the above studies were collected and were considered as mutants. The mutants obtained after UV exposure were labeled as UV1, UV2, UV3, UV4, so on. Mutants derived after treatment with ethyl methane sulphonate (for 5 and 10 min duration) were labeled as EMS1, EMS2, EMS3, so on. All these mutants were screened on CMC plates for their cellulolytic activity based on the formation of clear zones because of the hydrolysis of CMC that was made visible with Congo red staining (Figure 10-3). A similar method with a clear zone of hydrolysis on Mandel's medium/suitable medium with CMC was followed for screening mutants for cellulase production.[34,36,37] Of all the mutants tested in this study, EMS5 displayed a clear zone of hydrolysis with the largest diameter (Figure 10-4). In view of the largest size of the clear zone, subsequent studies were carried out with EMS5 mutant. Growth of hyper producing mutant EMS5 for seven days on a combination of rice bran and wheat bran in equal proportions in solid state fermentation yielded tires of 40, 18, and 30 U/gDS as against 30, 13, and 22 U/gDS by the wild type with respect to FPase, CMCase, and β-glucosidase. Thus, EMS5 enhanced the yields of cellulolytic enzymes over the wild type by 30–50%. Similarly, most of the *T. reesei* mutants such as QM9414[18] and RUTC- 30[19] useful for the commercial production of cellulases were derived from the wild strain *T. reesei* QM6a,that was originally isolated from the US army laboratories[38] through UV and chemical mutagenesis. Catabolite de-repression by resistance to the anti metabolite—2-Deoxyglucose[39], increased the yields of the cellulolytic enzymes by 2–5 folds. Systematic improvement of the production strains by mutagenesis and screening resulted in selective mutant strains producing high levels of extracellular protein of over 40 g/liter with approximately half of the protein being the main cellulase and cellobiohydrolase I.[22] Production of cellulolytic enzymes by EMS5 on combination of rice bran and wheat bran in solid state fermentation in the present study was further optimized for nutrient factors, physical factors, and biological factors. Among the carbon sources tested in this study, CMC and cellulose supported a higher secretion of extracellular enzymes FPase, CMCase, and β-glucosidase than glucose (Figure 10-5).

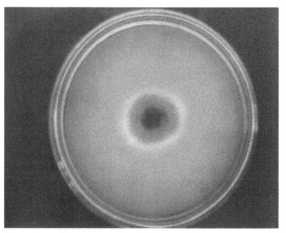

FIGURE 10-3   Screening of mutants on CMC plates with Congo red staining.

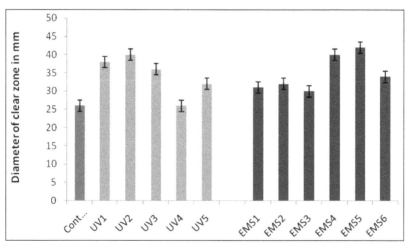

*Data represented in the figure are the means of two replicates with deviation.

FIGURE 10-4   Size of clear zone formed by the mutants on CMC plates.

Similarly, the influence of supplementation of different nitrogen sources to solid state fermentation medium at 1% on cellulase production by EMS5 was examined (Figure 10-6). The yield of FPase was high (49.50 U/gDS) in the presence of urea and the activities of the other individual components of cellulase such as CMCase (24.20 U/ml), and β-glucosidase (35.60 U/g) were higher in the presence of urea (Figure 10-6).

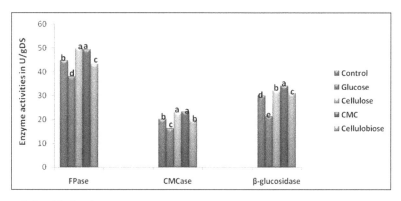

*Control devoid of carbon source.

** Bars with the same letter are not statistically significant according to the DMR test.

FIGURE 10-5   Effect of various carbon sources on the production of cellulase by the mutant EMS5.

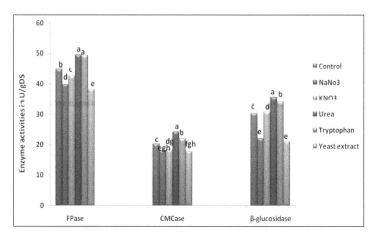

*Control devoid of nitrogen amendment

** Bars with same letter are not statistically significant according to DMR test

FIGURE 10-6   Effect of various Nitrogen sources on production of cellulase from mutant EMS5

The effectiveness of the nitrogen source in supporting cellulase production along with the secretion of extra cellular protein by EMS5 mutant strain decreased in the following order: urea > tryptone > $NaNO_3$ > yeast extract > KNO_3. Cultivation of EMS5 at 30 °C temperature for seven days secreted higher titres of cellulase enzyme in comparison with culturing at 25 °C and 40 °C on solid substrate fermentation medium (Figure 10-7). The fermented bran grown with EMS5 at 30 °C contained higher yields of FPase, CMCase, and β-glucosidase to the tune of 48.71, 22.91, and 32.10 U/gDS, respectively. Lower enzyme activities of FPase (32.20 U/gDS), CMCase (11.24 (U/gDS), and β-glucosidase (20.12 U/gDS) were recorded in the fermented bran incubated at 40 °C. Among different spore densities tested in this study, solid medium inoculated with spore density of 2×10^8 of EMS5 produced higher yields of cellulase (Figure 10-8). Solid state medium with an initial moisture content of 80% was optimal for maximum extracellular cellulase production (Figure 10-9). Cultivation of *Myceliophthora thermophila* M77 on a combination of soybean bran and sugarcane bagasse in 1 : 9 ratio in solid state fermentation at 80% moisture content gave a maximum yield of 11 FPU/gDS which was 4.4-folds higher than the production on pure wheat bran[40] (Figure 10-9).

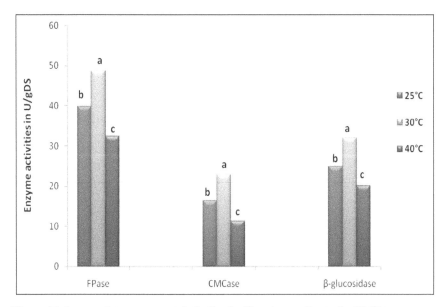

*Bars with the same letter are not statistically significant according to the DMR test.

FIGURE 10-7   Effect of temperature on the production of cellulase by the mutant EMS5 in SSF.

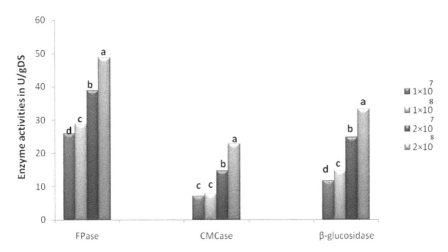

*Bars with the same letter are not statistically significant according to the DMR test.

FIGURE 10-8   Effect of spore density on the production of cellulase by the mutant EMS5 in SSF.

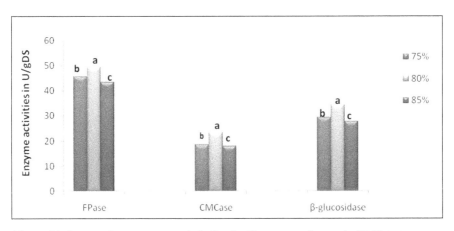

*Bars with the same letter are not statistically significant according to the DMR test.

FIGURE 10-9   Effect of moisture content on the production of cellulase by the mutant EMS5.

10.4   CONCLUSION

Chemical mutagen - EMS at concentrations employed in this study was more lethal to the spores of *A. niger* than UV radiation as reflected by the survival percentage of the spores. The percentage survival of the spores with EMS treatment was 10–20% as against 10–90% with the UV irradiation. The mutant—EMS5 was a hyper cellulase producing mutant with 30–50% more yields over the wild type in solid state fermentation on a combination of rice bran and wheat bran in equal ratio. The mutant EMS5 produced maximum titer of FPase (35.42 U/g), CMCase (14.81 U/g), and 26.42 (U/g) of β-glucosidase after seven days of growth in solid state fermentation. Cellulose, urea, a temperature of 30 °C, spore density of 2×10^8, and moisture content of 80% were the optimal conditions for the maximum secretion of the cellulolytic enzymes.

KEYWORDS

- *A. niger*
- Cellulases
- CMCase
- Congo red
- EMS
- FPase
- Lignocelluloses

REFERENCES

1. Bakare, M. K.; Adewale, I. O.; Ajayi, A.; and Shonukan, O. O. Purification and characterization of cellulase from the wild-type and two improved mutants of *Pseudomonas fluorescens. Afr. J. Biotechnol.* **2005,** *4*, 898–904.

2. Howard, R.; Masoko, P.; and Abotri, E. Enzyme activity of a *Penicillium Chrysosporium* Cellobiohydrolase (CBH!) expressed as a heterologous protein from *E. coli. African J. Biotechnol.* **2003,** *2* (9), 296–300.

3. Fan, L. T.; Gharpuray, M. M.; and Lee, Y. H. *Cellulose Hydrolysis Biotechnology Monographs;* Springer: Berlin, 1987, p 57.

4. Chahal, P. S.; Chahal, D. S.; and Andre, G. Cellulase production profile of *Trichoderma reesei* on different cellulosic substrates at various pH levels. *J. Ferment. Bioeng.* **1992,** *74*, 126–128.

5. Reczey, K.; Szengyel, Z.; Eklund, R.; and Zacchi, G. Cellulase production by *T. reesei. Biores. Technol.* **1996,** *57*, 25–30.

6. Wu, Z.; and Lee, Y. Y. Inhibition of the enzymatic hydrolysis of cellulose by ethanol. *Biotech. Lett.* **1997**, *19*, 977–979.

7. Solomon, B. O.; Amigun, B.; Betiku, E.; Ojumu, T. V.; and Layokun, S. K. Optimization of cellulase production by *Aspergillus flavus Linn* isolate NSPR 101 grown on bagasse. *JNSChE.* **1999**, *16*, 61–68.

8. Aristidou, A.; and Penttila, M. Metabolic engineering applications to renewable resource utilization. *Curr. Opin. Biotechnol.* **2000**, *11*, 478–483.

9. Akin-Osanaiye, B. C.; Nzelibe, H. C.; and Agbaji, A. S. Production of ethanol from *Carica papaya* (pawpaw) agro waste: Effect of saccharification and different treatments on ethanol yield. *Afr. J. Biotechnol.* **2005**, *4* (7), 657–659.

10. Yano, S.; Ozaki, H.; Matsuo, S.; Ito, M.; and Wakayama, M. Production, purification and characterization of D-aspartate oxidase from the fungus *Trichoderma harzianum* SKW-36. *Adv. Biosci. Biotechnol.* **2012**, *3* (1), 7–13.

11. Lynd, L. R.; Wyman, C. E.; and Gerngross, T. U. Biocommodity engineering. *Biotechn. Prog.* **1999**, *15*, 777–793.

12. Bhat, M. K. Cellulases and related enzymes in biotechnology. *Biotechnol. Adv.* **2000**, *18*, 355–383.

13. Lynd, L. R.; Weimer, P. J.; VanZyl, W. H.; and Pretorius, I. S. Microbial cellulose utilization: fundamentals and biotechnology. *Microbial. Mol. Biol. Rev.* **2002**, *66*, 506–577.

14. Pandey, A.; Selvakumar, P.; Soccol, C. R.; and Nigam, P. Solid State fermentation for the production of industrial enzymes. *Current Sci.* **1999**, *77*, 149–162.

15. Schulein, M. Protein engineering of cellulases. *Biochim. Biophys. Acta.* **2000**, *1543*, 239–252.

16. Wither. S. G. Mechanism of glycosyl transferase and hydrolases. *Carbohydr. Poly.* **2001**, *44*, 325–337.

17. Wilson, D. B. Studies of *Thermobifida fusca* plant cell wall degrading enzymes. *Chem. Rec.* **2004**, *4*, 72–82.

18. Mandels, M.; Weber, J.; and Parizek, R. Enhanced cellulase production by a mutant of *Trichoderma viride*. *Appl. Microbiol.* **1971**, *21*, 152–154.

19. Montenecourt, B. S.; Sheir-Neiss, G. I; Ghosh, A.; and Gosh, B. K. Mutational approaches to enhance synthesis and secretion of cellulase. In *Proc. International Symposium on Ethanol from Biomass*; Duckworth, H. E., and Thompson, E. A., Eds.; Royal Society of Canada: Ottawa, 1983, 309–414.

20. Chaudhary, K.; Mittal, S. L.; and Tauro, P. Control of cellulose hydrolysis by fungi *Biotechnol. Lett.* **1985**, *7*, 455–56.

21. Morikawa, Y.; Kawamori, M.; Ado, Y.; Shinsha, Y.; Oda, Y.; and Takasawa, S. Improvement of cellulase production in *Trichoderma reesei*. *Agric. Boil. Chem.* **1985**, *49*, 1869–1871.

22. Durand, H.; Clanet, M.; and Tiraby, G. Genetic improvement of *Trichoderma reesei* strains for large scale cellulase production. *Enzyme Microb. Technol.* **1988**, *10*, 341–345.

23. Esterbauer, H.; Steiner, W.; Labudova, I; Hermann, A.; and Hayn, M. Production of *Trichoderma* cellulase in laboratory and pilot scale. *Bioresource Technol.* **1991**, *36*, 51–65.

24. Rajoka, M. I.; Bashir, A.; Hussain, S. R. S.; and Malik, K. A. Gamma ray induced mutagenesis of *Cellulomonas biazotea* for improved production of cellulases. *Folia. Microbiol.* **1998**, *43* (1), 15–22.

25. Kotchoni, O. S.; Shonukan, O. O.; and Gachomo, W. E. *Bacillus pumilus BPCRI6* a promising candidate for cellulase production under conditions of catabolite repression. *African J. Biotechnol.* **2003**, *2* (6), 140–146.

26. Zaldivar, M.; Velasquez, J. C.; Contreras, I.; and Perez, L. M. *Trichoderma aureoviridae* 7-121, a mutant with enhanced production of lytic enzymes: its potential use in waste cellulose degradation and/or biocontrol. *Elect. J. Biotechnol.* **2001,** *4* (3), 160–168.

27. Nanmori, T.; Shinke, R.; Aoki, K.; and Nishara, N. fl-Amylase production by a rifampin-resistant, asporogenous mutant from *Bacillus cereus BQ10-SI. Agric. Biol. Chem.* **1983,** *47*, 609–611.

28. Narasimha, G.; Babu, B. V. A. K.; and Rajasekhar Reddy, B. Cellulolytic activity of fungal cultures isolated from soil contaminated with effluents of cotton ginning industry. *J. Env. Biol.* **1999,** *20*, 235–239.

29. Teather, R. M.; and Wood, P. J. Use of congo red–polysaccharide interactions in enumeration and characterization of cellulolytic bacteria from the bovine rumen. *Appl. Environ. Microbiol.* **1982,** *43*, 777–780.

30. Wood, T. M.; and Bhat, M. K. Methods for measuring cellulase activities. In *Methods in Enzymology*; Wood, W., and Kellog, S. J., Eds.; Academic Press: New York, 1988; Vol. 160, pp 87–112.

31. Mandels, M.; and Weber, J. *Cellulases and Its Application. Advances in Chemistry Series*; Gould, R. F., Ed.; American Chemical Society: Washington, DC., 1969, Vol. 95, pp 391–414.

32. Miller, G. L. Use of dinitrosalicylic acid reagent for determination of reducing sugars. *Anal. Chem.* **1959,** *31*, 426–429.

33. Herr, D. Secretion of cellulases and β-glucosides by *Trichoderma viridae* TTCC-1433 in submerged cultures on different substrates. *Biotechnol. Bioeng.* **1979,** *21*, 1361–1363.

34. Shafique, S.; Bajwa, R.; and Shafique, S. Strain improvement in *Trichoderma viride* through mutation for over expression of cellulase and characterization of mutants using random amplified polymorphic DNA (RAPD). *Afr. J. Biotechnol.* **2011,** *10* (84), 19590–19597.

35. Veerapagu, M.; Jeya, K. R.; and Ponmurugan, K. Mutational effect of *Penicillium chrysogenum* on Antibiotic Production. *Advan. Biotech.* **2008,** 16–19.

36. Shahbazi, S.; Ispareh, K.; Karimi, M.; Hamed, A.; and Ebrahimi, M.ali. Gamma and UV radiation induced mutagenesis in *Trichoderma reesei* to enhance cellulases enzyme activity. *Intl. J. Farm. Alli. Sci.* **2014,** *3* (5), 543–554.

37. Elakkiya, P.; and Muralikrishnan, V. Isolation and mass multiplication of solid state fermentation for cellulase producing fungi *Trichoderma viride. Gold. Res. Thou.* **2014,** *4*, 1–6.

38. Mandels, M.; and Reese, E. T. Induction of cellulase in *Trichoderma viride* as influenced by carbon sources and metals. *J. Bacteriol.* **1957,** *73*, 269.

39. Peterson, R.; and Nevalainen, H. *Trichoderma reesei* RUT-C30–thirty years of strain improvement. *Microbiology* **2012,** *158*, 58–68.

40. Kilikian, B. V.; Afonso, L. C.; Souza, T. F. C.; Ferreira, R. G.; and Pinheiro, I. R. Filamentous fungi and media for cellulase production in solid state cultures. *Braz. J. Microb.* **2014,** *45* (1), 279–286.

CHAPTER 11

SCREENING, QUANTIFICATION, AND PURIFICATION OF CELLULASES FROM SOIL ACTINOMYCETES

PAYAL DAS[1], RENU SOLANKI[1], and MONISHA KHANNA[1*]

[1]*Acharya Narendra Dev College, University of Delhi, Govindpuri, Kalkaji, New Delhi – 110019, India.*

Corresponding author: Dr. Monisha Khanna. E-mail: monishaandc@ gmail.com

CONTENTS

Abstract 220
11.1 Introduction 220
11.2 Materials and Methods 221
11.3 Results 226
11.4 Discussion 242
11.5 Conclusion 243
Acknowledgments 244
Keywords 244
References 244

ABSTRACT

Actinomycetes are a group of gram positive soil microbes, widely known as producers of various extracellular enzymes such as peptidase, ligninase, xylanase, and amylase. These enzymes have gained increased attention because of their widespread applications in several industries. In an attempt to discover new cellulase producing strains, isolates from diverse ecological habitats were subjected to primary screening. The diameter of the zones of hydrolysis ranged from 17–33 mm in the case of cellulose and 6–13 mm in the case of carboxymethyl cellulose (CMC). Colonies 194, 51, and 157 representing different habitats and showing substantial zones of clearance were selected for secondary screening. The crude extract was partially purified by ammonium sulphate precipitation and dialysis followed by concentration. The enzyme activity in crude and in partially purified samples was determined. The activity ranged from 5.4–12.13 IU/ml/min (in crude) and 12.04–21.5 IU/ml/min (in partially purified). Colony 194 showing maximum cellulase activity was subjected to sodium dodecyl sulphate-polyacrylamide gel electrophoresis (SDS-PAGE) profiling. Phenotypic and taxonomic characterization of selected isolates was done by polyphasic approach. It was confirmed from 16S rRNA gene sequence analyses that the isolates belonged to the genus *Streptomyces*.

11.1   INTRODUCTION

Enzymes are complex proteins, produced by all living beings.[1-3] Extracellular enzymes are secreted by micro-organisms including bacteria, fungi, and nematodes to break down complex chemical structures.[4] Soil is a rich source of complex organic matter. Those micro-organisms which cannot transport complex molecules inside their cytoplasm depend on the action of extracellular enzymes for the breakdown of these molecules into useful nutrients.[2,3] The secretion pattern of extracellular enzymes from Gram positive and negative bacteria shows a difference as they either follow the type I pathway through Chaperons or type II protein secretion pathway through Secretory (Sec) or Twin arginine translocation pathway (Tat) system for the transport of enzymes from the cytoplasmic space to the outside.[4] Once the extracellular enzymes have been secreted in the outside environment, they may be degraded and denatured. However, some of these enzymes are glycosylated, and have disulfide bonds which provide stability at various temperature and pH conditions. The excessive demand for energy has resulted in the utilization of widely available alternative energy sources which would be cost effective.[5] One such bio-residue is the lignocellulosic biomass generated from different industries,[6] consisting of renewable biopolymers such as cellulose, hemicellulose, and lignin.[7] Enzymatic degradation of substrates is of great biological as well as economic importance. Cellulose, the major component of biomass, consisting of β-1,4 glucose units linked by β-1,4-D-glycosidic bond, is commonly degraded by the enzyme system, cellulose.[5–13] This system constitutes a group of enzymes, namely, endoglucanase, which randomly attacks the internal β-1-4 linkages; exoglucanase acts on the nonreducing end of the cellulose chain, resulting in the formation of cellobiose units, and β-glucosidase which breaks cellobiose units into glucose.[14]

Actinomycetes are a group of filamentous Gram positive bacteria, widely distributed in different habitats and are known as degraders of organic matter and producers of antibiotics[15–18] and extracellular enzymes such as cellulases,[8,19,20] chitinases,[9] xylanases,[10] peptidases, proteases, amylases,[11] pectinases, hemicellulases, and keratinases.[12] The high cellulase producing bacterial genera include *Streptomyces, Bacillus, Pseudomonas, Thermonospora, Actinosynnema*, and *Nocardiopsis*.[13,21–23]

11.2 MATERIALS AND METHODS

11.2.1 ISOLATION OF ACTINOMYCETES

Soil samples were collected from diverse ecological habitats for the isolation of actinomycetes (Table 11-1). Dilutions of dried samples were prepared in saline and plated on different actinomycete specific mediums. Various actinomycete specific media were used for the isolation, such as arginine glycerol agar,[24] glycerol asparagine agar,[25] organic agar Gause 2,[26] starched casein agar,[27] and yeast extract malt extract agar.[25] Single actinomycete colonies were purified by several rounds of restreaking on Yeast extract-Malt extract (YM) agar plates and stored at −20 °C/−80 °C as 20% glycerol stocks.[19]

TABLE 11-1   Number of Isolates from Different Ecological Habitats

S. No.	Habitat	Total No. of Colonies
1.	**AGRICULTURAL SOILS**	
	Agricultural soil, Dhanaura, Uttar Pradesh	86
	Agricultural soil, Yamuna Bank, Delhi	37
	Agricultural soil, Nainital	13
	Agricultural soil, Kashipur	17
2.	**INDUSTRIAL SOILS**	
	Sugar Plant, Dhanaura, Uttar Pradesh	7
	Chemical Plant, Faridabad	10
3.	**LANDFILL SOILS**	
	Dumping site, Sarai Kale Khan, Delhi	28
4.	**RIVER/LAKE SOILS**	
	Yamuna Bank, Delhi	6
	Lake soil, Purana Quila, Delhi	14
5.	**DIVERSITY PARK SOILS**	
	Diversity Park, Sarai Kale Khan, Delhi	8
	Great Himalayan National Park, Teerthan Valley, Himachal Pradesh	42
	Great Himalayan National Park near a narrow spring, Teerthan Valley, Himachal Pradesh	38
6.	**SEA/BEACH SOILS**	
	Catamaran Beach Hotel, Colombo, Sri Lanka	20
	Havelock Islands, Andaman and Nicobar Islands	76
	Carbon Island, Andaman and Nicobar Islands	98
7.	**FOREST SOILS**	
	Killingpong 4000 ft, Kolkata	44
	Pine Forest, Teerthan Valley, Himachal Pradesh	102

11.2.2 PRIMARY SCREENING OF ISOLATES FOR THE PRODUCTION OF CELLULASE ENZYME

For analyses of the isolates for the production of cellulase enzyme, the isolates along with the positive controls and some known enzyme producers were spot inoculated on basal agar medium containing (gL^{-1})

$(NH_4)_2SO_4$ 2.64, KH_2PO_4 2.38, $K_2HPO_4.3H_2O$ 5.65, $MgSO_4.7H_2O$ 1, 1 ml of trace salt solution (composition gL^{-1}, $CuSO_4.5H_2O$ 6.43, $FeSO_4.7H_2O$ 1.1, $MnCl_2.4 H_2O$ 7.9, $ZnSO_4.7 H_2O$ 1.5), agar 15, pH 6.8–7[19,28] supplemented with 2% cellulose, and carboxymethyl cellulose (CMC) as the substrate, respectively.[8] The plates were incubated at 28 °C for 7 days to allow secretion of cellulase. A clear zone of hydrolysis around the colonies indicated cellulase production[19,29,30] which was measured by subtracting the inoculum size from the total zone diameter.[19,30] Congo red (0.1%) staining was done followed by counterstaining with 1M NaCl for 15–20 min to visualize the zone of hydrolysis in the case of CMC.[8] The experiment was repeated three consecutive times in each case, and the average zone diameter (mm) was calculated. Based on the average, standard deviation was also determined.

11.2.3 PCR AND SEQUENCING

For amplification of 16S rRNA gene, genomic DNA from the colonies was isolated by standardized methods for actinomycetes.[19,31,32] Two sets of universal primers were used namely: 8F (5'-AGAGTTTGATCCTGGCTCAG-3') and 1492R (5'-TACGGTTACCTTGTTACGACTT-3'); 27F (5'-AGAG TTTGATCCTGGCTCA-3'), and 1542R (5'-AAGGAGGTGATCCAG CCGCA-3') (SIGMA). The respective amplicons were purified with Bangalore Genei kit and were sequenced by the Micro Seq[R] 16S rRNA gene sequencing kit (Applied biosystems, USA) and Applied Biosystems 3100 Avant™ Genetic Analyzer Sequencer. After analysing the sequences with sequence analyser tools, the resultant sequences were of 1437 nt for isolate no. 51, 1450 nt for isolate no. 157, and 1432 nt for isolate no. 194. The sequences were then manually aligned with the known sequences of *Streptomyces* available on Eztaxon database.[33] Using Clustal_X 1.81 and MEGA 4.1 softwares,[18,34–36] phylogenetic trees were constructed by neighbor-joining method.[37] Bootstrap re-sampling method with 1000 replicates were selected[38,39] and *Actinomadura hibisca* JCM 9627[T] (from Japan Collection of Micro-organisms) (AF163115) (as per GenBank Accession Number prefixes) was selected as an outgroup for rooting the evolutionary trees.[18,19,40] The 16S rRNA gene sequences of colonies 51, 157, and 194 were submitted for accession numbers to GenBank nucleotide database of the National Centre for Biotechnology Information (NCBI).

11.2.4   BACTERIAL CHARACTERIZATION

Phenotypic and taxonomic characterization of selected cellulase isolates was studied as per protocols of the International *Streptomyces* project.[25,31,35] Spore morphology of the isolates 51, 157, and 194 was observed by phase contrast microscope (Nikon E600).[35]

11.2.5   SECONDARY SCREENING OF ENZYME ACTIVITY BY SUBMERGED FERMENTATION PROCESS

On basis of the results during qualitative screening, the isolates showing appreciable zones of clearance were inoculated in 50 ml of 148G medium (g/L^{-1}) containing glucose 22, beef extract 4, bacto peptone 5, yeast extract 0.5, tryptone 3, NaCl 1.5(pH 7.5). CFUs/ml for each culture were calculated.[19] Standard inoculum having an average viable count of 10^5–10^7 CFUs/ml was transferred to the production medium (200 ml) and incubated at 28 °C for 5–6 days on a rotary shaker (New Brunswick Scientific, Excella E24R) at 200 rpm.[4,29]

11.2.6   PROTEIN CONTENT IN CRUDE CULTURE BROTH

Protein content in the crude enzyme was determined by Lowry's method[41] with bovine serum albumin (BSA) as a standard.

11.2.7   ENZYME ASSAY IN CRUDE CULTURE BROTH

Crude supernatant was obtained by centrifuging the well grown, 5 days old cultures in the production medium.[13] Quantitative analysis of cellulase enzyme activity was done using the dinitrosalicyclic (DNS) method.[19,42,43] A quantity of 1.5 ml of 2% cellulose prepared in 0.05M sodium citrate buffer (pH 4.8) was added to 0.5 ml of crude culture filtrate and incubated at 40 °C for 30 min. After incubation, the reaction was terminated by adding 3 ml of DNS reagent and subsequent incubation at 100 °C for 5 min. Reducing sugars were estimated spectrophotometrically (UV Vis Elico SI-159) at 540 nm[14] and using glucose as the standard.[44–46] One unit of enzyme activity was defined as the amount of enzyme that released 1 µM of glucose/ml/min.[29] The enzyme activity (IU/ml/min) can be calculated either by the formula − Concentration of glucose (µg/ml)* × reaction volume/Molecular weight of glucose × reaction time × volume of enzyme (*where, Concentration of glucose = Actual optical density (OD)/Slope from the graph),[19,46,47] or Enzyme activity (IU/ml/min) − Concentration of glucose (µg/ml)* × reaction volume/Molecular weight of glucose × reaction time × volume of enzyme × dilution factor (*where, Concentration of glucose = Actual OD/Slope from the graph). The experiment was repeated three consecutive times, for checking the reproducibility of the result. The average enzyme activity (IU/ml/min) and standard deviation were calculated.

11.2.8 PURIFICATION OF THE ENZYME BY AMMONIUM SULPHATE SATURATION AND DIALYSIS METHOD

Crude culture extract of 250 ml was saturated by sequential ammonium sulphate precipitation (20–100%) and the mixture was kept for protein precipitation each time at 4 °C, overnight.[4,13,19,29] The precipitate was collected by centrifugation at 8000 rpm, for 20 min, 4 °C, and dissolved in 50 mM Tris HCl buffer (pH 8.0). This was dialyzed against the same buffer using dialysis membrane with a molecular cutoff of 5 KDa (HiMedia), overnight at 4 °C with continuous stirring. Dialysis was followed by concentration up to 5 ml using the Amicon ultra centrifugation tubes (membrane cut-off of 3 kDa) at a speed of 3000 rpm for 2 h at 4 °C. Estimation of enzyme activity and protein content was done in partially purified sample.

11.2.9 PROTEIN CONTENT IN PARTIALLY PURIFIED PRODUCTS

Protein content in the samples was determined as mentioned earlier.

11.2.10 ENZYME ASSAY IN PARTIALLY PURIFIED PRODUCTS

After estimating the protein contents in partially purified sample of each isolates, the protein content was equalized in all the samples by using 50 mM Tris-Cl (pH 8.0). Protein content and enzyme activity was measured in the equalized protein samples by DNS method using BSA and glucose standard curve, respectively.

11.2.11   SDS-PAGE PROFILING OF ISOLATES

SDS-PAGE profiling was done to observe the protein profile of isolate no 194. Gel plates were set by placing spacers between the plates. A solution of 10% resolving gel [water - 4 ml, Tris-Cl (pH 4.8) - 2.5 ml, acrylamide:bisacrylamide - 3.3 ml, SDS (10%) - 100 µl, Ammonium persulfate (APS)- 100 µl, Tetramethyl ethylenediamine (TEMED) - 8 µl] was prepared. The resolving gel solution was filled into the gap between the glass plates to 3/4th level using a pipette. A layer of water was added on the top of the resolving gel to make it horizontal. The gel was left for polymerization at room temperature for 1 h. The top layer of water was discarded after 1 h. A solution of 5% stacking gel [water - 2.96 ml, tris-Cl (pH 6.8) - 1.25 ml, acrylamide:bisacrylamide - 650 µl, SDS(10%) - 50 µl, APS - 75 µl, TEMED - 8 µl] was prepared and added over the resolving gel. Wells were formed in the gel by inserting a comb and left for 20–30 min to polymerize. The comb was removed after complete polymerization of the stacking gel. The glass plates were set in the buffer tank. Reservoir (running) buffer was poured into the inner chamber until the buffer reached the required level in the outer chamber. The protein samples were mixed with lamelli dye (loading dye) and heated at 96 °C for 5 min. Samples of 80–100 µl were loaded in the wells of the stacking gel along with the marker (3 µl). A voltage of 80 V was applied to the electrophoresis unit. After running, the gel was removed from the glass plates and was kept in the staining solution (0.1% Coomasie brilliant blue R250, 50% methanol, 10% glacial acetic acid, and water), overnight. Next day, several rounds of destaining were done in the destaining solution (40% methanol, 10% glacial acetic acid, and water) for 2–3 h and then differentiated in water for 2 h. The protein bands were observed.[48,49]

11.3   RESULTS

11.3.1   ISOLATION OF ACTINOMYCETES

The soil samples were collected from various ecological habitats (Table 11-1, Figure 11-1). More than 600 bacterial colonies were isolated using actinomycete specific media. Among the various media used, it has been found that arginine glycerol agar, glycerol asparagine agar, and organic agar Gause 2 were found to be most effective for the isolation of actinomycetes followed by starch casein agar and yeast extract malt extract agar (Table 11-2).

TABLE 11-2 Comparison of the Efficacy of Different Media for the Selective Isolation of Actinomycetes

Medium	Appearance of Actinomycete Colonies
Arginine Glycerol (AG) agar	+++
Glycerol asparagine agar	+++
Organic Gause agar	+++
Starch casein agar	++
Yeast extract malt extract agar	+

+++ large number of colonies++ moderate number of colonies.

+ less number of colonies.

Agricultural soil, Yamuna Bank, Delhi, India

Agricultural soil, Dhanaura, Uttar Pradesh, India

Dumping site, Sarai Kale Khan, Delhi, India

Sugar plant, Dhanaura, Uttar Pradesh, India

FIGURE 11-1 Some collection sites for soil samples.

11.3.2 PRIMARY SCREENING OF ISOLATES

It is evident from the results of preliminary screening, that the extent of degradation of cellulose varied from isolate to isolate (Table 11-3). Among the tested isolates, around 150 isolates showed cellulase activity. Colonies 4, 51, 157, 196, 222, 186, 194, and 169 represented varied habitats. *Streptomyces albogriseolus* (NRRLB 1305), *S. subrutilus* (NRRLB 12377), *S. mexicanus* (NRRLB 24196), *S. albidoflavus* (NRRLB 16746), *S. venezuelae* (ISP 5230), *S. stramineus* (NRRLB 12292), and *S. coelicolor* (NRRLB 16638) showed appreciable cellulase production. The control cultures were from Northern Regional Research Laboratory (NRRL) culture collection (also known as National Center for Agricultural Utilization Research).

TABLE 11-3   Primary Screening Results for the Cellulase Enzyme

Isolates	Activity
AGRICULTURAL SOIL, DHANAURA	
Colony no. 4	+++
Colony no. 23	−
Colony no. 43	++
Colony no. 51	+++
Colony no. 157	+++
AGRICULTURAL SOIL, YAMUNA RIVER	
Colony no. 85	++
Colony no. 101	+
Colony no. 138	++
Colony no. 196	+++
Colony no. 222	+++
DUMPING SITE, SARAI KALE KHAN	
Colony no. 102	−
Colony no. 112	++
Colony no. 136	++
Colony no. 186	+++
Colony no. 194	+++
SUGAR PLANT, DHANAURA	
Colony no. 130	−
Colony no. 169	++
CHEMICAL PLANT, FARIDABAD	
Colony no. 184	−
Colony no. 202	−
Streptomyces albogriseolus (NRRL B-1305) (cellulase control)	+++
Streptomyces mexicanus (NRRL B-24196) (xylanase control)	+++

TABLE 11-3   *(Continued)*

Isolates	Activity
Streptomyces thermocoprophilus (NRRL B-24314) (xylanase control)	–
Streptomyces albidoflavus (NRRL B-16746) (chitinase control)	+++
Streptomyces venezuelae (ISP 5230) (chitinase control)	+++
Streptomyces stramineus (NRRL B-12292) (phosphatase control)	+++
Streptomyces coelicolor (NRRL B-16638) (phosphatase control)	+++

+++ High activity.

++ Moderate activity.

+ Low activity.

– No activity.

It was found that the diameter of the zone of hydrolysis produced by *S. albidoflavus* (NRRL B-16746), isolate no. 194, isolate no. 51, isolate no. 157, and *S. albogriseolus* were 33 mm, 31 mm, 26 mm, 18 mm, and 17 mm in the case of cellulose and 6 mm, 13 mm, and 8 mm for isolates 51, 157, and *S. albogriseolus* in the case of CMC as a substrate, respectively (Figure 11-2, Table 11-4, Graphs 11-1 and 11-2). When the result of the plate assay obtained both in the case of cellulose and CMC was compared, it was observed that the diameter of zone of hydrolysis in the case of CMC was less as compare with cellulose; as a result, cellulose was chosen as a substrate for further analysis. On the basis of the results of the primary screening, colonies 51, 157, 194, *S. albidoflavus*, and *S. albogriseolus* (control) which showed substantial cellulase activity and represented diverse ecological habitats were selected for further analyses.

(a) *Streptomyces albidoflavus* (NRRL B-16746) culture showing zone of clearance (size - 33 mm) on the basal medium with cellulose

(b) Colony no. 194 culture showing zone of clearance (size - 31 mm) on the basal medium with cellulose

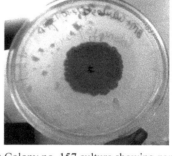

(c) Colony no. 51 culture showing zone of clearance (size - 26 mm) on the basal medium with cellulose

(d) Colony no. 157 culture showing zone of clearance (size - 18 mm) on the basal medium with cellulose

(e) *Streptomyces albogriseolus* (NRRL B-1305) culture showing zone of clearance (size - 17 mm) on the basal medium with cellulose

(f) Colony no. 157 culture showing zone of clearance (size - 13 mm) on the basal medium with CMC

(g) Colony no. 51 culture showing zone of clearance (size - 6 mm) on the basal medium with CMC

(h) *Streptomyces albogriseolus* (NRRL B-1305) culture showing zone of clearance (size - 8 mm) on the basal medium with CMC

FIGURE 11-2   Plates showing zone of clearance of the isolates because of the production of cellulase (in the case of cellulose and CMC).

TABLE 11-4 Clear Zone Produced by Isolates Because of the Production of Cellulase

S. No.	Strains	Clear Zone Diameter (mm) (with Cellulose)	Clear Zone Diameter (mm) (with CMC)
1.	NRRLB 16746, *Streptomyces albidoflavus* (chitinase control)	33	No zone
2.	Colony 194 (dumping site, Sarai Kale Khan, Delhi)	31	No zone
3.	Colony 51 (agricultural soil, Dhanaura, Uttar Pradesh)	26	6
4.	Colony 157 (agricultural soil, Dhanaura, Uttar Pradesh)	18	13
5.	Control- NRRLB 1305, *Streptomyces albogriseolus* (cellulase control)	17	8

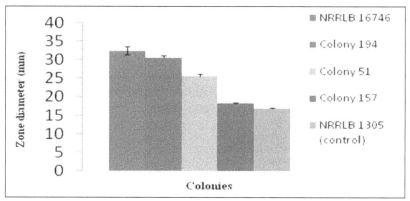

GRAPH 11-1 Comparison of the activity of different cellulase producing isolates (in the case of cellulose as the substrate; standard deviation is also shown).

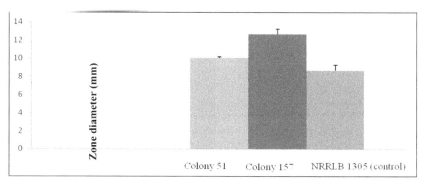

GRAPH 11-2 Comparison of the activity of different cellulase producing isolates (in the case of CMC as the substrate; standard deviation is also shown)

11.3.3   IDENTIFICATION OF ISOLATES

The 16S rRNA gene sequence comparison of the isolates 194, 51, and 157 with the sequences of their respective phylogenetic relatives, given in the database, showed that they belong to the genus *Streptomyces* and were assigned accession numbers KJ934595, KJ995861, KJ934594 (as per GenBank Accession Number prefixes) respectively.

Rooted phylogenetic trees were constructed based on neighbor joining method, separately for isolates 51, 194, and 157 indicating that these were included in distinct clades in their respective trees (Figures 11.3–11.5). Isolate 51 showed 100% similarity with its phylogenetic relative *Streptomyces griseochromogenes* NBRC 13413[T] (AB184387; NBRC- NITE Biological Resource Center, Department of Biotechnology, National Institute of Technology and Evaluation, Kisarazu, Chiba, Japan),[50] isolate 157 showed 100% similarity with *Streptomyces rochei* NBRC 12908[T] (AB184237),[51] *Streptomyces enissocaesilis* NRRL B-16365[T] (DQ026641; DQ is as per GenBank Accession Number prefixes),[26] and *Streptomyces plicatus* NBRC 13071[T] (AB184291; AB is as per DDBJ Accession Number prefixes),[52] and isolate 194 showed 96.20% similarity with *Streptomyces albidoflavus* DSM 40455[T] (Z76676; DSMZ-Deutsche Sammlung von Mikroorganismen und Zellkulturen GmbH, Braunschweig, Germany)[53].

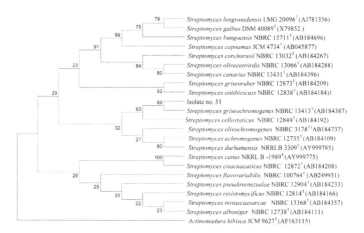

FIGURE 11-3   Rooted phylogenetic tree based on 16S rRNA gene sequences, showing the relationship between isolate no. 51 (KJ995861) and related representative species of the genus *Streptomyces*. The sequence of the 16S rRNA gene of *Actinomadura hibisca* JCM 9627[T] (AF163115) was used as an outgroup. The tree was generated using the neighbor-joining method (Clustal_X version 1.81 and MEGA version 4.1) and includes bootstrap percentages based on the analysis of 1000 resampled datasets.

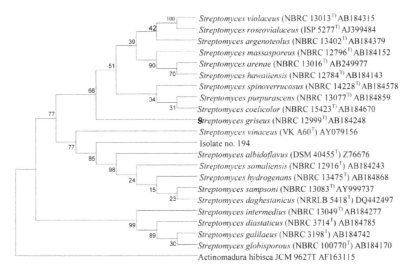

FIGURE 11-4   Rooted phylogenetic tree based on 16S rRNA gene sequences, showing the relationship between isolate no. 194 (KJ934595) and related representative species of the genus *Streptomyces*. The sequence of the 16S rRNA gene of *Actinomadura hibisca* JCM 9627^{T} (AF163115) was used as an outgroup. The tree was generated using the neighbor-joining method (Clustal_X version 1.81 and MEGA version 4.1) and includes bootstrap percentages based on the analysis of 1000 resampled datasets.

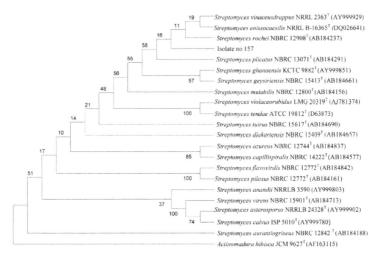

FIGURE 11-5   Rooted phylogenetic tree based on 16S rRNA gene sequences, showing the relationship between isolate no. 157 (KJ934594) and related representative species of the genus *Streptomyces*. The sequence of the 16S rRNA gene of *Actinomadura hibisca* JCM 9627^{T} (AF163115) was used as an outgroup. The tree was generated using the neighbor-joining method (Clustal_X version 1.81 and MEGA version 4.1) and includes bootstrap percentages based on the analysis of 1000 resampled datasets.

11.3.4 DESCRIPTION OF THE ISOLATES

The isolates were phenotypically characterized by analyses of spore chain morphology and of culture morphology on different ISP media, utilization of sugars, and hydrolysis of compounds. The results indicated that all the isolates belonged to the group of Gram positive actinomycetes, with the ability to form both vegetative and aerial mycelia. However, they differed in morphological and biochemical characters (Table 11-5) which could be because of the adaptation of isolates to their respective habitats.

Isolate 51 (KJ995861) showed cream/yellow/beige/brown substrate mycelium, white/gray aerial mycelium, and abundant sporulation on most ISP media. It was found to utilize fructose, mannitol, sucrose, meso-inositol, arabinose, raffinose, showed high caseinase activity and degrades Tween 80, but starch moderately. The isolate had a moderate metabolism of urea (Table 11-5). The spore chain morphology was Retinaculiaperti (Figure 11-6).

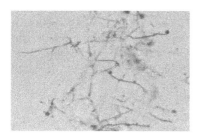

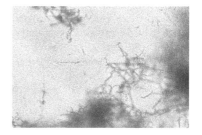

Colony 51: Hooks, loops, or spirals with one to two turns (retinaculiaperti)

Colony 157: Hooks, loops, or spirals with one to two turns (retinaculiaperti)

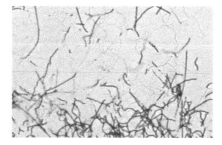

Colony 194: Straight to flexuous (rectiflexibles)

FIGURE 11-6 Spore chain morphology of the strains.

Isolate no. 157 (KJ934594) showed yellow/beige substrate mycelium, white/gray aerial mycelium and abundant sporulation on most ISP media.

TABLE 11-5 Comparison of Morphological and Biochemical Characteristics of the Isolates with their Respective Phylogenetic Relatives

Characteristics	Isolate no. 51	Streptomyces griseochromogenes (NBRC 13413) (Fukunaga 1955)	Isolate no. 157	Streptomyces rochei (NBRC 12908T) (Kavitha and Vijayalakshmi, 2007)	Streptomyces enissocaesilis (NRRL B-16365T) (Gause 1986)	Streptomyces plicatus (NBRC 13071T) (Pridham et al. 1958)	Isolate no. 194	Streptomyces albidoflavus (DSM 40455T) (Rossi Doria 1891) Waksman and Henrici 1948)
Color of: Substrate mycelium	Cream (ISP 1,3) Yellow (ISP 2, 5,7) Beige (ISP 5,7) Brown (ISP 6)	Grayish-yellow-orange brown (ISP 2) Grayish-yellow-light olive brown/gray greenish yellow (ISP 3,4,5)	Yellow (ISP 1,2,7) Beige (3,4,6,)	Light to dark yellow (ISP 2)	Brown (ISP 3) Black grayish brown (ISP 5)	N.R.	Yellow (ISP 1, 2,6) Yellow-Black (ISP 4,7) Cream (ISP 3) Beige (ISP 5)	N.R.
Aerial mycelium	White (ISP 1,2, 3,4,5) Gray (ISP 7) No aerial mycelium on ISP 6	Gray (ISP 2,3,4,5)	White (ISP 1, 3,4) Grey (ISP 2,5,6,7)	Gray-white (ISP 2)	Poorly developed (ISP 3) Absent or poorly developed, whitish (ISP 5)	Gray (ISP 2)	White (ISP 1,2, 3, 5,6) Gray (ISP 4) Yellow-White (ISP 7)	White/Gray (ISP 2,5) No sporulation observed in any other ISP media
Growth	Good (ISP 2, 4, 5,7) Moderate (ISP 3) Poor (ISP 1,6)	N.R.	Good (ISP 2,3,4,7) Moderate (ISP 3,6,7)	Good (ISP 3,4,5,7)	N.R.	N.R.	Good (ISP 2,3) Moderate (ISP 4,5,7) Poor (ISP 1,6)	N.R.
Sporulation	Good (ISP 2,4,5,7) Moderate (ISP 3) Poor (ISP 1,6)	N.R.	Good (ISP 2,5,6,7) Moderate (ISP 1,3,4)	N.R.	N.R.	N.R.	Good (ISP 2) Moderate (ISP 4,5,7) Poor (ISP 1,3,6)	N.R.

TABLE 11-5 *(Continued)*

	No diffusible pigment on ISP 1,2,3,4,5,7 Brown (ISP 6)	No pigment on ISP 2,3,4,5 Melanoid pigment on ISP 1,6,7	No diffusible pigment on any ISP media	No diffusible pigment	Brownish, weak (ISP 3) No diffusible pigment (ISP 5)	N.R.	No diffusible pigment on any ISP media	No diffusible pigment on any ISP media
Production of diffusible pigment								
Spore chain	Retinaculiaperti	Spirales to retinaculiaperti	Retinaculiaperti	Spirales	Retinaculiaperti	Spirales	Rectiflexibles	Rectiflexibles
Utilization of:								
D-fructose	+	+	–	+	+	+	+	+
Mannitol	+	+	Weak	+	+	+	Weak	+
Xylose	Weak	N.R.	+	+	+	+	+	+
Sucrose	+	+	+	+	–	–	+	+
Meso-inositol	+	+	+	N.R.	+	+	–	–
L-arabinose	+	+	+	–	+	–	+	+
L-rhamnose	Weak	–	+	–	–	+	+	–
Raffinose	+	+	+	+	–	–	+	–
Tween	+	N.R.	+	N.R.	N.R.	N.R.	+	N.R.
Hypoxanthine	–	N.R.	+	N.R.	N.R.	N.R.	+	N.R.
Urea	+	N.R.	+	N.R.	N.R.	N.R.	+	N.R.
Casein	+	N.R.	+	–	N.R.	N.R.	+	N.R.
Starch	+	N.R.	+	+	N.R.	N.R.	+	N.R.
Production of extracellular enzymes	Cellulase, Xylanase, Chitinase		Cellulase, Xylanase, Chitinase	Amylases, Asparaginases, Chitinase, Cellulase	Cellulase, Xylanase, Lipase, β-amylase	Chitinase, Endoglycosidase H	Cellulase, Xylanase, Chitinase	Cellulase, pectinase, isomerase

It was found to utilize xylose, sucrose, meso-inositol, and arabinose. High caseinase activity and degrades Tween 80 and starch moderately. The isolate had a good metabolism of urea and moderate metabolism of Hypoxanthine (Table 11-5). The spore chain morphology was retinaculiaperti (Figure 11-6).

Isolate no. 194 (KJ934595) showed yellow/yellow-black/cream/beige substrate mycelium, white/gray/yellow-white aerial mycelium, and abundant sporulation on many ISP media. It was found to utilize fructose, xylose, sucrose, arabinose, and raffinose, showed good caseinase activity and degraded Tween 80 and starch moderately. The isolate had a good metabolism of urea and hypoxanthine (Table 11-5). The spore chain morphology was rectiflexibles (Figure 11-6).

11.3.5 SECONDARY SCREENING FOR ENZYME ACTIVITY BY SUBMERGED FERMENTATION PROCESS

Because of the appreciable cellulase activity observed in *S. albidoflavus* (NRRL B-16746), the isolates 194, 51, and 157 along with the control strain were selected for subsequent secondary screening. Standard inoculum having an average viable count of 10^5 to 10^7 CFUs/ml was transferred to the production medium (BSA + cellulose).

11.3.6 PROTEIN CONTENT IN CRUDE CULTURE BROTH (IN 250 ML VOLUME)

Protein content in the crude culture extract was calculated. It was found to increase in all the culture extracts (Table 11-6).

TABLE 11-6 Cellulase Enzyme Activity and Protein Content in Crude Culture Extract (in 250 ml Volume)

S. No.	Cultures	Protein Content (mg/ml)	Enzyme Activity (IU/ml/min)	Protein Content After Equalization (mg/ml)	Enzyme Activity (IU/ml/min) (After Equalization)
1.	NRRLB 16746	0.88	12.11	0.49	12.13
2.	Colony 194	0.71	10.46	0.49	10.4
3.	Colony 51	0.51	7	0.48	7.3
4.	Colony 157	0.49	5.42	0.48	5.7
5.	NRRLB 1305 (control)	0.45	5.4	0.48	5.4

11.3.7   CELLULASE ACTIVITY IN CRUDE CULTURE BROTH (IN 250 ML VOLUME)

Cell free culture extract was used as a source of crude cellulase enzyme. Cellulase enzyme activity units were calculated as mentioned earlier.

The highest enzyme activity was shown by *S. albidoflavus* (NRRL B-16746) (12.13 IU/ml/min) followed by colony 194 (10.4 IU/ml/min), colony 51, and colony 157 and *S. albogriseolus* (NRRL B-1305) (Table 11-6, Graph 11-3).

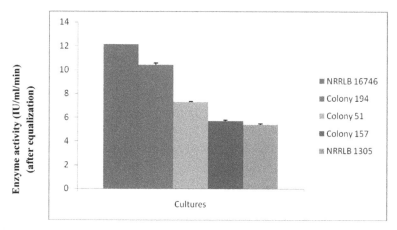

GRAPH 11-3   Comparison of the enzyme activity (in crude) of different cellulase producing isolates (standard deviation is also shown).

11.3.8   PURIFICATION OF CELLULASE ENZYME BY AMMONIUM SULPHATE SATURATION AND DIALYSIS METHOD

Crude culture extracts were partially purified by sequential ammonium sulphate precipitation followed by dialysis and concentration.[45] At 60% saturation, maximum protein precipitation was observed in colony 51 and at 80% in colonies 157 and 194, respectively.

11.3.9   PROTEIN CONTENT IN PARTIALLY PURIFIED SAMPLES OF THE CELLULASE ENZYME

Protein content was determined in the partially purified protein samples as per the earlier mentioned procedure (Table 11-7). The protein content increased in all the culture extracts after purification. Maximum protein content was found in NRRLB 16746, followed by colony 194, NRRLB 12377, colony 51, colony 157, and NRRLB 1305. For the comparison of enzyme activity in partially purified extract, protein contents in all the samples were equalized.

TABLE 11-7 Protein Content and Cellulase Enzyme Activity in Partially Purified Protein Samples (in 5 ml of Concentrated Volume)

S. No	Cultures	Protein Content After Purification (mg/ml)	Enzyme Activity (IU/ml/min)	Protein Content After Equalization (mg/ml)	Enzyme Activity (IU/ml/min) (After Equalization)
1.	NRRLB 16746	1.08	21.24	0.58	21.5
2.	Colony 194	0.9	19.08	0.59	19.11
3.	Colony 51	0.60	14.12	0.58	14.19
4.	Colony 157	0.52	12.38	0.56	12.5
5.	NRRLB 1305 (control)	0.51	12.08	0.52	12.04

11.3.10   ENZYME ASSAY IN PARTIALLY PURIFIED SAMPLES (30 ML AFTER DIALYSIS, 5 ML AFTER CONCENTRATION)

During primary screening, NRRL B-16746 (*S. albidoflavus*) showed a maximum zone of clearance followed by colonies 194, 51, and 157. The results of the primary screening were confirmed during secondary screening. The enzyme activity was found to be maximum in NRRL B-16746 followed by colony 194, colony 51, colony 157, and NRRL B-1305 both in the case of crude extract as well as in the partially purified extract of the samples (Table 11-9).

The strain NRRL B-16746 (21.5 IU/ml/min) showed a maximum cellulase activity followed by colony 194 (KJ934595) (19.11 IU/ml/min), colony 51 (KJ995861) (14.19 IU/ml/min), colony 157 (KJ934594) (12.5 IU/ml/min), and NRRL B-1305 (12.04 IU/ml/min) in 1 ml of concentrated samples (Table 11-8, Graph 11-4).

TABLE 11-8   Comparison of Cellulase Enzyme Activity of the Crude Culture Extracts and Partially Purified Samples

S. No	Cultures	Enzyme Activity in Crude Culture Extracts (IU/ml/min) (in 250 ml Volume; After Protein Equalization)	Enzyme Activity in Partially Purified Samples (IU/ml/min) (in 5 ml of the Concentrated Volume; After Protein Equalization)
1.	NRRLB 16746	12.13	21.5
2.	Colony 194	10.4	19.11
3.	Colony 51	7.3	14.19
4.	Colony 157	5.7	12.5
5.	NRRLB 1305 (control)	5.4	12.04

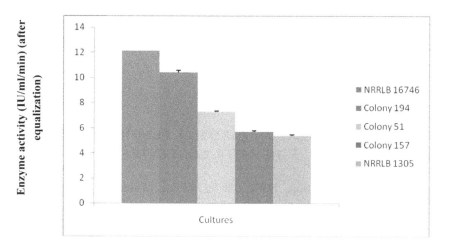

GRAPH 11-4   Comparison of the enzyme activity (in partially purified) of different cellulase producing isolates (standard deviation is also shown).

The results of the primary screening were confirmed during secondary screening. During primary and secondary screening, NRRL B-16746 (*S. albidoflavus*) showed a maximum activity in plate and broth followed by colonies 194, 51, and 157 and NRRL B-1305 (both in the case of crude extract as well as in partially purified extract of the samples) (Table 11-9).

TABLE 11-9 Comparison of Zone Size and Enzyme Activity of the Cultures

S. No.	Cultures	Clear Zone Diameter (mm)	Enzyme Activity in Crude Culture Extract (IU/ml/ min) (in 250 ml Volume)	Enzyme Activity in Partially Purified Samples (IU/ml/ min) (in 5 ml of the Concentrated Volume)
1.	NRRLB 16746	33	12.13	21.5
2.	Colony 194	31	10.4	19.11
3.	Colony 51	26	7.3	14.19
4.	Colony 157	18	5.7	12.5
5.	NRRLB 1305 (control)	17	5.4	12.04

11.3.11   SDS-PAGE PROFILING OF ISOLATES

The crude and partially purified sample of colony 194 was run on denaturing SDS–PAGE gel to view the total protein profile of the extracts (Figure
11-7). With the help of SDS-PAGE, it was confirmed that the samples were
partially purified as the number of protein bands obtained in dialyzed and
concentrated samples was less when compared with the crude extract of the
culture as shown in Figure 11-7. The band(s) corresponding to the cellulase
enzyme in the samples could be further identified by western blotting using
anticellulase antibodies and its molecular weight could be determined.

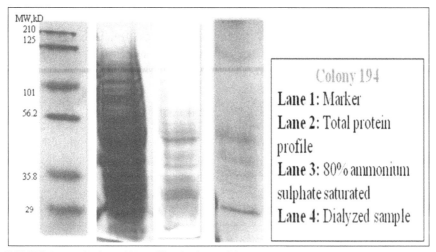

FIGURE 11-7   SDS-PAGE profile of colony 194.

11.4 DISCUSSION

Actinomycetes are well known as producers of extra cellular enzymes.[5,31,40,47,48] In the course of our study, actinomycetes from diverse ecological habitats were screened for the production of extra cellular cellulase. Actinomycetes have been isolated by researchers from various ecological environments for identifying potential cellulase producers.[17,28] During primary screening, out of the 646 bacterial colonies, a total of 150 isolates showed cellulase activity. Mohanta,[8] identified 18% of the isolates as cellulase producing organisms. Selvam et al.[12] screened 56 actinomycete cultures from marine sediments for the production of cellulase enzyme. Deepthi et al.[20] isolated actinomycetes from mangrove soils for the identification of cellulase producing strains.

The diameters of the clearance zone produced by isolate nos. 51, 157, 194, *S. albidoflavus* (NRRL B-16746), and *Streptomyces albogriseolus* (NRRL B-1305) because of the production of cellulase in the present study were 26 mm, 18 mm, 31 mm, 33 mm, and 17 mm, respectively in the case of cellulose and 6 mm, 13 mm, and 8 mm for colonies 51, 157, and NRRL B-1305, respectively in the case of CMC. Based on the results of the primary screening, isolate nos. 51, 157, 194, *S. albidoflavus* (NRRL B-16746) along with *Streptomyces albogriseolus* (NRRL B-1305, control) were selected for quantitative analyses. El sersy[30] tested actinomycetes for cellulase activity and found that the sizes of the zone of clearance in *Streptoverticillum moroo-kaense*, *Streptomyces globosus*, *Streptomyces alanosinicus*, *Streptomyces ruber*, *Streptomyces gancidicus,* and *Nocardiopsis aegyptia* were 10 mm, 10 mm, 15 mm, 25 mm, 15 mm, and 10 mm, respectively. Nayaka and Vidyasagar[54] tested 57 thermotolerant actinomycetes isolates from different habitats for the production of different enzymes, and found that isolate no. VSAC 1-57 produced the zone of clearance in the range of 19–52 mm.

The protein content and enzyme activity (IU/ml/min) in 250 ml crude extracts was determined. The enzyme assay was done by dinitrosalicyclic acid (DNS) method (Table 11-6). Golinska and Dahm[55] estimated cellulase (CMCase) activity in twenty strains of *Streptomyces* isolated from soil and found activity in the range of 0.011–0.1539 units/ml. FPase and CMCase activities was determined by Mohanta[8] in nine actinomycete isolates from Bhitarkanika National Park, Odisha. Cellulase activity using whatman filter paper as the substrate ranges from 0.266–0.734 U/ml and in case of CMCase it ranges from 0.501–1.381 U/ml.

The crude supernatant was subjected to partial purification by 80% ammonium sulphate saturation, dialysis followed by concentration. Amore et al.[56] precipitated the secreted proteins from *Streptomyces* spp using ammonium sulphate. Partially purified protein samples were used for the estimation of protein content and for the assay of enzyme activity (Tables 11-7 and 11-8, Graph 11-2). Yassein et al.[13] purified the enzyme samples produced by *Streptomyces* spp. by ammonium sulphate precipitation followed by dialysis, and then using DEAE cellulose and Sephadex columns. The enzyme obtained after purification was 38.85-fold purified. The enzyme activity recorded in the crude enzyme extract was 3.8 U/50 ml, 2.5 U/25 ml after ammonium sulphate precipitation and dialysis and 0.876 U/ml in 3 ml and 0.517 U/ml in 2 ml after purification by DEAE-cellulose column and Sephadex column, respectively.

SDS-PAGE result showed multiple bands in the case of the crude supernatant of colony 194. The number of bands was found to be decreased in the dialyzed and concentrated sample as the sample was partially purified.

11.5 CONCLUSION

The present work aims to isolate cellulolytic actinomycetes from varied ecological habitats. For this, the isolates were subjected to plate assay using both cellulose and CMC as the substrate and based on the results, cellulose was chosen as a substrate for further study and isolates 51, 157, 194 representing diverse habitats and NRRLB 16746 (*S. albidoflavus*) along with the positive control NRRLB 1305 (*S. albogriseolus*) were selected for secondary screening. The enzyme activity and protein content in the crude culture extract was determined. The cell free extract was then subjected to partial purification and enzyme activity, and the protein content was again determined in the partially purified extract after protein equalization. In the partially purified samples, both activity and protein content were found to be increased as compared with the crude extracts of the cultures. The protein profile of the cultures was observed using SDS-PAGE. The isolates were morphologically and biochemically characterized. The 16S rRNA gene sequencing and phylogenetic analyses showed that the isolates 51 (KJ995861), 157 (KJ934594), and 194 (KJ934595) belonged to the genus *Streptomyces*. Further work needs to be done for the purification of cellulase enzyme by column chromatography, determining the enzyme activity and protein content in the purified culture extracts followed by protein profiling on SDS-PAGE.

ACKNOWLEDGMENTS

The author Payal Das acknowledges UGC (University Grants Commission), Government of India, for granting financial assistance for the research work. Infra-structural facilities provided by Acharya Narendra Dev College are gratefully acknowledged.

KEYWORDS

- **Actinomycetes**
- **Ammonium sulphate precipitation**
- **Dialysis**
- **Enzyme activity**
- **Primary screening**
- **Secondary screening**

REFERENCES

1. Haggag, K.; Ragheb, A. A.; EL-Thalouth, I. A.; Nassar, S. H.; and Sayed, H. E. L. A review article on enzymes and their role in resist and discharge printing styles. *Life Sci. J.* **2013,** *10, 1646*–1654.
2. Baldrian, P.; Snajdr, J.; Merhautova, V.; Dobiasova, P.; Cajthaml, T.; and Valaskova, V. Responses of the extracellular enzyme activities in hardwood forest to soil temperature and seasonality and the potential effects of climate change. *Soil Biol. Biochem.* **2013,** *56,* 60–68.
3. Baldrian, P. Distribution of extracellular enzymes in soils, spatial heterogeneity and determining factors at various scales. *Soil Sci. Soc. Am. J.* **2014,** *78,* 11–18.
4. Shanmugapriya, S.; Saravana, P. S.; Krishnapriya; Manoharan, M.; Mythili, A.; and Joseph, S. Isolation, screening and partial purification of cellulase from cellulase producing bacteria. *Int. J. Adv. Biotechnol. Res.* **2012,** *3,* 509–514.
5. Wushke, S.; Levin, D. B.; Cicek, N.; and Sparling, R. Characterization of enriched aerotolerant cellulose-degrading communities for biofuels production using differing selection pressures and inoculum sources. *Can. J. Microbiol.* **2013,** *59,* 679–83.
6. Moreno, A. D.; Ibarra, D.; Alvira, P.; Tomas-Pejo, E.; and Ballesteros, M. A review of biological delignification and detoxification methods for lignocellulosic bioethanol production. *Crit. Rev. Biotechnol.* **2015,** *35,* 342–354. DOI: 10.3109/07388551.2013.878896.
7. Naggar, E. L. N.; Sherief, A. A.; and Hamza, S. S. Bioconversion process of rice straw by thermotolerant cellulolytic *Streptomyces viridiochromogenes* under solid-state fermentation conditions for bioethanol production. *Afr. J. Biotechnol.* **2011,** *10,* 11998–12011.

8. Mohanta, Y. K. Isolation of cellulose degrading actinomycetes and evaluation of their cellulolytic potential. *Bioengineering Bioscience* **2014**, *2*, 1–5.

9. Sowmya, B.; Gomathi, D.; Kalaiselvi, M.; Ravikumar, G.; Arulraj, C.; and Uma, C. Production and purification of chitinase by *Streptomyces sp.* from soil. *J. Adv. Sci. Res.* **2012**, 3, 25–29.

10. Kamble, R. D.; and Jadhav, A. D. Isolation, purification, and characterization of xylanase produced by a new species of *Bacillus* in solid state fermentation. *Int. J. Microbiol.* **2012**, 1–8. DOI: 10.1155/2012/683193.

11. Sonia, M. T.; Hafesh, B.; Abdennaceur, H.; and Ali, G. Studies on the ecology of actinomycetes in an agricultural soil amended with organic residues II, Assessment of enzymatic activities of Actinomycetales isolates. *World J. Microbiol. Biotechnol.* **2011**, *27*, 2251–2259.

12. Selvam, K.; Vishnupriya, B.; and Yamuna, M. Isolation and description of keratinase producing marine actinobacteria from South Indian Coastal Region. *Afr. J. Biotechnol.* **2013**, *12*, 19–26.

13. Yassien, M. A. M.; Jiman-Fatani, A. A. M.; and Asfour, H. Z. Production, purification and characterization of cellulase from *Streptomyces sp. Afr. J. Microbiol.* **2014**, *4*, 348–354.

14. Kuhad, R. C.; Gupta, R.; and Singh, A. Microbial cellulases and their industrial applications. *Enzyme Res.* **2011**, 1–10. DOI: 10.4061/2011/280696.

15. Khanna, M.; Solanki, R.; and Lal, R. Selective isolation of rare actinomycetes producing novel antimicrobial compounds. *Int. J. Adv. Biotechnol. Res.* **2011**, *2*, 357–375.

16. Solanki, R.; Khanna, M.; and Lal, R. Review article entitled "Bioactive compounds from marine actinomycetes". *India J. Microb.* **2008**, *48*, 410–431.

17. Solanki, R.; Lal, R.; and Khanna, M. Antimicrobial activities of actinomycetes from diverse ecological habitats in Delhi and its adjoining states. *India J. Microb. World* **2011**, *13*, 233–240.

18. Solanki, R.; Das, P.; and Khanna, M. Metabolic profiling of actinomycetes having antimicrobial properties. *Int. J. Adv. Biotechnol. Res.* **2013**, *4*, 444–459.

19. Das, P.; Solanki, R.; and Khanna M. Isolation and screening of cellulolytic actinomycetes from diverse habitats. *Int. J. Adv. Biotechnol. Res.* **2014**, *5*, 438–451.

20. Deepthi, M. K.; Sudhakar, M. S.; and Devamma, M. N. Isolation and screening of *Streptomyces* sp. from Coringa mangrove soils for enzyme production and antimicrobial activity. *Int. J. Pharmaceut. Chem. Biol. Sci.* **2012**, *2*, 110–116.

21. Anderson, I.; Abt, B.; Lykidis, A.; Klenk, H. P.; Kyrpides, N.; and Ivanova, N. Genomics of aerobic cellulose utilization systems in actinobacteria. *PLoS One* **2012**, *7*, 1–10.

22. Cheng, C. L.; and Chang, J. S. Hydrolysis of lignocellulosic feedstock by novel cellulases originating from *Pseudomonas* sp. CL3 for fermentative hydrogen production. *Bioresour. Technol.* **2011**, *102*, 8628–8634.

23. Rastogi, G.; Bhalla, A.; Adhikari, A.; Bischoff, K. M.; Hughes, S. R.; Christopher, L. P.; and Sani, R. K. Characterization of thermostable cellulases produced by *Bacillus* and *Geobacillus* strains. *Bioresour. Technol.* **2010**, *101*, 8798–8806.

24. El-Nakeeb, M. A.; and Lechevalier, H. A. Selective isolation of aerobic actinomycetes. *Appl. Microbiol.* **1963**, *11*, 75–77.

25. Shirling, E. B.; and Gottlieb, D. Methods for characterization of *Streptomyces species*. *Int. J. Syst. Bacteriol.* **1966**, *1*, 313–340.

26. Gause, G. F.; Preobrazhenskaya, T. P.; Sveshnikova, G. V.; Terekhova, L. P.; and Maksimova, T. S. *A Guide for Determination of Actinomycetes*; Nauka: Moscow, Russia, 1983.

27. Kuester, E.; and Williams, S. T. Selection of media for isolation of *Streptomycetes*. *Nature* **1964**, *202*, 928–929.

28. Gautam, S. P.; Bundela, P. S.; Pandey, A. K.; Jamaluddin; Awasthi, M. K.; and Sarsaiya, S. Diversity of cellulolytic microbes and the biodegradation of municipal solid waste by a Potential Strain. *Int. J. Microbiol.* **2012**, 1–8. DOI: 10.1155/2012/325907.

29. Shaikh, N. M.; Patel, A. A.; Mehta, S. A.; and Patel, N. D. Isolation and screening of cellulolytic bacteria inhabiting different environment and optimization of cellulase production. *Univ. J. Environ. Res. Technol.* **2013**, *3*, 39–49.

30. El-Sersy, N. A.; Abd-Elnaby, H.; Abou-Elela, G. M.; Ibrahim, H. A. H.; and El-Toukhy, N. M. K. Optimization, economization and characterization of cellulase produced by marine *Streptomyces ruber*. *Afr. J. Biotechnol.* **2010**, *9*, 6355–6364.

31. Khanna, M.; and Solanki, R. *Streptomyces antibioticalis*, a novel species from sanitary landfill soil. *Indian J. Microbiol.* **2012**, *52*, 605–611.

32. Lal, R.; Lal, S.; Grund, E.; and Eichenlaub, R. Construction of a hybrid plasmid capable of replication in *Amycolatopsis mediterranei*. *Appl. Environ. Microbiol.* **1991**, *57*, 665–671.

33. Chung, J.; Lee, J. H.; Jhung, Y.; Kim, M.; Kim, B. K.; and Lim. Y. W. Ez/Taxon, a web based tool for the identification of prokaryotes based on 16S ribosomal RNA gene sequences. *Int. J. Syst. Evol. Microbiol.* **2007**, *57*, 2259–2261.

34. Charbonneau, D. M.; Mouelhi, F. M.; Boissinot, M.; Sirois, M.; and Beauregard, M. Identification of thermophilic bacterial strains producing thermotolerant hydrolytic enzymes from manure compost. *Indian J. Microbiol.* **2012**, *52*, 41–47.

35. Prasad, P.; Tanuja; and Bedi, S. Characterization of a novel thermophilic cellulase producing strain *Streptomyces matensis* strain St-5. *Int. J. Curr. Microbiol. App. Sci.* **2014**, *3*, 74–88.

36. Tamura, K.; Dudley, J.; Nei, M.; and Kumar, S. MEGA 4, molecular evolutionary genetic analysis (MEGA) software version 4.0. *Mol. Biol. Evol.* **2007**, *24*, 1596–1599.

37. Saitou, N.; and Nel, M. The neighbour-joining method, a new method for reconstructing phylogenetic trees. *Mol. Biol. Evol.* **1987**, *4*, 406–425.

38. Kaur, J.; Verma, M.; and Lal, R. *Rhizobium rosettiformans sp. nov.,* isolated from hexachlorocyclohexane (HCH) dump site in India, and reclassification of *[Blastobacter] aggregatus* Hirsch et al. [1985] as *Rhizobium aggregatum* comb. *nov. Int. J. Syst. Evol. Microbiol.* **2011**, *61*, 1218–1225.

39. Malhotra, J.; Anand, S.; Jindal, S.; Raman, R.; and Lal, R. *Acinetobacter indicus* sp. nov. isolated from hexachlorocyclohexane dumpsite. *Int. J. Syst. Evol. Microbiol.* **2012**, *62*, 2883–2890. DOI: 10.1099/ijs.0.037721-0.

40. Tomita, K.; Nishio, M.; Saitoh, K.; Yamamoto, H.; Hoshino, Y.; Ohkuma, H.; Konishi, M.; Miyaki, T.; and Oki, T. Pradimicins A, B and C, new antifungal antibiotics. I. Taxonomy, production, isolation and physico-chemical properties. *J. Antibiot.* **1990**, *43*, 755–762.

41. Lowry, O. H.; Rosenbrough, N. J.; Farr, A. L.; and Randall, R. J. Protein measurement with the Folin phenol reagent. *J. Biol. Chem.* **1951**, *193*, 265–275.

42. Ghosh, T. K. Measurement of cellulose activities. *Pure Appl. Chem.* **1987**, *59*, 257–68.

43. Sadhu, S.; and Maiti, T. K. Cellulase production by bacteria, A review. *Brit. Microbiol. Res. J.* **2013**, *3*, 235–258.

44. Miller, G. L. Use of dinitrosalicylic acid reagent for determination of reducing sugar. *Anal. Chem.* **1959**, *31*, 426–428.

45. Ponnambalam, A. S.; Deepthi, R. S.; and Ghosh, A. R. Qualitative display and measurement of enzyme activity of isolated cellulolytic bacteria. *Biotechnol. Bioinf. Bioeng.* **2011,** *1*, 33–37.

46. Rathnan, R. K.; and Ambili, M. Cellulase enzyme production by *Streptomyces sp.* Using fruit waste as substrate. *Aust. J. Basic Appl. Sci.* **2011,** *5*, 1114–1118.

47. Azzeddine, B.; Abdelaziz, M.; Estelle, C.; Mouloud, K.; Nawel, B.; Nabila, B.; Francis, D.; and Said, B. Optimization and partial characterization of endoglucanase produced by *Streptomyces sp.* B-PNG23. *Arch. Biol. Sci.* **2013,** *65*, 549–558.

48. Laemmli, U. K. Cleavage of structural proteins during the assembly of the head of bacteriophage T4. *Nature* **1970,** *277*, 680–685.

49. Murugan, M.; Srinivasan, M.; Sivakumar, K.; Sahu, M. K.; and Kannan, L. Characterization of an actinomycete isolated from the estuarine finfish, *Mugil cephalus* Lin. (1758) and its optimization for cellulose production. *J. Sci. Ind. Res.* **2007,** *66*, 388–393.

50. Fukunaga, K.; Misato, T.; Ishii, I.; and Asakawa, M. Blasticidin, a new antiphytopathogenic fungal substance I. *Bull. Agric. Chem. Soc. JPN* **1955,** *19*, 181–188.

51. Kavitha, S.; and Vijayalakshmi, M. Studies on cultural, physiological and antimicrobial activities of *Streptomyces rochei. J. Appl. Sci. Res.* **2007,** *3*, 2026–2029.

52. Pridham, T. G.; Hesseltine, C. W.; and Benedict, R. G. A guide for the classification of *Streptomycetes* according to selected groups; placement of strains in morphological sections. *Appl. Microbiol.* **1958,** *6*, 52–79.

53. Rong, X.; Guo, Y.; and Huang, Y. Proposal to reclassify the *Streptomyces albidoflavus* clade on the basis of Multilocus sequence analysis and DNA-DNA hybridization, and taxonomic elucidation of *Streptomyces griseus* subsp. *solvifaciens. Syst. Appl. Microbiol.* **2009,** *32*, 314–322.

54. Nayaka, S.; and Vidyasagar, G. M. Occurrence and extra cellular enzyme potential of actinomycetes of a thermotolerant, northern region of Karanataka, India. *Int. Multidiscip. Res. J.* **2012,** *2*, 40–44.

55. Golinska, P.; and Dahm, H. Enzymatic activity of actinomycetes from the genus *Streptomyces* isolated from the bulk soil and rhizosphere of the *Pinus sylvestris. Dendrobiology* **2011,** *65*, 37–46.

56. Amore, A.; Pepe, O.; Ventorino, V.; Birolo, L.; Giangrande, C.; and Faraco, V. Cloning and recombinant expression of a cellulase from the cellulolytic strain *Streptomyces* sp. G12 isolated from compost. *Microb. Cell Fact.* **2012,** *11*, 1 12

.

CHAPTER 12

KINETIC STUDIES OF ALKALI PRETREATED SWEET SORGHUM BAGASSE USING CELLULASE

L. SAIDA[1*] and K. VENKATA SRI KRISHNA[1]

[1]*Center for Biotechnology, Institute of Science and Technology, Jawaharlal Nehru Technological University Hyderabad, Hyderabad, Telangana – 500085, India.*

*Corresponding author: L. Saida. E-mail: lavudisaida@jntuh.ac.in

CONTENTS

Abstract 250
12.1 Introduction 250
12.2 Materials and Methods 252
12.3 Results and Discussion 255
12.4 Conclusion 261
Keywords 261
References 261

ABSTRACT

Kinetics study of cellulase produced over pretreated sweet sorghum bagasse as a limiting substrate was taken up for understanding its dynamic nature on crystalline material hydrolysis. The influence of substrate concentration, pH, and temperature on the production of reducing sugars from crystalline material was studied. The kinetic parameters (maximal velocity; V_{max}, and half-saturation constant; Km) were determined from the initial velocities by Michaelis–Menten equation. Michaelis-Menten constant, Km and maximum reaction rate, V_{max} for the cellulase on sweet sorghum bagasse were found to be 5.005 and 1.168 g/L min, respectively. The optimum pH and temperature were 5 and 50 °C, respectively.

12.1   INTRODUCTION

The growing demands for CO_2-neutral transportation fuels and the desire to achieve a reduced dependence on fossil resources have been the major driving forces for the substantial increase in the amounts of bioethanol produced by fermentation of lignocelluloses. Cellulosic materials obtained from agro residues, municipal solid wastes, and energy crops represent an abundant source of biomass.[1] Search for cheap substrate and economical process setups is still in vogue to make bioethanol production a viable answer to meet the growing demand for renewable fuels. Many reports suggested that pre-treatment of lignocelluloses make fermentability condition favorable. To make cellulosic biomass favorable for fermentation, enzymatic hydrolysis plays a key role. The search for viable and cheap substrate can be answered by sweet sorghum, which is drought-tolerant, grows very tall, and the stalks contain a high volume of fermentable sugars. First generation biofuel is being produced by the fermentation of juice extracted from sweet sorghum by setting up a pilot scale plant in various countries like Brazil, and USA.[2] After the extraction of syrup from sweet sorghum, the spent is obtained which can be used for second generation biofuel production by adapting additional processing steps.

Spent obtained from the stalk of sweet sorghum is referred to as bagasse which is classified as lignocellulosics. Sweet sorghum bagasse is pretreated, to decrease the crystallinity, increase the porosity, and enhance the receptiveness for enzymatic treatment for obtaining simple sugars. This step followed by fermentation, accelerates ethanol production and its recovery processes. Crystallinity indexes of sweet sorghum bagasse for untreated and pretreated material were 0.83 and 0.73, respectively.[3] Solid residue obtained after the pretreatment and sugar extraction of sweet sorghum stalks was rich in cellulose (40.4%, w/w) and hemicellulose (35.5%, w/w) fractions.[4] The group of non cellulosic carbohydrates detected in the enzymatic hydrolysate of sweet sorghum bagasse (SSB) included xylose, arabinose, and galactose, because arabinoxylan comprises the major hemicellulosic component of this agricultural by-product, and is accompanied by various types of galactans.[5] The use of sweet sorghum bagasse for alcohol production is based on the efficient pretreatment method employed to carry out saccharification and fermentation efficiently by using yeast.[6]

Enzymatic hydrolysis of cellulase depends on many factors such as physical properties of substrate (composition, crystallinity, degree of polymerization, etc.), enzyme synergy (origin, composition, etc.), mass transfer (substrate adsorption, bulk and pore diffusion, etc.), and intrinsic kinetics. Most of these said effects show up concurrently, and therefore, cannot be distinguished from each other. An accurate assessment of intrinsic kinetics requires a pure form of cellulosic substrates unhindered by mass transfer resistances or physical factors of the substrate. The enzyme-substrate interaction varies from one enzyme-substrate complex to another. The formation of enzyme substrate complex is usually by weak Van der Waals forces and hydrogen bonding. The substrate binds to a specific site of the enzyme called as the active site. As the size of the substrate is very small when compared with the size of the enzyme, it easily fits into any region of the enzyme to form the complex.[7] Park reported that cellulase enzyme loses its apparent activity during hydrolysis because of the light absorption of the enzyme on the cellulose surface.[8]

Cellulases include three main types of enzymes, namely, endoglucanases, cellobiohydrolases or exoglucanases, and β-glucosidases. These enzymes can either be free (mostly in aerobic microbes) or grouped in a multicomponent enzyme complex (cellulosome) found in anaerobic cellulosome bacteria.[9] Cellulases from different sources have also been reported to show similar modes of action. The enzymatic hydrolysis of the glycosidic bonds takes place through general acid catalysis involving two carboxylic acids. Endoglucanases form intermediate compounds with the cellulose chains and hydrolyse them at random, giving rise to less polymerized chains and soluble reducing sugars.[10] The indiscriminate action of endoglucanases progressively increases the accessibility of cellulose chain ends, in this manner, increasing the specific surface area of the substrate for the exocellulase activity. The endoglucanase attacks the β-1,4 glycosidic bonds within the amorphous regions of the cellulose chains.[11] The products of this attack are oligosaccharides of various lengths and subsequently, new chain reducing ends. Exoglucanases degrade the crystalline cellulose most efficiently acting in a progressive mode and binding to the reducing or non reducing ends of the cellulose polysaccharide chains, releasing either glucose (glucohydrolases) or cellobiose (cellobiohydrolases) as the major products.[12]

The kinetics of the enzymatic hydrolysis of cellulose was studied by a number of scientists.[13,14] The kinetics of hydrolysis depends of the source of the enzyme, nature of the substrate, inhibitory effects of the intermediates or the end products, water content, pH, additional substances such as lignin, temperature, etc.[15] Studies carried on the product formation kinetics of SSB revealed that the effect of pretreatment severity, particle size, and substrate loading play an important role in the hydrolysis step. The aim of the present study was to study the effects of pH, temperature, and substrate concentration during the hydrolysis of SSB on cellulase and to determine experimentally the undeterminable constants such as energy of activation, Arrhenius constant, K1 and K2, V_{max}, Km, and identify optimal conditions.

12.2   MATERIALS AND METHODS

12.2.1   MATERIALS

The sweet sorghum bagasse (SSB) used in our study was procured from the Punjab Agricultural University, Ludhiana, Punjab, India. The SSB used in this study was composed of 40.6% cellulose, 29.04% hemicelluloses, 20.1% lignin, and 5.6% ash. Celluclast (C-2730), Novozyme 188 (C-6105), and Pectinase (P-2611) were procured from Sigma-Aldrich (St. Louis, MO, USA). Instruments like laminar air flow, orbital shaker incubator, ultra high pressure incubator, High Performance Liquid Chromatography (HPLC), UV-Vis spectrophotometer, and water bath were used for the experimentation.

12.2.2   EFFECT OF TEMPERATURE AND THE ESTIMATION OF KINETIC PARAMETERS

Alkali pre-treated SSB with 15% substrate concentration (w/v) was suspended in 0.1 M citrate buffer (pH 5.0) in the capped polycarbonate flasks. The flasks were autoclave-sterilized for 15 min, cooled, and supplemented with cellulase by a loading of 20 IU/mL, filtered through 0.45 μm PVDF membranes

(Millipore India Pvt. Ltd., Bangalore, India). Hydrolysis was performed at variable temperature conditions i.e., 40 °C, 50 °C, 60 °C, 70 °C, and 80 °C at 120 rpm in an incubator shaker. Samples were withdrawn after every 10 min and centrifuged for obtaining a supernatant for sugar profile analysis by HPLC (Dionex Corporation, CA, USA). A Shodex SP-0810 column (300 mm × 7.8 mm) fitted with a SP-G guard column (Waters Inc., USA) was used to analyze glucose in the enzymatically hydrolyzed samples. Degassed deionised water was used as a mobile phase at a flow rate of 1.0 mL/min. The column oven and refractive index detector were maintained at 80 °C and 55 °C, respectively. The samples were diluted, centrifuged, and filtered through Phenomenex 0.45 μm Regenerated Cellulose (RC) membranes. The peaks were detected by refractive index detector and quantified on the basis of the area and retention time of the sugar standards. Further, the rate of the reaction was deduced from product formation versus time course plot taking the slope. Cellulase temperature dependence kinetic parameters were tested with respect to the model proposed by Arrhenius.[16] The Arrhenius constant (A) and Ea were calculated by plotting the graph between temperature versus denaturation constant of the reaction using the curve fitting tool (Matlab® 7.1).

$$k = Ae^{-Ea/RT}$$

Here, A refers to Arrhenius constant, Ea refers to the activation energy, R represents the universal gas constant, k represents rate of the reaction and T is the temperature.

12.2.3 EFFECT OF PH AND THE ESTIMATION OF KINETIC PARAMETERS

In our present study to find the effect of pH on kinetic parameter, the alkali pre-treated SSB concentration was maintained at 15% (w/v) with sodium citrate buffer at pH values of 3, 4, 5, 6, and 7 by the addition of acid/base, in capped polycarbonate flasks. The flasks were autoclave-sterilized for 15 min, cooled, and supplemented with a cellulase loading of 20 IU/mL, filtered through 0.45 μm PVDF membranes (Millipore India Pvt. Ltd., Bangalore, India). Hydrolysis was performed at 50 °C at 120 rpm in an incubator shaker. The samples were withdrawn after every 10min and centrifuged to obtain the supernatants for sugar profiling by HPLC. Cellulase pH dependent kinetic parameters were tested with respect to the pH model developed by Henderson–Hasselbalch equation.[17] The equilibrium constants K_1 and K_2 were calculated by plotting the graph between pH and velocity of the reaction using the curve fitting tool (Matlab® 7.1).

$$V_{max} = \frac{ke_0}{\left[1+\left(\dfrac{h^+}{K_1}\right)+\left(\dfrac{K_2}{h^+}\right)\right]}$$

Here, h^+ represents pH, e_0 = total enzyme concentration, and k is the rate of the reaction.

12.2.4   EFFECT OF SUBSTRATE CONCENTRATION AND KINETIC PARAMETERS ESTIMATION

Alkali pre-treated SSB was suspended in 0.1 M sodium citrate buffer (pH 5.0) in capped polycarbonate flasks. The addition of buffer was done in such a way so as to maintain the substrate concentration at 10%, 12%, 14%, 16%, and 18% (w/v). The flasks were autoclaved for 15 min, cooled, and supplemented with the cellulase having 20 IU/mL activity by filtering through 0.45 μm PVDF membranes (Millipore India Pvt. Ltd., Bangalore, India). Hydrolysis was carried out at 50 °C for 96 h at 120 rpm in an incubator shaker. Samples were withdrawn after every 12 h and centrifuged to obtain the supernatants for sugar profiling with HPLC. Cellulase over substrate utilization kinetic parameters were tested with respect to Michaelis Menten model.[18] The V_{max} and Km values were calculated by plotting the graph between the substrate versus velocity of the reaction using curve fitting tools (Matlab® 7.1).

$$v = \frac{V_{max}[s]}{Km+[s]}$$

Here, Km refers to Michaelis Menten constant, S for substrate (units), V_{max} implies maximum rate achieved at maximum substrate concentration, v for reaction rate.

12.3 RESULTS AND DISCUSSION

12.3.1 EFFECT OF TEMPERATURE ON ENZYME KINETICS

In the present study, the effect of temperature on the rate of the reaction was investigated at constant substrate concentration and pH. To study the kinetics at variable temperature, sugar profiling was done after every 10 min and the data obtained is shown in (Table 12-1). We had observed that at different temperatures, the rate of the reaction got altered. Arrhenius proposed that the rate of the reaction depends upon the frequency of collisions among the reactant molecules.

TABLE 12-1 Sugar Profile Report at Variable Temperatures and Calculated Rate of Reaction (k) of Each Group

		Temperature (°C)				
		40	**50**	**60**	**70**	**80**
Time (Min)	10	69.37	70.03	100.4	142.5	36.5
	20	83.08	84.07	111.9	126.19	31.71
	30	92.28	96.62	120.3	118.13	30.22
	40	103.89	106.78	128.25	114.3	24.87
	50	119.6	113.64	138.91	110.09	28.24
	60	125.2	137.89	160.38	106.85	26.33
	70	138.08	129.83	146.67	101.97	22.18
	80	140.2	128.5	145.52	101.5	21.5
	k	**0.00784**	**0.007468**	**0.006911**	**0.006071**	**0.00551**

After acquiring the necessary activation energy, the enzyme acts as a catalyst in the conversion of the substrate to the product.[19] With an increase or decrease in temperature, the degree of freedom gets altered, which results in a change in the kinetic behavior.[20] In our study, we found 50 °C to be the optimal temperature for maximal activity and our results are in agreement with earlier work reported.[21] In our study, the optimal temperature indicates that the rate of the reaction was maximum i.e., the enzyme cellulase attained optimal Km and V_{max} values, and was stable i.e., the active site was in a good configuration for the uptake of the substrate and liberated out the product quickly. In general, most of the cellulase requires an optimum temperature of 50 °C for reaching a peak of the transition state, whereas in our study it was 50 °C. At 50 °C, the enzyme attained maximum reactivity which implied that it participated in a reaction with higher collisions. Our studies are in agreement with the earlier reports[22,23] where 50 °C was the optimum temperature. After developing the plot, we found the experimentally undeterminable constants such as the Arrhenius constant to be 2.515 and that activation energy of 4.83 kcal/mol and 20.2 kJ.mol was required for the enzyme to be a part of the reaction. Typical standard free energies of activation (15–70 kJ M^{-1}) gave rise to an increase in the rate by factors between 1.2 and 2.5 for every 10 °C rise in the temperature. In our study, we found that 20.2 kJ/mol was required for the activation. Experiments were conducted beyond 70 °C; however, it was reported by earlier workers[24] that denaturation of cellulase occurs at 75 °C. Our results were found to closely resemble the results declared by earlier workers,[25] who reported that 5.1 kcal/mol of activation energy was required for cellulase to hydrolyse amorphous cellulose. On the whole, we could conclude that cellulase followed the Arrhenius equation as the experimental values obtained were in correlation with the theoretical model (Figure 12-1), which were validated based on

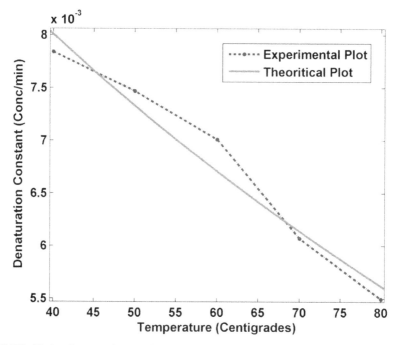

FIGURE 12-1   Comparative graph between experimental data obtained and theoretical model developed using Arrhenius equation.

the coefficient of determination (R^2) and sum of squares of the error (SSE). As the value of regression coefficient (R^2) was found to be more than 0.90 and SSE value was less than 0.001, it was confirmed that the plot obtained was in agreement with the Arrhenius model.[26]

12.3.2 EFFECT OF pH ON ENZYME KINETICS

In the present study, the effect of pH on the kinetic behavior of cellulase towards the hydrolysis of SSB was studied at constant substrate concentrations and a temperature of 50 °C. To study kinetics at a variable pH, sugar profiling was done after every 10 min and the data obtained is shown in (Table 12-2). It was reported that slightest change in pH would affect the affinity of the enzyme toward the substrate.[27]

TABLE 12-2 Sugar Profile Report at Variable pH and Calculated Rate of Reaction (k) of Each Group

		pH				
		3	**4**	**5**	**6**	**7**
Time (Min)	0	0	0	0	0	0
	10	33.76	76.61	84.21	69.49	29.45
	20	42.51	85.15	92.75	81.54	37.15
	30	51.12	99.76	103.14	92.56	46.68
	40	63.4	108.15	116.19	102.7	57.92
	50	77.79	119.45	128.24	113.75	64.16
	60	90.12	138.47	145.15	129.45	76.61
	70	78.15	122.15	131.29	110.19	65.55
	k	**1.4314**	**2.3328**	**2.4819**	**2.1758**	**1.2251**

We observed that the rate of the reaction got altered at different pH conditions, as catalytically active structural part of the enzyme might change its conformation upon the variation of pH as it directly affects the ionic strength of the solution.[28] From our studies, we found that a pH of 5.0 was best suited for the hydrolysis of SSB as the rate of the reaction was maximum. By using the curve fitting tool, Michaelis pH function constants K1 and K2 were calculated as 0.0864 and 238.8, respectively. In the earlier studies, cellulase from fungi grown on media composed of wheat straw, wheat straw holocellulose, cellulose, and xylan was found to have an optimum pH value of 5.5.[29] However, cellulase isolated from different microorganisms also demonstrated different pH optima.[30] From the graph, we observed that the alkaline arm and the acidic arm were quite steep, as the change in pH alters the net charge on the protein and thereby affects the enzyme activity (Figure 12-2). From the plot, it could also be inferred that the enzyme kinetics did not get affected as the experimental data showed co-linearity with the theoretical model (Figure 12-2). On the whole, we could conclude that the crude enzyme had followed Michaelis pH function, which we had validated based on the coefficient of determination (R^2) and SSE. In the present investigation, the R^2 value was found to be 0.97 which is more than 0.90 and the SSE value of 0.0003047 which is less than 0.001, enabling us to confirm that our results were in agreement with the Michaelis pH function.[26]

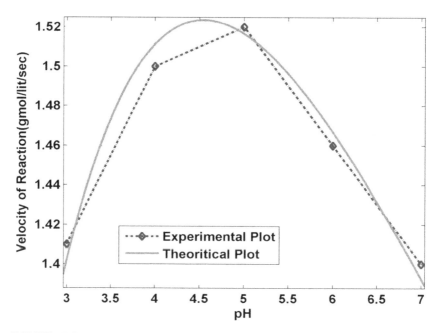

FIGURE 12-2   Comparative graph between the experimental data obtained at optimized condition and theoretical values.

12.3.3 EFFECT OF SUBSTRATE CONCENTRATION ON THE ENZYME KINETICS

In the present study, the effect of the substrate concentration on the catalytic activity of cellulase was studied at a constant temperature of 55 °C and a pH of 5.0. Here, the substrate concentration varied from 10% w/v to 20% w/v with a 2% increment to know the optimal turnover number. To study kinetics at a variable substrate concentration, sugar profiling was done after every 12 h and the data obtained is shown in (Table 12-3). We observed a significant increase in the hydrolysis rate with an increase in the substrate concentration from 10–16%. Thereafter, the hydrolysis rate decreased. Earlier workers reported that increasing the substrate concentration from 4–16% improved the hydrolysis rate.[30] An increase in rate of the reaction implied that the concentration of the substrate availability at the enzyme active site got improved and other dependent parameters such as pH, temperature, nature of the substrate, structure of the substrate had favored the enzyme to adsorb appropriately and bound to the substrate in the stable form to carry out the hydrolysis reaction.[31]

TABLE 12-3   Sugar Profile Report at Variable Substrate Concentration and Calculated Rate of the Reaction (k) of Each Group

		Substrate Concentration (%)					
		10	12	14	16	18	20
Time (h)	12	23.6	28.49	29.33	30.21	30.28	30.48
	24	28.63	32.6	33.05	36.02	36.39	36.79
	36	31.12	36.17	37.32	42.63	47.15	47.87
	48	36.46	40.42	43.86	48.86	53.01	53.95
	60	37.09	36.37	38.78	42.41	44.21	45.35
	k	0.7614	0.8241	0.8723	0.8930	0.9118	0.9183

In our study, Km value was determined for SSB at half of the V_{max} using Michaelis–Menten equation. The maximum rate of the reaction of 1.168 g/L/min and half saturation constant Km of 5.005 were determined for crude enzyme in the current study using Matlab® 7.1 (Figure 12-3). From the plot, it is observed that the enzyme followed Michaelis Menten kinetics without any significant deviation when compared with theoretical model developed using Matlab® 7.1.[18] The gradual rise in the rate of the reaction shows that concentration of the substrate made a significant impact to show first order rate of the reaction. It is reported that cellulase has a very high affinity toward microcrystalline cellulose such as Whatman paper for hydrolysis. In our study, SSB hydrolysis by cellulase reached an optimal affinity at 16% though it had a higher crystallinity index when compared with microcrystalline substrates. Thus, we infer that the enzyme kinetics did not get affected as experimental data had shown co-linearity with the theoretical model. On the whole, we concluded that the crude enzyme followed Michaelis pH function, which was validated based on the coefficient of determination (R^2) and SSE. In the present investigation, the R^2 value was found to be 0.9638 which was more than 0.90 and the SSE value was 0.00065 which was less than 0.001, which enabled us to confirm that our results were in agreement with the Michaelis Menten equation[26] and similar to that of the results obtained earlier.[32]

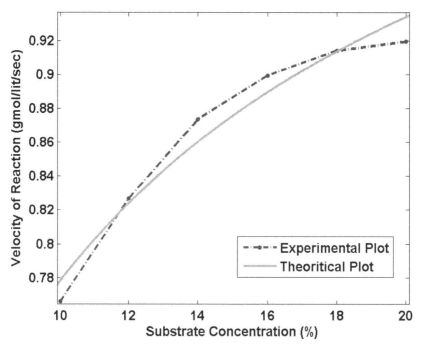

FIGURE 12-3 Comparative graph plotted between experimental data of the substrate conversion by crude enzyme obtained at optimized condition and theoretical values based on Michaelis Menten equation.

12.4 CONCLUSION

In the present investigation, kinetics studies conducted over the hydrolysis of crystalline substance revealed that an enzyme gets saturated with higher concentration of the substrate up to a certain range. It was seen that the hydrolysis rate increased with an increase in the temperature up to certain range, thereafter, the hydrolysis rate decreased, which implies quick denaturation. On the whole, cellulase performed well within the range suggested by earlier reporters. From our observation, slight first order effect was seen with the range of 40–60 °C. Yet, the overall performance was found to be satisfactory which implied that the enzyme had altered in its dynamic nature within the above mentioned range. From our observation, we found that the enzyme had increased in its tendency toward the adsorption of the substrate to carry out hydrolysis between the range of pH 4 to 5; thereafter, a decrease in the hydrolysis rate was observed which conveyed that cellulase showed a negative affinity toward the substrate beyond the range of pH 4 to 5. We conclude that cellulase has an affinity toward alkali pretreated SSB and can perform hydrolysis at a lower rate compared with pure amorphous substrates.

KEYWORDS

- **Hydrolysis**
- **Kinetics**
- **Sweet Sorghum Bagasse**

REFERENCES

1. Lin, Y.; and Tanaka, S. Ethanol fermentation from biomass resources: Current state and prospects. *App. Microbio. Biotechnol.* **2006,** *69*, 627–642.
2. Almodares, A.; and Hadi, M. R. Production of bioethanol from sweet sorghum: A review. *Afr. J. Agr. Res.* **2009,** *4*, 772–780.
3. Goshadrou, A.; Karimi, K.; and Taherzadeh, M. J. Bioethanol production from sweet sorghum bagasse by Mucor hiemalis. *Ind. Crop Prod.* **2011,** *34*, 1219–1225.
4. Dogaris, I.; Vakontios, G.; Kalogeris, E.; Mamma, D.; and Kekos, D. Induction of cellulases and hemicellulases from *Neurospora crassa* under solid-state cultivation for bioconversion of sorghum bagasse into ethanol. *Ind. Crop Prod.* **2009,** *29*, 404–411.

5. Goto, Y.; Nonaka, I.; and Horai, S. A new mtDNA mutation associated with mitochondrial myopathy, encephalopathy, lactic acidosis and stroke-like episodes (MELAS). *Biochim. Biophys. Acta.* **1991,** *1097*, 238–240.

6. Saha, B. C.; Iten, L. B.; Cotta, M. A.; and Wu, Y. V. Dilute acid pretreatment, enzymatic saccharification and fermentation of wheat straw to ethanol. *Process Biochem.* **2005,** *40*, 3693–3700.

7. Shuler, M. L.; and Kargi, F. *Bioprocess Engineering: Basic Concepts*; Prentice Hall: Upper Saddle River, NJ, 2002.

8. Park, J. W.; Takahata, Y.; Kajiuchi, T.; and Akehata, T. Effects of nonionic surfactant on enzymatic-hydrolysis of used newspaper. *Biotechnol. Bioeng.* **1992,** *39*, 117–120.

9. Bayer, E. A.; Lamed, R.; and Himmel, M. E. The potential of cellulases and cellulosomes for cellulosic waste management. *Current Opinion in Biotechnol.* **2007,** *18*, 237–245.

10. Bravo, V.; Paez, M. P.; Aoulad, M.; Reyes, A.; and Garcia, A. I. The influence of pH upon the kinetic parameters of the enzymatic hydrolysis of cellobiose with Novozym 188. *Biotechnol. Progr.* **2001,** *17*, 104–109.

11. Mosier, N. S.; Hall, P.; Ladisch, C. M.; and Ladisch, M. R. Reaction kinetics, molecular action, and mechanisms of cellulolytic proteins. In *Recent Progress in Bioconversion of Lignocellulosics;* 1999, 23–40.

12. Lynd, L. R. Overview and evaluation of fuel ethanol from cellulosic biomass: Technology, economics, the environment, and policy. *Annu. Rev. Energ. Env.* **1996,** *21*, 403–465.

13. Gonzàlez, G.; Caminal, G.; De Mas, C.; and Lopez-Santín, J. A kinetic model for pretreated wheat straw saccharification by cellulase. *J. Chem. Technol. and Biotechnol.* **1989**, 44, 275–288.

14. Wang, G. S.; Post, W. M.; Mayes, M. A.; Frerichs, J. T.; and Sindhu, J. Parameter estimation for models of ligninolytic and cellulolytic enzyme kinetics. *Soil Bio. Biochem.* **2012,** *48*, 28–38.

15. Helmreich, E.; and Cori, C. F. The effects of Ph and temperature on the kinetics of the phosphorylase reaction. *Proc. Nat. Aca. Sci. Uni. St. Ame.* **1964,** *52*, 647–654.

16. Laidler, K. J. The development of the Arrhenius equation. *J. Chem. Edu.* **1984,** *61*, 494.

17. Bailey, J. E.; and Ollis, D. F. Biochemical engineering fundamentals. *Biochemi. Edu.* 1986, 129–132.

18. Michaelis, L. and Menten, M. L. Die kinetik der invertinwirkung. *Biochem. Z.* **1913,** *49*, 352.

19. Child, M. S. *Molecular Collision Theory*; Courier Dover Publications: New York, 1974.

20. Wilkinson, G. N. Statistical estimations in enzyme kinetics. *Biochem J.* **1961,** *80*, 324–332.

21. Fadel, M. Production physiology of cellulases and ß-glucosidase enzymes of *Aspergillus niger* grown under solid state fermentation conditions. *J. Bio. Sci.* **2000,** *1*, 401–411.

22. Saxena, A.; Garg, S. K.; and Verma, J. Simultaneous saccharification and fermentation of waste newspaper to ethanol. *Bioresource Technol.* **1992,** *42*, 13–15.

23. Iqbal, H. M. N.; Ishtiaq, A.; Zia. M. A.; and Irfan, M. Purification and characterization of the kinetic parameters of cellulase produced from wheat straw by *Trichoderma viride* under SSF and its detergent compatibility. *Adv. Biosci. Biotechnol.* **2011**, 2, 149–156.

24. Baker, J. O.; Tatsumoto, K.; Grohmann, K.; Woodward, J.; Wichert, J. M.; Shoemaker, S. P.; and Himmel, M. E. Thermal denaturation of *Trichoderma reesei* cellulases studied by differential scanning calorimetry and tryptophan fluorescence. *App. Biochemi. Biotechnol.* **1992,** *34*, 217–231.

25. Li, L.; Flora, R.; and King, K. Individual roles of cellulase components derived from *Trichoderma viride*. *Arch. Biochem. Biophys.* **1965,** *111*, 439–447.
26. Engle, R. F.; and Granger, C. W. J. Co-integration and error correction: representation, estimation, and testing. *Econometric Soci.* **1987,** 251–276.
27. Trivedi, B.; and Danforth, W. H. Effect of pH on the kinetics of frog muscle phosphofruc-tokinase. *J. Biol. Chem.* **1966,** *241*, 4110–2.
28. Segel, I. *Enzyme Kinetics;* John Wiley and Sons: New York, 1975, Vol. 258, pp 1826–1832.
29. Lowe, S. E.; Theodorou, M. K.; and Trinci, A. P. Cellulases and xylanase of an anaerobic rumen fungus grown on wheat straw, wheat straw holocellulose, cellulose, and xylan. *Applied Envi. Microbiol.* **1987,** *53*, 1216–23.
30. Deng, S. P. and Tabatabai, M. A. Cellulase activity of soils. *Soil Bio. Biochem.* **1994,** *26*, 1347–1354.
31. Zheng, Y.; Pan, Z.; Zhang, R.; and Jenkins, B. M. Kinetic modeling for enzymatic hydro-lysis of pretreated creeping wild rye grass. *Biotechnol. Bioengg.* **2009,** *102*, 1558–69.
32. Carrillo, F.; Lis, M. J.; Colom, X.; Lopez-Mesas, M.; and Valldeperas, J. Effect of alkali pretreatment on cellulase hydrolysis of wheat straw: Kinetic study. *Process Biochem.* **2005,** *40*, 3360–3364.

CHAPTER 13

SECRETION OF LIGNINOLYTIC ENZYMES BY THE WHITE ROT FUNGUS *STEREUM OSTREA* IMMOBILIZED ON POLYURETHANE CUBES UNDER THE INFLUENCE OF CHLORPYRIFOS

B. S. SHANTHI KUMARI[1], KANDERI DILEEP KUMAR[1], K. Y. USHA[1], A. RAMYA[1], and B. RAJASEKHAR REDDY[1*]

[1]*Department of Microbiology, Sri Krishnadevaraya University, Anantapur, Andhra Pradesh – 515591, India.*

Corresponding author: B. Rajasekhar Reddy.
E-mail: rajasekharb64@gmail.com

CONTENTS

Abstract 266
13.1   Introduction 266
13.2   Materials and Methods 267
13.3   Results and Discussion 269
13.4   Conclusion 273
Acknowledgments 273
Keywords 274
References 274

ABSTRACT

Ligninolytic potential of white rot fungus *Stereum ostrea* on solid support polyurethane cubes (6 and 8 cubes) in liquid Koroljova medium under the influence of chlorpyrifos at 20 ppm concentration for 10 days was assessed in the present study. The growth and secretion of extracellular proteins by *S. ostrea* only on 8 sponge cubes were comparable with that of *Stereum ostrea* in the free state. Secretion of laccase by immobilized (8 cubes) and the free culture of *S. ostrea* occurred to the same extant of 42 U/ml on the 8th day of incubation. However, the yields of ligninolytic enzymes - laccase (LAC), manganese peroxidase (MnP), and lignin peroxidase (LiP) by immobilized culture with 8 cubes were higher than the titre of the respective enzymes by *Stereum ostrea* in the free state on the 10th day of incubation. The yields of laccase, MnP, and LiP by 8 cubes of immobilized culture were 62, 18.5, and 0.47 U/ml as against 15.0, 4.75, and 0.195 U/ml by *S. ostrea* in the free state. Reusability of immobilized culture for different applications has been discussed in the present study.

13.1   INTRODUCTION

Lignocellulosic biomass derived from plants is abundantly available and its supply is renewable. Annually about 1×10^{10} metric tons of lignocellulosic biomass is produced worldwide.[1,2] Ligninocellulosic biomass is made up of cellulose, hemicellulose, and a cementing material called lignin. Recycling of lignocellulosic biomass in carbon cycling is a major event that takes place in nature.

After cellulose, lignin is the second abundant biopolymer of lignocelluloses accounting for 15–25% of the total wood dry weight,[3] where it protects cellulose against the hydrolytic attack by saprophytic and pathogenic microbes. It is a non-carbohydrate aromatic hetero polymer composed of phenylpropanoid units of three precursor aromatic alcohols including coniferyl, sinapyl, and *p*-coumaryl alcohol.[4] These precursors form the guaiacyl-(G), syringyl-(S), and *p*-hydroxyphenyl (H) subunits in the lignin molecule, respectively,[5] linked through a variety of *non hydrolysable* C–C and C–O–C bonds.[5]

Lignin degradation plays a major role in carbon recycling in the ecosystem as well as in the conversion of plant biomass for the second generation bio fuel (ethanol) production because of the structural complexity recycling of this compound in the carbon cycle complicated in the biosphere. Lignin removal from lignocellulose materials is facilitated by lignin-degrading organisms. Even though different microorganisms including bacteria and fungi are involved in lignin degradation, the most rapid and efficient lignin degraders are basidiomycetes.[6,7] White rot fungi are the only organisms capable of mineralizing lignin more rapidly and extensively into CO_2 and H_2O.[5] White rot fungi bring about lignin decay through an oxidative process that is thought to involve secreted extracellular enzymes such as lignin peroxidase (LiP) (Enzyme Commission (EC) .1.11.1.14), Manganese dependent peroxidase (MnP) (EC.1.11.1.13), and Laccase (Lac) (EC.1.10.3.2)[8–12] and facilitating access to other polysaccharide components in lignocellulosic biomass for the utilization by other organisms.

Environmental pollution caused by pesticides and their degradation products is a major ecological problem.[13] It has been documented that organophosphorus pesticides (OP) constitute the largest group of pesticides used globally which account for about 38% of the total pesticides used worldwide.[14] Chlorpyrifos is a broad-spectrum insecticide which is considered as one of the most frequently used chlorinated xenobiotic organophosphorus pesticides in the agro forestry system.[15] Its massive application has led to the contamination of water and soil, and disruption of biogeochemical cycles.[16] In addition, its residues have been detected in various ecological systems.[17]

But, the extensive use and accumulation of this pesticide in the environment may affect the growth of several beneficial microorganisms in the environment including white rot fungi and disturb ecological functions in the biosphere. The influence of organophosphorus pesticides on white rot fungi was not assessed and virtually, information is lacking on the impact of organophosphate pesticides on the secretion of ligninolytic enzymes by the white rot fungi. The impact of chlorpyrifos on the growth and secretion of ligninolytic enzymes with respect to the white rot fungus—*Stereum ostrea* has not been assessed. The present study focuses on the secretion of ligninolytic enzymes by the white rot fungus *Stereum ostrea* immobilized on polyurethane cubes under the influence of chlorpyrifos.

13.2   MATERIALS AND METHODS

13.2.1   MICROORGANISM

Stereum ostrea used in this study was kindly supplied by Prof. M. A. Singaracharya, Department of Microbiology, Kakatiya University, Telangana, India, and was isolated from wood logs. This culture was maintained on Koroljova-Skorobogat medium.[18]

13.2.2   EXPERIMENTAL DESIGN

To reuse the culture for recycling purpose and to test the influence of chlor-pyrifos on the secretion of lignolytic enzymes by immobilized *S. ostrea*, 6 or 8 pre-weighed (polyurethane) sponge cubes (1 cm^3) were soaked in ethanol for 5 min and washed 2–3 times with sterile distilled water and finally placed in 50 ml of Koraljova broth in 250 ml Erlenmeyer flasks. The flasks, with the broth devoid of the cubes, served as the controls. There were flasks of three categories; all the flasks received chlorpyrifos from the stock solution of chlorpyrifos EC grade at only one concentration of - 20 ppm level. All the flasks were inoculated with the culture of *S. ostrea*. All the flasks were incubated at 30 °C and at a speed of 160 rpm into the incubator cum shaker (Scigenics Orbitek, Chennai, India).

13.2.3   ANALYTICAL ASSAYS

At regular intervals, the growing culture of *Stereum ostrea* in the flasks was filtered through Whatman filter no. 1 for the collection of fungal biomass on the pre-weighed filter and the culture filtrate in a suitable container. The fungal biomass on the filter was dried in an oven at 60 °C and the growth was expressed in terms of the dry weight of fungal biomass per flask. Activities of the extracellular ligninolytic enzymes in the culture filtrate after centrifu-gation were measured following standard protocols.

13.2.3.1   LACCASE ENZYME ASSAY

Laccase enzyme assay was determined according to the method of Das et al.[19] The assay solution contained 100 mM acetate buffer (1.2 ml), 10 mM guaicol (0.4 ml), and an aliquot of the enzyme source. The color change was monitored at 470 nm ($\epsilon = 6740$) for 3 min. The enzyme activity was expressed in terms of IU (international units) where one unit of the enzyme was defined as the amount of enzyme that oxidized one micromole of the substrate per minute.

13.2.3.2   MANGANESE PEROXIDASE ASSAY

Manganese Peroxidase (MnP) activity was determined following the method of Bonnen et al.[20] The assay medium contained 1mM guaicol (0.4 ml) and

1mM MnSO4 (0.5 ml) in 10mM citrate buffer with a pH of 5.5 (0.5 ml). The reaction was initiated by the addition of 50 μM H_2O_2 (0.2 ml). After an incubation at room temperature for 10 min, the change in absorbance because of the oxidation of guaicol was monitored at 460 nm for 3 min. The enzyme activity was expressed in terms of IU (International Units) where one unit of the enzyme was defined as the amount of enzyme that oxidized one micromole of substrate per minute.

13.2.3.3 LIGNIN PEROXIDISE ASSAY

Lignin Peroxidase (LiP) enzyme assay was based on the method of Tien and Kirk.[5] The assay medium contained 0.25 M tartaric acid (0.5 ml), 10 mM veratryl alcohol (0.5 ml), 5 mM H_2O_2 (0.5 ml). The absorbance was monitored at 310 nm for 3 min after the addition of the enzyme source. The enzyme activity was expressed in terms of IU where one unit of the enzyme was defined as the amount of enzyme that oxidized one micromole of substrate per minute.

13.2.3.4 PROTEIN ESTIMATION

Appropriate dilution of the culture filtrate from the submerged culture was used for the estimation of soluble protein content according to Lowry et al.[19] Bovine serum albumin was used as the protein standard.

13.3 RESULTS AND DISCUSSION

S. ostrea in liquid Koroljova medium was amended with chlorpyrifos at only 20 ppm concentration and ligninolytic potential of the culture was assessed in the present study because chlorpyrifos at 20 ppm significantly stimulated the secretion of ligninolytic enzymes by S. ostrea in a previous experiment. The growth of S. ostrea on sponge cubes was comparable to the growth of S. ostrea in the free (control) state (Table 13-1). Secretion of extracellular protein by S. ostrea in immobilized state and free state was comparable (Table 13-1).

TABLE 13-1   Growth and Secretion of Extracellular Protein by *Stereum ostrea* Immobilized on Sponge Cubes in the Medium in the Presence of Chlorpyrifos at 20 ppm level

Incubation Period (In Days)	Biomass in g/50ml			Extracellular Protein in µg/ml		
	Control	Sponge Cubes		Control	Sponge Cubes	
		6 cubes	8 cubes		6 cubes	8 cubes
2	-	-	-	-	-	-
4	0.27[a]	0.16[c]	0.20[b]	660[a]	519[a]	576[a]
6	0.35[a]	0.40[a]	0.37[a]	916[a]	858[b]	942[a]
8	0.50[a]	0.46[a]	0.45[b]	1036[a]	992.5[a]	1041[a]
10	0.75[a]	0.70[a]	0.68[b]	1067[a]	1001[a]	1198[a]

Values are the means of triplicates.

Means in each row, followed by the same superscript letter (a, b and c) are not significantly different ($P \leq 0.05$) from each other according to the DMR test.

Control: Devoid of sponge cubes (free state).

A comparison of the cultures of *S. ostrea* immobilized on an inert substrate (polyurethane) and a free state culture showed differences between the production of extracellular ligninolytic activities comparable with those observed in submerged cultures (Figures 13-1–13-3). Though the results on the yields of laccase by the culture of *Stereum ostrea* in immobilized (6 or 8 cubes) state were inconsistent, an observation of equal or higher production of ligninolytic enzymes by immobilized culture in comparison with the control was made. The culture, both in the immobilized (8 cubes) and in the free state gave production of laccase to the same extent of 42 U/ml on the 8th day of incubation. But on the 10th day of incubation, there was an increased production of all the three ligninolytic enzymes in 8 and 6 cubes used culture as compared with the production of all the three ligninolytic enzymes in the free state culture. On the whole, immobilized culture with 8 cubes gave yields of all the three ligninolytic enzymes higher than that of the respective enzyme by the culture in the free state.

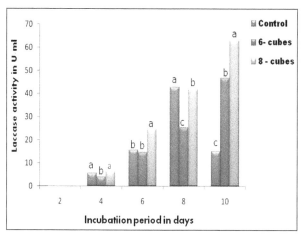

FIGURE 13-1   Secretion of laccase enzyme by *Stereum ostrea* immobilized on sponge cubes in the medium in the presence of chlorpyrifos at 20 ppm level.

The mean bars for each sampling interval, followed by the same letter are not significantly different ($P \leq 0.05$) from each other according to the DMR test.

Control: Devoid of amendment pesticide chlorpyrifos and sponge cubes.

Six and 8 cubes: Polyurethane cubes added into the medium.

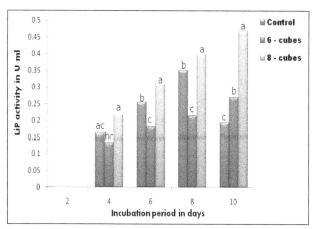

FIGURE 13-2   Secretion of manganese peroxidase enzyme by *Stereum ostrea* immobilized on sponge cubes in the medium in the presence of chlorpyrifos at 20 ppm.

The mean bars for each sampling interval, followed by the same letter are not significantly different ($P \leq 0.05$) from each other according to the DMR test.

Control : Devoid of amendment pesticide chlorpyrifos.

Six and 8 cubes: Polyurethane cubes added into the medium.

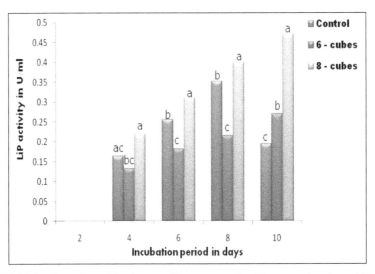

FIGURE 13-3   Secretion of lignin peroxidase enzyme by *Stereum ostrea* immobilized on sponge cubes in the medium in the presence of chlorpyrifos at 20 ppm level.

The mean bars for each sampling interval, followed by the same letter are not significantly different ($P \leq 0.05$) from each other according to the DMR test.

Control : Devoid of amendment pesticide chlorpyrifos.

Six and 8 cubes: Polyurethane cubes added into the medium.

The adsorption of mycelia of the fungal culture onto the sponge cubes in the present study resulted in immobilization of the fungal culture and functioned in the normal way of secretion of extracellular proteins including ligninolytic enzymes. Similarly, the use of sponge cubes with the same culture for immobilization was demonstrated earlier.[21] According to this study, *Stereum ostrea* in the immobilized state gave higher yields of laccase than the same culture in the free state in the presence of dye.[22] The use of different carriers for the immobilization of *Phanerochaete chrysosporium* revealed that polystyrene was as an optimum carrier on the basis of the adsorption of cells. Production of MnP and biomass with this immobilized culture was achieved on the 6th day of incubation in comparison with the corresponding free culture. The white rot fungus *P. chrysosporium*, immobilized on mineral kissiris, produced LiP of 174 U/l and 500 U/l within 7–9 days.[23]

Changes in pH took place in the medium upon the growth of *S. ostrea* in the presence or absence of the cubes (Table 13-2). The pH of the medium decreased from an initial value of 5.5.

TABLE 13-2   Changes in pH in the Medium upon the Growth of *Stereum ostrea* on Sponge Cubes in the Presence of Chlorpyrifos at 20 ppm level

Incubation Period (In Days)	pH		
	Control	Sponge Cubes	
		6 cubes	8 cubes
2	Nd	Nd	Nd
4	5.29[b]	5.46[a]	5.44[a]
6	5.20[c]	5.44[a]	5.35[b]
8	4.67[c]	5.19[a]	4.89[b]
10	4.20[b]	4.80[a]	4.30[b]

Values are the means of triplicates.

Means in each row, followed by the same superscript letter (a, b and c) are not significantly different ($P \leq 0.05$) from each other according to the DMR test.

Control: Devoid of sponge cubes (free state).

Nd: Not determined.

13.4   CONCLUSION

The performance of *Stereum ostrea* immobilized on sponge cubes under the influence of chlorpyrifos at 20 ppm level during growth was comparable with the same culture in the free state under the influence of chlorpyrifos in terms of growth, secretion of extracellular protein contents, and ligninolytic enzymes. Increased productions of all the three ligninolytic enzymes were shown in 8 cubes immobilized *Stereum ostrea* culture medium compared with the control and 6 cubes of immobilized culture. There was a decrease in the pH of the medium amended with chlorpyrifos in immobilized 6 and 8 sponge cubes and the control during the course of the growth of *Stereum ostrea*.

ACKNOWLEDGMENTS

The authors express gratitude to the University Grant Commission, New Delhi for providing financial assistance in the form major research project (UGC-MRP) awarded to Prof. B. Rajasekhar Reddy and BSR fellowships to Kanderi Dileep Kumar and A. Ramya, and are also thankful to the Department of Science and Technology (DST)-Inspire fellowship programme for providing financial assistance to K. Y. Usha. They sincerely express gratitude to Prof. M.A. Singaracharya, Department of Microbiology, Kakatiya University for providing the fungal culture - *Stereum ostrea*.

KEYWORDS

- **Chlorpyrifos**
- **Immobilization**
- **Ligninolytic enzymes (LiP, MnP, LAC)**
- **Polyurethane cubes**
- ***Stereum ostrea***

REFERENCES

1. Alvira, P.; Tomas-Pejo, E.; Ballesteros, M.; and Negro, M. J. Pretreatment technologies for an efficient bioethanol production process based on enzymatic hydrolysis: A review. *Bioresour. Technol.* **2010,** *101*, 4851–4861.
2. Sanchez, O. J.; and Cardona, C. A. Trends in biotechnological production of fuel ethanol from different feedstocks. *Bioresour. Technol.* **2008,** *99*, 5270–5295.
3. Dashtban, M.; Schraft, H.; Syed, T. A.; and Qin, W. Fungal biodegradation and enzymatic modification of lignin. *Int. J. Biochem. Mol. Biol.* **2010,** *1*, 36–50.
4. Wei, H.; Xu, Q.; Taylor, L. E.; Baker, J. O.; Tucker, M. P.; and Ding, S. Y. Natural paradigms of plant cell wall degradation. *Curr. Opin. Biotechnol.* **2009,** *20*, 330–338.
5. Martinez, A. T.; Speranza, M.; Ruiz-Duenas, F. J.; Ferreira, P.; Camarero, S.; Guillén, F.; Martínez, M. J.; Gutiérrez, A.; and del Río, J. C. Biodegradation of lignocellulosics: Microbial, chemical, and enzymatic aspects of the fungal attack of lignin. *Int. Microbiol.* **2005,** *8*, 195–204.
6. Kirk, T. K.; and Farrel, R. L. Enzymatic "combustion": The microbial degradation of lignin. *Ann. Rev. Microbio.* **1987,** *141*, 465–505.
7. Jurado, M.; Martinez, A. T.; Saparrat, M. C. N.; and Martinez, M. J. Application of white rot fungi in transformation, detoxification of agricultural wastes. *Com. Biotech.* **2011,** *6*, 595–603.
8. Tien, M.; and Kirk, T. K. Lignin peroxidase of *Phanerochaete chrysosporium. Methods Enzymol.* **1988,** *161*, 238–249.
9. Martinez, A. T. Molecular biology and structure-function of lignin-degrading heme peroxidases. *Enzyme Microbiol. Technol.* **2002,** *30*, 425–444.
10. Collins, P. J.; and Dobson, A. Regulation of laccase gene transcription in *Trametes versicolor. Appl. Environ. Microbiol.* **1997,** *63* (9), 3444–3450.

11. Hammel, K. E. Fungal degradation of lignin; In *Driven by Nature: Plant Litter Quality and Decomposition*; Cadisch, G., and Giller, K. E., Eds.; CAB International: United Kingdom, 1997; pp 33–45; Chapter 2.

12. Dashtban, M.; Schraft, H.; Syed, T. A.; and Qin, W. Fungal biodegradation and enzymatic modification of lignin. *Int. J. Biochem. Mol. Biol.* **2010**, *1*, 36–50.

13. Guliy, O. I.; Ignatov, O. V.; Makarov, O. E.; and Zgnatov, V. V. Determination of organophosphorus aromatic nitro insecticides and *p*-nitrophenol by microbial-cell respiratory activity. *Biosen. Bioelectro.* **2003**, *18*, 1005–1013.

14. Singh, B. K.; and Walker, A. Microbial degradation of organophosphorus compounds. FEMS *Microbiol. Rev.* **2006**, *30*, 428–471.

15. Maya, K.; Singh, R. S.; Upadhyay, S. N.; and Dubey, S. K. Kinetic analysis reveals bacterial efficacy for biodegradation of chlorpyrifos and its hydrolyzing metabolite TCP. *Process Biochem.* **2011**, *46*, 2130–2136.

16. Chishti, Z.; Hussain, S.; Arshad, K. R.; Khalid, A.; and Arshad, M. Microbial degradation of chlorpyrifos in liquid media and soil. *J. Environ. Man.* **2013**, *114*, 372–380.

17. Xue, N.; Xu, X.; and Jin, Z. Screening 31 endocrine-disrupting pesticides in water and surface sediment samples from Beijing Guanting reservoir. *Chemosphere* **2005**, *61*, 1594–1606.

18. Koroljova-Skorobogatko, O. V.; Stepanova, E. V.; Gavrilova, V. P.; Morozova, O. V.; Lubimova, N. V.; Dzchafarova, A. N.; Jaropolov, A. I.; and Makower, A. Purification and characterization of the constitutive form of laccase from the basidiomycete *Coriolus hirsutus* and the effect of inducers on laccase synthesis. *Biotechnol. Appl. Biochem.* **1998**, *28*, 47–54.

19. Das, N.; Sengupta, S.; and Mukerjee, M. Importance of laccase in vegetative growth of *Pleurotus Florida*. *Appl. Environ. Microbiol.* **1997**, *63* (10), 4120–4122.

20. Bonnen, A. M.; Anton, L. H.; and Orth, A. B. Lignin-degrading enzymes of the commercial button mushroom, *Agaricus bisporus*. *Appl. Environ. Microbiol.* **1994**, *60*, 960–965.

21. Pallavi, H. Decolorosation of dyes by fungi. Thesis, Department of Microbiology, Sri Krishnadevaraya University, India, 2011.

22. Urek, R.; and Pazarlioglu, N. Purification and partial characterization of manganese peroxidase from immobilized *Phanerochaete chrysosporium*. *Process Biochem.* **2004**, *39*, 2061–2068.

23. Ghasemzadeh, R.; Kargar, A.; and Lotfi, M. Decolorization of synthetic textile dyes by immobilized white-rot fungus. In *International Conference on Chemical, Ecology and Environmental Sciences (ICCEES'2011)*; Pattaya; 2011, 434–438.

CHAPTER 14

STRUCTURE AND GAS DIFFUSION PATH ANALYSIS OF HYDROGENASE ENZYMES BY HOMOLOGY MODELLING

NIVEDITA SAHU[1,2*], ANIRUDH NELABHOTLA[1], and PRADHAN NITYANANDA[2]

[1]CSIR-Indian Institute of Chemical Technology, Hyderabad, Telangana – 500007, India.

[2]AcSIR- CSIR-Indian Institute of Chemical Technology, Hyderabad, Telangana – 500007, India.

*Corresponding author: Nivedita Sahu. E-mail: nivedita@iict.res.in

CONTENTS

Abstract 278
14.1 Introduction 278
14.2 Materials and Methods 282
14.3 Results and Discussion 283
14.4 Conclusion 290
Acknowledgments 290
Keywords 290
References 290

ABSTRACT

Hydrogenase is an enzyme that catalyzes the reversible oxidation of molecular hydrogen (H_2). Here, the hydrogenase enzyme structures of *Chlamydomonas reinhardtii*, *Clostridium pasteurianum*, and *Desulfovibrio vulgaris* were taken as reference. Homology structures of the hydrogenase enzymes taken from *Clostridium acetobutylicum*, *Chlorella variabilis*, and *Scenedesmus obliquus* were developed. BLAST search was performed and the templates with a minimum of 35% sequence similarity were considered for homology modelling. Structure analysis was performed after minimizing the energies of the nine (3*3) structures. For analyzing the gas diffusion path and hydrogen bonds, *C. pasteurianum* (CpI) was taken as the reference. It has been seen that the path was formed around the H-cluster which was where the hydrogen molecule was produced. It was blocked by the oxygen molecule in its path reducing the efficiency of the enzyme. With this information we selected *C. variabilis* as the next model organism along with *C. reinhardtii* for our further research to study the exact mechanism of gas diffusion in the enzyme.

14.1   INTRODUCTION

Hydrogen (H_2) is most promising in the next generation fuel evolution, with many technical, economic, and environmental benefits making it a potential future. The energy content per unit weight for hydrogen (142 kJ/g) is the highest with no known fuel near to its competition. It can be transported for domestic/industrial consumption through conventional means making it an invincible fuel.[1,2] Among the various candidates, H_2 is regarded as the most promising future energy carrier as its energy content is higher by 2.75 times compared with hydrocarbon fuels (gasoline) and produces only water upon combustion.[3,4] The world has now accepted H_2 as an environmentally protective resource of renewable energy.

Nature has created biological reactions that use sunlight for the oxidation of water, and iron/nickel containing enzymes that use electrons for the generation of H_2.[5] Biological processes are catalysed by microorganisms in an aqueous environment at ambient temperature and atmospheric pressure. These techniques are well suited for decentralized energy production in small-scale installations in locations where biomass or wastes[6] are available, thus reducing the energy input and expenses for transport.[2] These processes are usually carried out by different anaerobic bacteria and/or algae.[7] The characteristics of these microorganisms widely differ from each other with respect to the substrates and process conditions.[8]

Hydrogenases catalyze the reversible redox reaction between H_2 and its component two protons and two electrons.[3] Creation of hydrogen gas occurs in the transfer of reducing equivalents produced during pyruvate fermentation to water. Water is decomposed into its component H^+ ions, electrons, and oxygen, which occurs in the presence of light in all photosynthetic organisms. Some algae and cyanobacteria have evolved a second step in the dark reactions in which protons and electrons are reduced to form H_2 gas by specialized enzymes in the chloroplast called hydrogenases.[5,9] They catalyze the $2H^+ + 2 e^- \leftrightarrow H_2$ reaction.[10] Fermentation in bioreactors with genetically modified algae have been put into practice.[11]

14.1.1 ALGAL SOURCES

Photosynthetic hydrogen production by green algae was discovered in the pioneering experiments of Gaffron and Rubin.[12] It is the light-activated simultaneous photoproduction of hydrogen and oxygen that is of primary interest in biophotolysis.[13,14] Light energy facilitates the oxidation of water molecules, the release of H^+ ions and electrons, and the endergonic movement of these electrons to ferredoxin.[15] The photosynthetic ferredoxin (PetF) serves as the physiological electron donor to the Fe-hydrogenase and thus, links the enzyme to the electron transport chain in the chloroplast of the green algae.[16]

The model green microalga *Chlamydomonas reinhardtii*, along with several species of unicellular green algae can operate under both aerobic photosynthesis and anaerobic metabolic conditions. The most significant part of its metabolic activity is that of "anaerobic oxygenic photosynthesis", in which the photosynthetically produced oxygen is utilized by the cell's own respiration, which in turn causes anaerobiosis in the culture and induction of the cellular "hydrogen metabolism" process.[5] After anaerobic induction for several hours in the dark, the hydrogenase activity could be measured for a short time when the algae were brought back into the light.[17,18] Although photosynthetic hydrogen production seemed to be a promising way for the biological conversion of sunlight into fuel, there are hardly any biotechnological applications.

Cyanobacteria and green algae are so far the only known organisms with both oxygenic photosynthesis and hydrogen production.[19] In cyanobacteria, H_2 production and nitrogen fixation are mostly coupled, whereas, photosynthetically generated electrons are utilized for H^+ reduction in unicellular green algae. One recent and interesting extension of photosynthesis research entails the development of methods for a sustained photo-biological H_2 gas production in green microalgae *C. reinhardtii*.[16,20–22] It couples an extremely O_2-sensitive enzyme [FeFe] hydrogenase to the photosynthetic electron transport pathway that generates O_2 during its normal function. The electrons are dissipated from the photosynthetic electron transport chain in the form of molecular H_2 which is enabled by the hydrogenase enzyme pathway.[23]

14.1.2   BACTERIAL SOURCES

Hydrogen can be produced from fermentation[24] modifying a microbial fuel cell using the end product acetate as a substrate. The theoretical efficiency for hydrogen gas production from waste water by fermentation was determined to be 12 mol H_2/mol-hexose and it was stated by Benemann[25] that if the H_2 gas conversion efficiency reaches to about 70%, it could have a potential and sustainable possibility to be used as a commercial fuel.

$$CH_3COO^- + 4H_2O \rightarrow 2HCO_3^- + H^+ + 4H_2 \uparrow$$

Hydrogen production by *Clostridium thermocellum* cultured on three insoluble cellulosic substrates, as well as on cellobiose was measured. It was observed that efficient H_2 production from cellulosic biomass by *Clostridium thermocellum* was acquired when the bioreactor was maintained at a pH of approximately 7.0 and avoiding the product inhibition by removing the gas products (H_2 and CO_2) maintain a lower H_2 partial pressure.[26] Carbon monoxide was oxidized to carbon dioxide by the biologically-mediated water-gas shift reaction, which simultaneously reduced water to hydrogen, making it a cost-effective technology for the production of hydrogen from synthesis gas.

The best hydrogen yield (0.0128 $kgH_2/kg_{biomass}$) by *Anabaena sp.* was achieved for $Ar + CO_2 + N_2$ gas atmosphere with high light intensity. This hydrogen yield was further increased by 8.1%, by using the recovered or residual cyanobacteria through a fermentative process. Therefore, the best value for H_2 production ratio versus energy consumption and CO_2 emissions was obtained for $Ar + CO_2 + 20\%$ N_2 gas atmosphere and medium light intensity conditions.[27]

The dark fermentation process converted the organic substrates into hydrogen in an anaerobic environment. Butyrate and acetate were the main co-products in dark fermentation.[28] Dark fermentation of organic compounds might be the most feasible one because the rates of hydrogen production and microbial growth were higher than those of photo fermentation and bio-photolysis.[9] The anaerobes responsible for fermentative hydrogen production can produce hydrogen all day long without light and can use various kinds of substrates such as refuse and food waste products.[4] The majority of microbial hydrogen production is driven by the anaerobic metabolism of pyruvate, formed during the catabolism of various substrates. The breakdown of pyruvate is catalyzed by one of two enzyme systems,[10] i.e., Pyruvate Formate Lyase (PFL) and Pyruvate Ferredoxin (flavodoxin) Oxido-Reductase (PFOR).

The photo fermentation process utilizes volatile fatty acids (VFAs) in addition to sugars as the substrate in the presence of light. Because VFAs are the co-products of hydrogen in dark fermentation, a two-stage process, involving photo fermentation after dark fermentation[15,29,30] is an attractive option for enhancing hydrogen production and reducing the chemical oxygen demand (COD) in the effluent from dark fermentation.[28] Photosynthetic efficiencies are lower under ideal (low) light conditions. It is also an assumption that photosynthetic efficiencies must be even lower under high-light (full sunlight) conditions assuming that the photo system of the photosynthetic bacteria is, like that of microalgae and cyanobacteria, optimized for low-light conditions.[10]

14.1.3 CURRENT RESEARCH

Oxygen suppresses both the expression of the hydrogenase genes and within minutes completely and irreversibly inactivates the hydrogenase.[31] Physically, it blocks the hydrogen molecule's way out of the enzyme which cannot be reversed. The major problem of the scientific community is on the high sensitivity of the hydrogenase enzyme toward molecular oxygen. The current research in the area of biohydrogen production concentrates on the molecular biology of the process and not just the production technology. The structure of the hydrogenase enzymes and the properties of amino acids involved in the oxygen sensitivity issue are being studied. Simulations for gas diffusion paths within the enzyme have given a great opportunity to scientists to explore genetic engineering techniques. The large scale productions of biohydrogen were found implemented using algal bioreactors.

In this chapter, we mainly discuss the algal sources of biohydrogen production and choose the next best model organism to carry out the experimental modules. The unicellular green alga, *Chlorella* spp. was selected as the model organism because of its universal availability and the ease to carry out research work. The alga can be cultured easily when the production levels are scaled up for its mass cultivation.

14.2   MATERIALS AND METHODS

For the experimental procedures, standard microorganisms like for algae *Chlamydomonas reinhardtii* and *Desulfovibrio vulgaris* and for bacteria *Clostridium pasteurianum* were selected. These organisms have the X-ray crystallographic structure of the hydrogenase enzymes submitted in the Protein Data Bank (PDB). Tools like BLAST and Clustal were used to find the other organisms which closely matched to the standard strains selected. An *in vitro* culture of *Chlorella* was made using artificial nutrient media for the production of biohydrogen.

14.2.1   *PROTEIN STRUCTURES*

The hydrogenase enzyme structures of *Chlamydomonas reinhardtii* (3LX4), *Clostridium pasteurianum* (3C8Y/CpI), and *Desulfovibrio vulgaris* (1HFE) were taken as the reference. Homology structures of the hydrogenases enzymes taken from *Clostridium acetobutylicum* (CA), *Chlorella variabilis* (CV), and *Scenedesmus obliquus* (SO) were developed. The BLAST search was performed and the templates with a minimum of 35% sequence similarity were considered for homology modelling. Structure analysis of all the sequences was performed after minimizing the energies of the nine (3*3) structures. Sequence alignment was done using Clustal 2.0 program online and few modifications (deletion of non-identical with long gap sequences) were done in the sequence using Swiss PDB Viewer.

14.2.2   *HOMOLOGY MODELING AND ANALYSIS*

The structures were then processed for energy minimization using Swiss PDB Viewer. *C. pasteurianum* was taken as the reference for obtaining the amino acids which built up the gas diffusion paths of the hydrogenase enzyme. They were further analyzed for the number of hydrogen bonds formed by the residues which were a part of the diffusion pathway. The work compares algal and bacterial hydrogenases using homology modeling and has given some critical insights into the mechanism of gas diffusion paths within the hydrogenase enzymes of CA, CV, and SO. Finally, all the structures were validated using PyMol analysis. Their RMSD values were calculated accordingly, to select the best model organism which could be easily accessible and comfortable to carry out large scale research work.

14.2.3 MICROALGAL CULTURE

A species of *Chlorella* was isolated from stagnant rainwater in the Indian Institute of Chemical Technology. Its pure culture was obtained by growing the culture on agar plates containing Modified Bristol Medium (MBM).[32] The media contained 2.94 mM of $NaNO_3$, 0.17 mM of $CaCl_2 \cdot 2H_2O$, 0.3 mM of $MgSO_4 \cdot 7H_2O$, 0.43 mM of K_2HPO_4, 1.29 mM of KH_2PO_4, and 0.43 mM of NaCl. Stocks of the salt solutions were prepared to the concentrations of 10X and were used for media preparation. The cells were inoculated and grown in an incubator maintained at 28 °C, pH 6.7, and with a continuous light source using Phillips Tube FL 20 W × 4 EA. The algae cultures were run in duplicates and revived every week using MBM.

14.3 RESULTS AND DISCUSSION

14.3.1 HOMOLOGY MODELLING

The Fast Alignment (FASTA) sequences of the enzymes were obtained from the BLAST search which showed a highest similarity of 71% among *C. pasteurianum* and *C. acetobutylicum*. The algal sequences had modest similarity among them and were perfectly suitable for homology modelling. Deletion of a few of the long gapped and non-identical sequences produced efficient Clustal multiple sequence alignment which helps in minimizing energy. The sequences were then processed for homology modelling using Swiss PDB Server. The energy was minimized using Swiss PDB Viewer post-modelling to the ranges of 30,000 KJ/mol. The structures were then compared with each other to obtain the gas diffusion paths. Let us discuss the importance of these paths and amino acids and how it affects hydrogen production with the help of a simple experiment of sulfur deprivation effect.

Sustained photoproduction of hydrogen by green algae can be achieved by reversible inactivation of photosynthetic water oxidizing activity, catalysed by photosystem II.[33] Sulfur is the main constituent in all electron acceptors and is also part of the enzymes that are responsible for growth and photosynthesis. So, when sulfur is deprived, photosynthesis is inhibited and the oxygen levels fall down. Because of respiration, the conditions become anaerobic and hydrogen production starts again. Thus, an S-deprived culture, transitions through five consecutive states: aerobic phase, oxygen consumption phase, anaerobic phase, hydrogen production phase, and termination phase. It is observed that the re-addition of sulphate to the S-deprived culture at the beginning of the sulfur deprivation period increases the rate and yield of the hydrogen produced.[34]

Lowering of the rate of photosynthesis to about the level of cellular respiration, enables the cell's own respiration to consume photosynthetically generated O_2. This allows [FeFe]-hydrogenase to continuously express and function. Presently, a balanced photosynthesis–respiration activity is the current research interest of many countries. The absence of sulfur in the growth medium caused a slow-down in the rate of photosynthesis which was compared to the rate of respiration,[20] resulting in necessary preconditions for H_2 evolution activity. This induces an internal anaerobiosis thereby permitting the expression of the HYDA1 gene and the activation of HydA1 enzyme.

During S-deprivation and H_2 production, the *C. reinhardtii* cells stop growth and down-regulate CO_2 assimilation[20,23] resulting in an inoperative photosynthetic electron sink. Thus, the hydrogenase pathway is activated, and leads to proton reduction and H_2 production. This becomes an alternative sink for photosynthetic electron transport. The latter stays active in the electron transport chain[35] starting at the plastoquinone (PQ) pool,[20] thereby dissipating the light energy by the cells producing adenosine tri- phosphate (ATP). A similar result was obtained by National Renewable Energy Laboratory (NREL, United States of America), Department of Energy (DOE, United States of America) stating algal photosynthesis and hydrogen production as sister processes.[28] They explain the process of water splitting and adoption of alternative reaction to lose the excess energy before the carbon fixation completes. A very special behavior was observed in the Rubisco-deficient *C. reinhardtii* strain CC-2803, which produces hydrogen in the absence, but also in the presence of sulfur.[36]

14.3.2   GAS DIFFUSION PATHS

There are two paths (A and B) which help the hydrogen molecule escape the enzyme structure after its production at the H-cluster and entering the cavity.[37] A schematic representation of the gas diffusion paths of the hydrogenase enzymes is shown in Figure 14-1. List of the residues of CpI that form the Paths A, B, the common central cavity for O_2 and H_2 transport in hydrogenase are defined by Cohen et al.[38] CpI, *C. reinhardtii* and three modelled structures with template as these two are represented in Figure 14-1(a), whereas *D. vulgaris* which is a small enzyme has only path A but not path B, and it is represented in Figure 14-1(b).

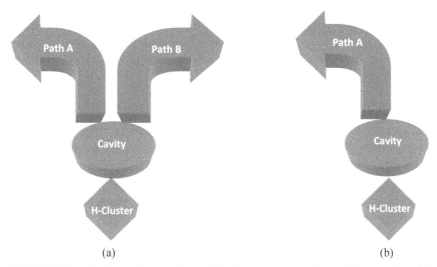

| (a) | (b) |

FIGURE 14-1   The figure denotes the graphical representation of gas diffusion paths of the hydrogenase enzymes of the respective organism (a) represents the paths in the structures which have paths A and B both i.e., modelled where CpI was used as the template (b) where the template was used as *D. vulgaris*.

Overcoming the O_2 sensitivity of the hydrogenase enzyme is the major focus of the photobiological hydrogen production research. The hydrogen production in CpI occurs at the well-known H-cluster, where the protons are reduced by the electrons gained from ferredoxin. The electrons reach the H-clusters by a series of Fe-S clusters, but how the protons reach the H-cluster is unknown. Near to the exit of the hydrogen gas molecules, the other gas molecules O_2 and CO bind to the CpI around the H-cluster and inactivate the enzyme. The O_2 pathways are found to be around large hydrophobic residues.[37] While the O_2-mediated deactivation of hydrogenase is in some cases beneficial for the host organism, it severely limits the practicality of using hydrogenase to produce H_2.[39]

Hydrogen production in CpI happens at the H-cluster, a metallic cluster embedded inside the CpI protein matrix. It is achieved by the reduction of H^+ ions from the external solution through the use of electrons acquired from a reduced carrier such as ferredoxin.[40,41] The H^+ ions (or protons) probably reach the H-cluster by the means of a putative, but yet unverified, proton path contained in the protein. The electrons are transferred to the embedded H-cluster through a series of accessory iron-sulfur clusters aligned in a chain between the H-cluster and one end of CpI. The CpI must allow the product to exit the protein, but it also allows in small gas molecules such as O_2 and CO to transit through the enzyme and reach the H cluster, which becomes inactivated upon their binding.[37]

14.3.3   STRUCTURAL ANALYSIS

Figure 14-2 is a graphical representation on the comparison of the number of hydrogen bonds formed by amino acids which build up the gas diffusion paths of the hydrogenase enzyme. The amino acids which are involved in the gas diffusion paths of hydrogenase enzymes in the respective organisms are mentioned below:

The list of amino acid residues in *D. vulgaris*; of the total 421 amino acids, those that form the cavity are: 149, 152, 153, 156, 176, 178, 203, 296, and 302. Path A: 151, 157, 160, 161, 164, 172, 174, 303, 306, 307, 310, 338, 340, 345, 346, 347, 371, and 372.

The list of amino acid residues in *C. reinhardtii*; of the total 457 amino acids, those that form the cavity are: 94, 97, 98, 101, 127, 129, 154, 250, and 256. Path A: 96, 102, 105, 106, 109, 123, 125, 257, 260, 261, 264, 292, 294, 344, 345, 346, 370, and 371. Path B: 97, 100, 101, 151, 157, 160, 161, 164, 424, 425, 428, 429, 436, 437, 440, and 441.

The list of amino acid residues in *C. variabilis*; of the total 717 amino acids, those that form the cavity are: 331, 334, 335, 338, 388, 390, 415, 509, and 515. Path A: 333, 339, 342, 343, 346, 384, 386, 516, 519, 520, 523, 551, 553, 569, 570, 571, 595, and 596. Path B: 334, 337, 338, 412, 418, 421, 422, 425, 649, 650, 653, 654, 661, 662, 665, and 666.

The list of amino acid residues in *C. acetobutylicum*; of the total 582 amino acids, those that form the cavity are: 271, 275, 276, 279, 297, 299, 324, 417, and 423. Path A: 273, 279, 282, 283, 286, 292, 294, 423, 426, 427,

430, 458, 460, 465, 466, 467, 491, and 492. Path B: 274, 277, 278, 320, 326, 329, 330, 333, 551, 552, 555, 556, 563, 564, 567, and 568.

The list of amino acid residues in *S. obliquus*; of the total 449 amino acids, those that form the cavity are: 119, 122, 123, 126, 152, 154, 179, 274, and 280. Path A: 121, 127, 130, 131, 134, 148, 150, 281, 284, 285, 288, 316, 318, 339, 340, 341, 365, and 366. Path B: 122, 125, 126, 176, 182, 185, 186, 189, 419, 420, 423, 424, 431, 432, 435, and 436.

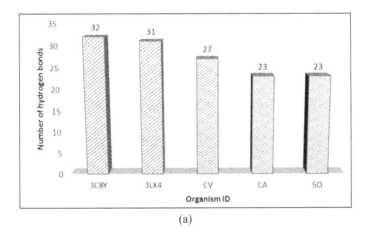

(a)

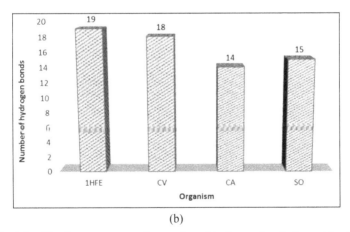

(b)

FIGURE 14-2 The figure compares the number of hydrogen bonds formed by the amino acids involved in the gas diffusion paths of modelled hydrogenase enzymes with their respective template enzymes. CV: *C. variabilis*; CA: *C. acetobutylicum*; and SO: *S. obliquus*. (a) Comparison of hydrogen bonds formed by the amino acids that build up the gas diffusion paths in the structures that contain the cavity, path A and path B. (b) Comparison of hydrogen bonds formed by the amino acids that build up the gas diffusion paths in the structures that contain the cavity and path A only.

The size and shape of the cavities and gas diffusion paths can be related to the number of hydrogen bonds formed by the amino acids building up the paths. The modelled structures showed a competitive number of bonds as in the standard structures which shows the reliability of homology modelling (Figure 14-2). The number of hydrogen bonds formed by *C. variabilis* (27 and 18; Figure 14-2 (a) and 14-2 (b), respectively) were highest of all the modelled enzymes. Figure 14-3 shows gas diffusion paths in each of the modelled organisms in shape and size of the intramolecular hydrophobic cavities leading to the [NiFe] active site of the regulatory hydrogenase in *Ralstonia eutropha* H16 which are crucial for oxygen insensitivity.[42] Bingham et al.[43] were successful in achieving reduced oxygen sensitivity in CpI derived hydrogenases combining the effects of three mutations. Forestier et al.[44] identified a HydA2 gene encoding 505 amino acid proteins that was 68% identical to the known HydA1 hydrogenase from *C. reinhardtii*.

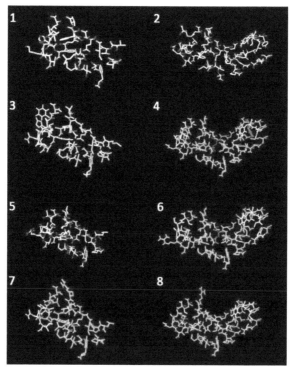

FIGURE 14-3 The figure shows gas diffusion paths within the proteins of the modelled organisms. (1) 1HFE; (2) 3LX4; (3) CA modelled based on 1HFE; (4) CA modelled based on 3C8Y/3LX4; (5) CV modelled based on 1HFE; (6) CV modelled based on 3C8Y/3LX4; (7) SO modelled based on 1HFE; (8) SO modelled based on 3C8Y/3LX4. CV: *C. variabilis*; CA: *C. acetobutylicum*, and SO: *S. obliquus*.

All the structures were then validated using PyMol analysis which revealed the Root Mean Square Deviation (RMSD) scores of all the modelled structures with respect to all the other structures. All the scores obtained were less than 3.00 units which according to Benkert et al.[45] prove that the modelling was efficient and reliable. The RMSD scores obtained for CV when valued against CpI was 0.08 and against *C. reinhardtii* and *D. vulgaris* were 0.08 and 0.09, respectively proving very close relationship with the standard organisms. *Clostridium acetobutylicum* when valued against CpI, *C. reinhardtii*, and *D. vulgaris* gave scores of 0.11, 0.1, and 0.09, respectively. Similarly, *Scenedesmus obliquus* when valued against CpI, *C. reinhardtii*, and *D. vulgaris* gave 0.08, 0.09, and 0.1 respectively. Hence, we conclude that the next best model organism that can be exploited for bio-hydrogen production through molecular modelling is the microalgae *Chlorella variabilis*.[46] It is followed by the other microalgae *Scenedesmus obliquus* and the bacteria *Clostridium acetobutylicum*. Figure 14-4 shows *Chlorella sp.* cultures being grown in MBM at 28 °C.

FIGURE 14-4 *Chlorella sp.* cultures grown in MBM at 28 °C (photo taken on 3/5/2015).

14.4   CONCLUSION

The study consolidates all the basic information with regard to the hydrogenase enzyme and its oxygen sensitivity which is one of the most primary reasons in efficient hydrogen production by algae and bacteria. By comparing a few major organisms which play a role in hydrogen production in a renewable manner, we try to provide information which helps scientists to choose the best model organism for biohydrogen research. We conclude that microalgae are dominant over bacteria as they are easy and economical to cultivate. Their nutrients limit to just salts which are cheaply available. There is no problem of contamination with microalgae as no glucose source is required for their growth.

ACKNOWLEDGMENTS

The authors gratefully acknowledge their technical support of Department of Psychopharmacology, National Institute of Mental Health and Neuro Sciences (NIMHANS), Bangalore. The authors would like to express their appreciation to Dr. Nityananda Pradhan, Mr. Shankarachariyar, Dr. Chittaranjan Andrade (NIMHANS) and Mr. Amit Kumar (CSIR-Indian Institute of Chemical Technology) for their guidance.

KEYWORDS

- BLAST
- H-Cluster
- Homology
- Hydrogenase
- Oxygen Sensitivity

REFERENCES

1. Sen, U.; Shakdwipee, M.; and Banerjee, R. Status of biological hydrogen production. *J. Sci. Ind. Res.* **2008**, *67*, 980–993.

2. Das, D.; and Veziroglu, T. N. Advances in biological hydrogen production processes. *Int. J. Hyd. Energy* **2008,** *33* (21), 6046–6057.

3. Kim, D. H.; and Kim, M. S. Hydrogenases for biological hydrogen production. *Bioresour. Technol.* **2011,** *102* (18), 8423–8431.

4. Jo, J. H.; Lee, D. S.; Park, D.; and Park, J. M. Biological hydrogen production by immobilized cells of *Clostridium tyrobutyricum* JM1 isolated from a food waste treatment process. *Bioresour. Technol.* **2008,** *99* (14), 6666–6672.

5. Hemschemeier, A.; Melis, A.; and Happe, T. Analytical approaches to photobiological hydrogen production in unicellular green algae. *Photosynth. Res.* **2009,** *102* (2–3), 523–540.

6. Kalia, V. C.; and Lal, S. Biological hydrogen production as a sustainable green technology for pollution prevention. In *VisionRI Nous: Discourse on Development in Asia, Pacific and Africa*; VisionRI: Haryana, 2005.

7. García, R. E.; Martínez, V. L.; Franco, J. I.; and Curutchet, G. Selection of natural bacterial communities for the biological production of hydrogen. *Int. J. Hyd. Energy* **2012,** *37* (13), 10095–10100.

8. Patel, S. K. S.; Kumar, P.; and Kalia, V. C. Enhancing biological hydrogen production through complementary microbial metabolisms. *Int. J. Hyd. Energy* **2012,** *2*, 1–14.

9. Levin, D. B.; Pitt, L.; and Love, M. Biohydrogen production, prospects and limitations to practical application. *Int. J. Hyd. Energy* **2004,** *29* (2), 173–185.

10. Hallenbeck, P. C.; and Benemann, J. R. Biological hydrogen production; fundamentals and limiting processes. *Int. J. Hyd. Energy* **2002,** *27*, 1185–1193.

11. Srirangan, K.; Pyne, M. E.; and Perry Chou, C. Biochemical and genetic engineering strategies to enhance hydrogen production in photosynthetic algae and cyanobacteria. *Bioresour. Technol.* **2011,** *102* (18), 8589–8604.

12. Gaffron, H.; and Rubin, J. Fermentative and photochemical production of hydrogen in algae. *J. Gen. Physiol.* **1942,** *26* (2), 219–240.

13. Lee, J. W.; and Greenbaum, E. A new oxygen sensitivity in photosynthetic H2 production. *Fuel Chem. Div. Repr.* **2002,** *47* (2), 761–765.

14. Greenbaum, E.; and Lee, J. Photosynthetic hydrogen and oxygen production by green algae. In *BioHydrogen*; Zaborsky, O., Benemann, J., Matsunaga, T., Miyake, J., and San Pietro, A., Eds.; Springer: US, 1998, 235–241.

15. Kapdan, I. K.; and Kargi, F. Bio-hydrogen production from waste materials. *Enzyme Microb. Technol.* **2006,** *38* (5), 569–582.

16. Melis, A.; and Happe, T. Update on hydrogen production: Hydrogen production. Green algae as a source of energy. *Plant Phisiol.* **2001,** *127*, 740–748.

17. Rosenbaum, M.; Schröder, U.; and Scholz, F. Utilizing the green alga *Chlamydomonas reinhardtii* for microbial electricity generation: A living solar cell. *Appl. Microbiol. Biotechnol.* **2005,** *68* (6), 753–756.

18. Wang, Y.; Ahmed, S.; Czernik, S.; Damle, A.; Vanderspurt, T.; and Hua, Y. *Hydrogen Production and Delivery: Summary of Annual Merit Review of the Hydrogen Production and Delivery Sub-Program*; U.S. Department of ENERGY: Washington, DC., 2010; pp 9–170.

19. Schütz, K.; Happe, T.; Troshina, O.; Lindblad, P.; Leitão, E.; Oliveira, P.; and Tamagnini, P. Cyanobacterial H2 production—a comparative analysis. *Planta* **2004,** *218* (3), 350–359.

20. Melis, A.; Zhang, L.; Forestier, M.; Ghirardi, M. L.; and Seibert, M. Sustained photobiological hydrogen gas production upon reversible inactivation of oxygen evolution in the green alga *Chlamydomonas reinhardtii*. *Plant Physiol.* **2000,** *122* (January), 127–136.

21. Ghirardi, M. L.; Zhang, L.; Lee, J. W.; Flynn, T.; Seibert, M.; Greenbaum, E.; and Melis, A. Microalgae: A green source of renewable H2. *Trends Biotechnol.* **2000,** *18*, 506–511.

22. Melis, A. Photosynthetic H2 metabolism in *Chlamydomonas reinhardtii* (unicellular green algae). *Planta* **2007,** *226* (5), 1075–1086.

23. Hemschemeier, A.; Jacobs, J.; and Happe, T. Biochemical and physiological characterization of the pyruvate formate-lyase Pfl1 of *Chlamydomonas reinhardtii*, a typically bacterial enzyme in a eukaryotic alga. *Eukaryot. Cell* **2008,** *7* (3), 518–526.

24. Niel, E. W. J. V.; Budde, M. A. W.; Haas, G. G. D.; Van Der Wal, F. J.; Claassen, P. A. M.; and Stams, A. J. M. Distinctive properties of high hydrogen producing extreme thermophiles, *Caldicellulosiruptor saccharolyticus* and *Thermotoga elÿi*. *Int. J. Hyd. Energy* **2002,** *27*, 1391–1398.

25. Benemann, J. Hydrogen biotechnology, progress and prospects. *Nat. Biotech.* **1996,** *14* (9), 1101–1103.

26. Levin, D. B.; Islam, R.; Cicek, N.; and Sparling, R. Hydrogen production by *Clostridium thermocellum* 27405 from cellulosic biomass substrates. *Int. J. Hyd. Ener.* **2006,** *31* (11), 1496–1503.

27. Ferreira, A. F.; Marques, A. C.; Batista, A. P.; Marques, P. A. S. S.; Gouveia, L.; and Silva, C. M. Biological hydrogen production by *Anabaena* sp. –yield, energy and CO2 analysis including fermentative biomass recovery. *Int. J. Hyd. Ene.* **2012,** *37* (1), 179–190.

28. Yasuhiro, F.; Yu-Jung, H.; Jhen-Wei, C.; Hung-Chun, L.; Liang-Ming, W.; Hsin, C.; Young-Chong, L.; and Chang, J. Material and energy balances of an integrated biological hydrogen production and purification system and their implications for its potential to reduce greenhouse gas emissions. *Bioresour. Technol.* **2011,** *102*, 8550–8556.

29. Ren, N.; Guo, W.; Liu, B.; Cao, G.; and Ding, J. Biological hydrogen production by dark fermentation: Challenges and prospects towards scaled-up production. *Curr. Opin. Biotechnol.* **2011,** *22* (3), 365–370.

30. Ozgur, E.; Mars, A. E.; Peksel, B.; Louwerse, A.; Yucel, M.; Gündüz, U.; Claassen, P. A. M.; and Eroğlu, İ. Biohydrogen Production from beet molasses by sequential dark and photofermentation. *Int. J. Hyd. Ene.* **2010,** *35* (2), 511–517.

31. Ghirardi, M. L.; Kosourov, S.; and Seibert, M. Cyclic photobiological algal H₂ -production. In *Proceedings of the 2002 U.S. DOE Hydrogen Program Review*; NREL: Golden, CO, 2001; p iii.

32. Yeh, K. L.; and Chang, J. S. Effects of cultivation conditions and media composition on cell growth and lipid productivity of indigenous microalga *Chlorella vulgaris* ESP-31. *Bioresour. Technol.* **2012,** *105*, 120–127.

33. Rupprecht, J.; Hankamer, B.; Mussgnug, J. H.; Ananyev, G.; Dismukes, C.; and Kruse, O. Perspectives and advances of biological H₂ production in microorganisms. *Appl. Microbiol. Biotechnol.* **2006,** *72* (3), 442–449.

34. Ghirardi, M. L.; Remick, R.; and Williams, T. Photobiological Production of Hydrogen. *Hydrog. Technol.* **2007,** NREL/FS-560-42285, 1–4.

35. Wykoff, D. D.; Davies, J. P.; Melis, and Grossman, A. R. The regulation of photosynthetic electron transport during nutrient deprivation in *Chlamydomonas reinhardtii*. *Plant Physiol.* **1998,** *117* (1327), 129–139.

36. Christine, A. The Anaerobic Life of the Photosynthetic Alga *Chlamydomonas Reinhardtii* Photofermentation and Hydrogen Production upon Das Anaerobe Leben Der Photosynthetischen Alga *Chlamydomonas Reinhardtii*. Ph.D. Dessertation, Ruhr University Bochum, August 2005, pp 1–141.

37. Cohen, J.; and Schulten, K. O_2 migration pathways are not conserved across proteins of a similar fold. *Biophys. J.* **2007**, *93* (10), 3591–3600.

38. Cohen, J.; Kim, K.; King, P.; Seibert, M.; and Schulten, K. Finding gas diffusion pathways in proteins: Application to O_2 and H_2 transport in CpI [FeFe]-hydrogenase and the role of packing defects. *Structure* **2005**, *13*, 1321–1329.

39. Seibert, M.; King, P.; Zhang, L.; Mets, L.; and Ghirardi, M. Molecular engineering of algal H_2-production. In *Proceedings of the 2002 U.S. DOE Hydrogen Program Review*; NREL: Golden, CO, 2002; *80401* (1993), pp 1–10.

40. Adams, M. W. W. The structure and mechanism of iron-hydrogenases. *Biochim. Biophys. Acta - Bioenerg.* **1990**, *1020* (2), 115–145.

41. Peters, J. W. Structure and mechanism of iron-only hydrogenases. *Curr. Opin. Struct. Biol.* **1999**, *9* (6), 670–676.

42. Buhrke, T.; Lenz, O.; Krauss, N.; and Friedrich, B. Oxygen tolerance of the H_2-sensing [NiFe] hydrogenase from *Ralstonia eutropha* H16 is based on limited access of oxygen to the active site. *J. Biol. Chem.* **2005**, *280* (25), 23791–23796.

43. Bingham, A. S.; Smith, P. R.; and Swartz, J. R. Evolution of an [FeFe] hydrogenase with decreased oxygen sensitivity. *Int. J. Hydro. Ene.* **2012**, *37* (3), 2965–2976.

44. Forestier, M.; King, P.; Zhang, L.; Posewitz, M.; Schwarzer, S.; Happe, T.; Ghirardi, M. L.; and Seibert, M. Expression of two [Fe]-hydrogenases in *Chlamydomonas reinhardtii* under anaerobic conditions. *FEBS J.* **2003**, *2758*, 2750–2758.

45. Benkert, P.; Schwede, T.; and Tosatto, S. C. E. QMEANclust: Estimation of protein model quality by combining a composite scoring function with structural density information. *BMC Struct. Biol.* **2009**, *9* (35), 1–17.

46. Nelabhotla, A. B. T.; Sahu, N.; and Pradhan, N. Microalgae as model organism for biohydrogen production : A study on structural analysis. *Int. J. Curr. Microbiol. Appl. Sci.* **2014**, *3* (7), 481–494.

PART IV
Microorganisms in the Environment: Role and Industrial Applications

CHAPTER 15

YEAST *SACCHAROMYCES CEREVISIAE* AS AN EFFICIENT BIOLOGICAL AGENT FOR THE DECOLORIZATION OF REACTIVE DYES USED IN TEXTILE INDUSTRY

B. BHIMA[1*], N. HANUMALAL[1], and B. CHANDRASEKHAR[1]

[1]*Department of Microbiology, University College of Science, Osmania University, Hyderabad, Telangana – 500007, India.*

Corresponding author: B. Bhima. Email: bhima.ou@gmail.com

CONTENTS

Abstract 298

15.1 Introduction...298

15.2 Materials and Methods 300

15.3 Results and Discussion..302

15.4 Conclusion 305

Keywords 306

References..306

ABSTRACT

In textile industries, reactive dyes are extensively used, which are recalcitrant xenobiotic compounds causing a serious damage to the environment. In our study, 85 yeasts were isolated from different sources collected from various regions of the Telangana State, India. The isolates were morphologically, culturally, and biochemically identified as *Saccharomyces* species, *Candida krusei*, and two unidentified strains. Among the 85 yeast strains, only 5 strains were able to degrade 2 textile reactive dyes, namely Reactive Red 4 and Reactive Green 19 at a concentration of 100 mgl⁻¹ within 36 h of incubation. The experiments were carried out to determine some parameters such as decolorization, pH, temperature, and initial dye concentration. *Saccharomyces* species (OHB11) was highly efficient toward two reactive dyes tested, Reactive Red 4 and Reactive Green 19, demonstrating the potential as a biological agent for the treatment of textile effluent. The decolorization was ranging from 70–90%. Complete decolorization was observed within 12 and 16 h depending on the inoculum size, temperatures (40 °C), and pH (5.0).

15.1   INTRODUCTION

Reactive dyes are xenobiotic compounds whose degradation in the environment is quite difficult. Reactive dyes are chemical dyes with numerous varieties of colors and for that reason, they have been broadly used by the textile industries. These dyes are characterized by the presence of one or more azo linkages (R1 − N = N − R2) and aromatic structures which account for the recalcitrant nature and hence are not biodegradable.[1,2] Textile industry is a very important segment of the Indian economy, with considerable growth in the past few years. As a consequence, there is an increase of an environmental pollution caused by the huge amount of dyes involved in the textile industries that are discharged in the form of liquid effluents. Around 10–15% of the dyes are released into the environment.[3] The textile industry is one of the major polluters in the world with the release of waste dyes and auxiliary products to the groundwater and river environments. They are toxic to aquatic life because of the presence of heavy metals, chlorides, etc., in effluents.[4] Recently, the existing physico-chemical methods for the removal of dyes from the effluents suffer strict issues such as low efficiency and high operation cost.[5] The biological effects of the reactive dyes before degradation have been shown to be toxic.[6] These compounds are carcinogenic and mutagenic.[7,8] In the bioaccumulation process, the first step is a rapid buildup of contaminants on the cell surface which is independent of temperature and metabolism. The second step depends upon metabolism which allows the accumulation of a large number of cells in the industrial effluents. Bioaccumulation takes place for the removal of different types of dyes. The cells need sufficient sources of nitrogen and carbon for absorption in the growth medium instead of the dyes. The use of microbial cells provides an advantage over dead or resting cells because the loss and cell growth takes place to avoid simultaneous and cell yield separately before the treatment of industrial effluents. Moreover, other components or issues in waste water can also be toxic to living cells such as extreme pH, high salt concentration, etc. If the problem of dye toxicity for microbial cell growth is solved by employing dye resistant microorganisms, the renewable system can operate continuously for extended periods.[9–12]

There are many valuable waste water treatments. As these technologies are extremely costly, biological treatments are conventionally applied. A few textile dyes are moderately resistant to microbial degradation, besides anaerobic microorganisms when degrading various dyes produce aromatic amines that are toxic and carcinogenic.[13] Therefore, in modern years, intensive research on fungal decolorization of textile waste water is being carried out. The use of fungi is a promising alternative for the replacement or additional treatment of industrial effluents.[14 16]

From the 1970s, efforts have been started for isolating dye-degrading bacteria. In 1977, *Bacillus subtilis* was first isolated, followed by *Aeromonas hydrophila* in 1978, and in 1980s *Bacillus cereus* was isolated. After that, several strains like *Pseudomonas* and *Aeromonas* were reported as dye-degraders.[17] In recent times, bacterial cultures which are competent of degrading reactive dyes were listed.[17–19] Aerobically and anaerobically, numerous studies have demonstrated the capability of bacteria in monoculture for the degradation of reactive dyes.[17] The selected microorganisms have showed effective decolorization of dyes and their adaptability in the textile effluents. *Bacillus* sp. which is highly alkali and thermo stable, was isolated from the wastewater drain of a textile industry and was found to be an effective biological agent to degrade reactive dye.[20] An endospore forming bacterial isolate was isolated from the textile industry for waste water treatment to decolorize the reactive dye remazol black B.[21] Many microbes have developed an enzyme system for the mineralization and decolorization of reactive dyes under certain environmental conditions.[22–24]

From the present study, we infer that the effective biological decolorization and degradation method of textile dyes especially Reactive Red 4 and Reactive Green 19 was possible by using an efficient microorganism i.e., a yeast isolate *Saccharomyces cerevisiae* for the treatment of textile effluents.

15.2 MATERIALS AND METHODS

15.2.1 ISOLATION AND CHARACTERIZATION OF YEAST

Eighty five yeast strains were isolated from different samples from various regions of the Telangana State, India. One yeast strain was procured from MTCC (Microbial Type Culture Collection, Chandigarh, India). These isolates were maintained on YEPD Broth (yeast extract 10 g L^{-1}, peptone 20 g L^{-1}, and dextrose 20 g L^{-1}), and YEPD Agar (22.5 g L^{-1}). The isolates were characterized and screened morphologically, microscopically, and biochemically. Finally, the isolates were characterized by sequencing their 26S rRNA. The characterized isolates were maintained on slants temporarily at 4 °C and preserved as glycerol stocks. For the dye decolorization studies, the growth medium used was (g L^{-1}) yeast extract, 0.2; glucose, 8; KH_2PO_4, 1; $(NH_4)_2SO_4$, 1; $MgSO_4 \cdot 7H_2O$, 0.2, and the pH was adjusted to 5.5–6.0. A loopful of culture was inoculated in 50 ml conical flask containing 10 ml medium. These flasks were incubated at 40 °C in an orbital shaker incubator (150 rpm) for 24 h.

15.2.2 REACTIVE DYES

Two reactive dyes, Reactive Red 4 ($C_{32}H_{19}CIN_8O_{14}S_4Na_4$) and Reactive Green 19 ($C_{40}H_{23}CI_2N_{15}Na_6O_{19}S_6$) were selected for the decolorization studies (Figure 15-1). The selected technical grade dyes were procured from MP Biomedicals, liquid-liquid chromatography (LLC), as they are used in the textile industry in different countries. Stock solutions of the dyes were prepared by 1000 mg L^{-1} concentration.

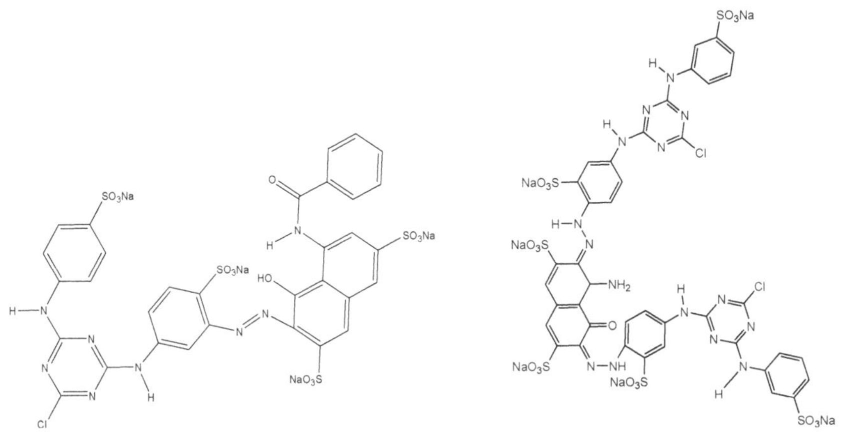

Reactive Red 4 ($C_{32}H_{19}ClN_8O_{14}S_4Na_4$)   **Reactive Green 19** ($C_{40}H_{23}Cl_2N_{15}Na_6O_{19}S_6$)

FIGURE 15-1 Chemical structures of dyes used in biodegradation experiments.

15.2.3 DYE DECOLORIZATION

The medium for dye decolorization with 100 mg L^{-1} dye concentration was taken. The dyes were filtered and sterilized separately in addition to the medium. 10 ml of sterile medium without the dye was inoculated with a loopful of yeast culture in 50 ml Erlenmeyer flasks and incubated for 24 h. Further, 1 ml of 24 h grown culture was inoculated into the medium with 100 mg L^{-1} dye. The flasks were incubated in a shaking incubator at 40 °C/150 rpm for 24 h. A flask containing the same medium without the inoculation of yeast culture was treated as the control.

15.2.4 OBSERVATION OF DYE DEGRADATION

After incubation, the culture medium was spinned at 5000 rpm and the supernatant was collected for the observation of dye decolorization in a spectrophotometer at 520 nm. Color disappearance was reported as 100%, by the following formula:

$$\text{Decolourisation } 100\% = \frac{(A_0 - A_t)}{A_0 \times 100}$$

where, A_0 and A_t were the absorbence of the dye solution initially and at cultivation time (t), respectively. All the decolorization experiments were carried out in triplicates.

15.2.5   IDENTIFICATION OF DYE DEGRADING YEAST

Isolated yeast strains with high dye decolorization properties were first identified according to their respective morphological and physiological characteristics.[17] In further studies, the isolated yeast strains were identified as reactive dye degrading yeast by decolorization of the dye and by comparing with the dye degrading MTCC strain (MTCC2665).

15.3   RESULTS AND DISCUSSION

15.3.1   ISOLATION AND SCREENING OF YEAST FOR DYE DECOLORIZATION

In the current study, Reactive Red 4 and Reactive Green 19 were selected for determining the decolorization ability from the 85 yeast strains. The reactive dyes were chosen as they are extensively used in the textile industry.[25] The color of the decolorized culture medium and the yeast cells were observed. Although these observations could not give the imminent rate of color removal, they accomplish at least partial information on the decolorization ability of the yeast strains tested (Table 15-1). The results of decolorization of Reactive Red 4 and Reactive Green 19 reveal that only four yeast isolates could decolorize the dye concentration of up to 100 mg L^{-1}. These strains had noticeable effect on the decolorization of Reactive Red 4 and Reactive Green 19. The laboratory isolates of yeast strains *Saccharomyces cerevisiae* OHB11 (Osmania Hanuma Bhima), *Saccharomyces italicus* OHB2, *Candida krusei* MTCC2665, and two unidentified strains (Table 15-1) showed the ability to decolorize the dye in high concentration. In all the cases of decolorization, the yeast cells absorbed the dye and appeared in color. Along with the five strains, another 12 yeast isolates (not listed), had affected dye decolorization, and even the cells of these strains absorbed a small amount of Reactive Red 4. This result was in accordance with the earlier reports.[26–28]

TABLE 15-1 Preliminary Observations on Decolorization Characteristics of Five Yeast Strains for Reactive Red 4 at 100 mg L⁻¹

S. No.	Strain Numbers	Strain Name	Color of the Medium	Color of the Yeast Cells	Color of the Yeast Cells in the Controls
1	OHB11	*Saccharomyces cerevisiae*	Pink	Red	White
2	OHB2	*Saccharomyces italicus*	Colorless	Red	White
3	2665	*Candida krusei*	Colorless	Red	White
4	K6	Unidentified	Pink	White	White
5	S18	Unidentified	Brown	Red	White

To screen out the best strain for decolorization of both the dyes Reactive Red 4 and Reactive Green 19 at a concentration of 100 mg L⁻¹, the results were compared. On comparison with Reactive Green 19, 90% Reactive Red 4 was decolorized by *Saccharomyces cerevisiae* (OHB11) (Figure 15-2). All the five strains had the ability of decolorizing 70% dye in 18 h. Strains like OBH11 and OBH2 were more active at removing 90% and 85% of Reactive Red 4 dye, respectively in 18 h of incubation.

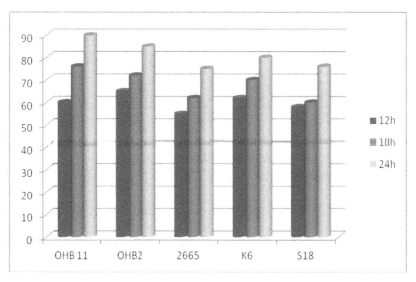

FIGURE 15-2 Comparison of the decolorization of Reactive Red 4 at 100 mg L⁻¹ by five yeast strains at different time intervals.

15.3.2   IDENTIFICATION OF YEAST ISOLATES

The isolated strains (OHB11 and OHB2) from different sources were identified by morphological, biochemical, and molecular characteristics (Table 15-2). The OBH11 was identified as *Saccharomyces cerevisiae* and OHB2 was identified as *Saccharomyces italicus*. Molecular level characterization was carried out by phylogenetic analysis based on 26S rRNA gene sequences and identified as being most closely related to *Saccharomyces* sp.OBH 11 and OBH 2, and the same were submitted to the NCBI data base.

TABLE 15-2   Biochemical Characterization of Yeast Isolates Based on their Sugar Fermentation Ability

S. No	Characterization	OHB11	OHB2	2665	K6	S18
1	Glucose	+	+	+	+	+
2	Sucrose	+	–	–	+	–
3	Maltose	+	+	–	+	–
4	Fructose	+	+	–	–	+
5	Cellobiose	–	–	+	–	–
6	D - Ribose	+	–	–	–	–
7	Lactose	–	–	–	–	+
8	Xylose	–	–	–	–	–

+ Positive for utilization.

– Negative for utilization.

15.3.3   EFFECT OF YEAST ON DIFFERENT DYE CONCENTRATIONS

Only Reactive Red 4 was taken for determining the effect of yeast on different dye concentrations. To evaluate the effect on different dye concentrations, five yeast strains were tested against different concentrations of the dye ranging from 100–1000 mg L^{-1} for 24 h (Figure 15-3). Both OHB11 and OHB2 were able to decolorize up to 80% of Reactive Red 4 at a concentration of 200 mg L^{-1} in 24 h. Above a concentration of 200 mg L^{-1}, the decolorization rate of the dye decreased periodically till 1000 mg L^{-1} concentration of the dye. Only about 20% decolorization was achieved by these strains at 1000 mg L^{-1}.

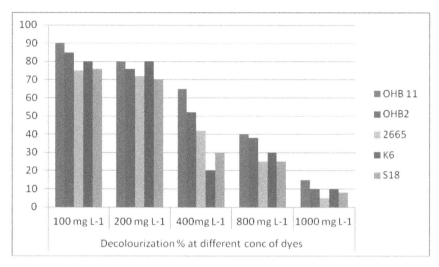

FIGURE 15-3   Effect of yeast strains on different concentrations of Reactive Red 4 dye.

15.4   CONCLUSION

In the present study, a total of 85 yeasts were isolated for the decolorization of reactive dyes. Among the 85 yeasts, 4 isolates namely, OBH11 (*Saccharomyces cerevisiae*), OBH2 (*Sacharomyces italicus*), and two unidentified strains were selected as efficient dye degraders. In summary, the yeast strains *Saccharomyces cerevisiae* (OHB11) and *Saccharomyces italicus* (OHB2) were capable of decolorizing the dyes, which shows a significant capability for removing multiple types of dyes, mostly the reactive dye types. A positive control experiment was carried out with *Candida krusei* (MTCC2665). The OBH11 strain showed 90% removal of Reactive Red 4 in 24 h, whereas OBH2 showed 85% compared with the positive control (MTCC2665) where it showed only 75% dye decolorization. The OBH11 and OBH2 strains were able to decolorize upto 40–65% dye at high concentrations of 400–800 mg L^{-1} in 24 h.

KEYWORDS

- **Decolorization**
- **Reactive dyes**
- *Saccharomyces cerevisiae*
- **Textile effluent**

REFERENCES

1. He, F.; Hu, W. R.; and Li, Y. Biodegradation mechanisms and kinetics of azo dye 4BS by a microbial consortium. *Chemosphere* **2004,** *57,* 293–301.
2. Tan, N. C. G.; Prenafeta-Boldu, F. X.; Opsteeg, J. L.; Lettinga, G.; and Field, J. A. Biodegradation of azo dyes in cocultures of anaerobic granular sludge with aerobic aromatic amine degrading enrichment cultures. *Appl. Microbiol. Biotechnol.* **1999,** *51,* 865–871.
3. Singh, H. *Mycoremediation: Fungal remediation;* Wiley-Interscience: New Jersey, 2006.
4. Clarke E. A.; and Anliker, R. Organic Dyes and Pigments. In *Hand-book of Environmental Chemistry, Part (A) Anthropogenic Compounds*; Springer: New York, 1980; 181.
5. Khehra M. S.; Saini H. S.; Sharma D. K.; Chadha, B. S.; and Chimni, S. S. Comparative studies on potential of consortium and constituent pure bacterial isolates to decolorize azo dyes. *Water Res.* **2005,** *39,* 5135–5141.
6. Ramsay, J. A.; and Nguyen, T. Decoloration of textile dyes by Trametes versicolor and its effect on dye toxicity. *Biotechnol. Lett.* **2002,** *24,* 1757–1761.
7. El-Rahim, W. M. A.; and Moawad, H. Enhancing bioremoval of textile dyes by eight fungal strains from media supplemented with gelatine wastes and sucrose. *J. Basic Microbiol.* **2003,** *43,* 367–375.
8. Novotny, C.; Dias, N.; Kapanen, A.; Malachova, K.; Vandrovcova, M.; Itavaara, M.; and Lima, N. Comparative use of bacterial, algal and protozoan tests to study toxicity of azo and anthraquinone dyes. *Chemosphere* **2006,** *63,* 1436–1442.
9. Renganathan, S.; Thilagaraj W. R.; Miranda, L. R.; Gautam, P.; and Velan, M. Accumulation of acid orange 7, acid red 18 and reactive black 5 by growing Schizophyllum commune. *Bioresour. Technol.* **2006,** *97,* 2189–2193.
10. Aksu, Z.; and Donmez, G. Combined effects of molasses sucrose and reactive dye on the growth and dye accumulation properties of *Candida tropicalis. Process Biochem.* **2005,** *40,* 2443–2454.
11. Dursun, A. Y.; Uslu, G.; Cuci Y.; and Aksu, Z. Bioaccumulation of copper (II), lead (II) and chromium (IV) by growing *Aspergillus niger. Process Biochem.* **2003,** *38,* 1647–1651.
12. Malik, A. Metal bioremediation through growing cells. *Environ. Int.* **2004,** *30,* 261–278.
13. Yesilada, O.; Cing, S.; and Asma, D. Decolourisation of the textile dye Astrazon Red FBL by *Funalia trogii* pellets. *Biores. Technol.* **2002,** *81,* 155–157.

14. Conneely, A.; Smyth, W. F.; and McMullan, G. Metabolism of the phthlocyanine textile dye remazol turquoise blue by *Phanerochaete chrysosporium*. *FEMS Microbiol. Lett.* **1999,** *179*, 333–337.

15. Dos Santos, A. Z.; Neto, J. M. C.; Tavares, C. R.; and da Costa, M. G. Screening of filamentous fungi for the decolorization of a commercial reactive dye. *J. Basic Microbiol.* **2004,** *44*, 288–295.

16. Fu, Y.; and Viraraghavan, T. Fungal decolorization of dye wastewaters: A review. *Biores. Technol.* **2001,** *79*, 251–262.

17. Banat, I. M.; Nigam, P.; Singh, D.; and Marchant, R. Microbial decolorization of textile-dyecontaining effluents: A review. *Bioresour. Technol.* **1996,** *58*, 217–227.

18. Pearce, C. I.; Lloyd, J. R.; and Guthrie, J. T. The removal of color from textile wastewater using whole bacterial cells: A review. *Dyes Pig.* **2003,** *58*, 179–196.

19. Forgacs, E.; Cserhati, T.; and Oros, G. Removal of synthetic dyes from wastewaters: A review. *Environ. Int.* **2004,** *30*, 953–971.

20. Maier, J.; Kandelbauer, A.; Erlacher, A.; Cavaco-Paulo, A.; and Gubitz, G. M. A new alkali thermostable azo reductases from *Bacillus* sp. Strain SF. *Appl. Environ. Microbiol.* **2004,** *70*, 837–844.

21. Meehan, C.; Banat, I. M.; McMullan, G.; Nigam, P.; Smyth, F.; and Marchant, R. Decolorization of remazol Black-B using a thermotolerant yeast, *Kluyveromyces marxianus* IMB3. *Environ. Inter.* **2000,** *26* (1–2), 75–79.

22. Khalid, A.; Arshad, M.; and Crowley, D. E. Accelerated decolorization of structurally different azo dyes by newly isolated bacterial strains. *Appl. Microbiol. Biotechnol.* **2008,** *78*, 361–369.

23. Hao, J. J.; Song, F. Q.; Huang, F.; Yang, C. L.; Zhang, Z. J.; Zheng, Y.; and Tian, X. J. Production of laccase by a newly isolated deuteromycete fungus *Pestalotiopsis* sp. and its decolorization of azo dye. *J. Ind. Microbiol. Biotechnol.* **2007,** *34*, 233–240.

24. Pandey, A.; Singh, P.; and Iyengar, L. Bacterial decolourisation and degradation of azo dyes. *Int. Biodeter. Biodeg.* **2007,** *59*, 73–84.

25. Hu, T. L. Removal of reactive dyes from aqueous solution by different bacterial genera. *Wat. Sci. Technol.* **1996,** *10*, 89–95.

26. Knapp, J. S.; and Newby, P. S. The microbiological decolorization of an industrial effluent containing a diazo-linked chromophore. *Water Res.* **1995,** *29*, 1807–1809.

27. Sani, R. K.; and Banerjee, U. C. Decolorization of triphenylmethane dyes and textile and dye-stuff effluent by *Kurthia* sp. *Enzyme Microb. Technol.* **1999,** *24*, 433–437.

28. Chen, K. C.; Wu, J. Y.; Liou, D. J.; and Hwang, S. C. Decolorization of the textile dyes by newly isolated bacterial strains. *J. Biotechnol.* **2003,** *101*, 57–68.

CHAPTER 16

METABOLITE PROFILING AND BIOLOGICAL ACTIVITIES OF EXTROLITES FROM *ASPERGILLUS TURCOSUS* STRAIN KZR131 ISOLATED FROM KAZIRANGA NATIONAL PARK, ASSAM, INDIA

C. GANESH KUMAR[1*], POORNIMA MONGOLLA[1,5], JAGADEESH BABU NANUBOLU[2], PATHIPATI USHA RANI[3], and KUMAR KATRAGUNTA[4]

[1]*Medicinal Chemistry and Pharmacology Division, CSIR-Indian Institute of Chemical Technology, Uppal Road, Hyderabad 500007, Telangana, India.*

[2]*Laboratory of X-ray Crystallography, CSIR-Indian Institute of Chemical Technology, Uppal Road, Hyderabad 500007, Telangana, India.*

[3]*Biology and Biotechnology Division, CSIR-Indian Institute of Chemical Technology, Uppal Road, Hyderabad 500007, Telangana, India.*

[4]*Natural Products Chemistry Division, CSIR-Indian Institute of Chemical Technology, Uppal Road, Hyderabad 500007, Telangana, India.*

[5]*Acharya Nagarjuna University, Guntur, Andhra Pradesh – 522510, India.*

*Corresponding author: Dr. C. Ganesh Kumar. E-mail: cgkumar@iict.res.in

CONTENTS

Abstract 311

16.1 Introduction 311

16.2 Materials and Methods 313

16.3 Results and Discussion 318

16.4 Conclusion 330

Acknowledgments 330

Keywords 332

References 332

ABSTRACT

In an ongoing survey for the bioactive potential of microorganisms isolated from soil and dung samples collected from the biodiversity hotspot, the Kaziranga National Park, Assam, India, an *Aspergillus turcosus* strain Kaziranga-131 (KZR131) was isolated from forest soil, exhibiting broad spectrum antimicrobial, antifeedant, and insecticidal activities. Bioactivity-guided purification resulted in the isolation of two compounds. Based on ^{1}H and ^{13}C NMR, Fourier transform infrared spectrum (FT-IR), and mass spectroscopic techniques, the chemical structures were elucidated as glio-toxin and acetylgliotoxin. The crystal structure for acetylgliotoxin has been established for the first time with the accession number Cambridge Crystallographic Data Center (CCDC) 1054601. Gliotoxin and acetyl-gliotoxin exhibited a broad spectrum antimicrobial activity against Gram-positive and Gram-negative bacteria, methicillin-resistant *Staphylococcus aureus* (MRSA), vancomycin-resistant Enterococci (VRE), and different *Candida* strains. Gliotoxin showed a dose-dependent antifeedant activity of 76.4% against *Spodoptera litura* and 66.4% against *Achaea janata*, respectively, at 60 µg cm^{-2} area of castor leaf. The LC_{50} (50% lethal concentration) values of gliotoxin on *S. oryzae, T. casteneum*, and *C. chinensis* after 96 h were 5.52, 3.57, and 4.25, respectively. However, acetylgliotoxin did not exhibit antifeedant and insecticidal properties. To our knowledge, this is the first report on extrolites produced by *Aspergillus turcosus* strain KZR131 exhibiting broad spectrum antimicrobial activity, and promising antifeedant and insecticidal activities against several agriculturally important pests.

16.1 INTRODUCTION

Approximately one-third of the global food production is destroyed annually by field and storage pests.[1] Despite the use of expensive and often environmentally hazardous control measures, insects remain the chief pests of crops and stored products.[2] Synthetic pesticides are currently the most effective means of pest control; however, the emergence of insect resistance[3] and other negative side effects has prompted the search for new alternatives. With regard to this, a renewed interest has been generated in the recent years on natural agroactive compounds to overcome the resistance and pollution problems accompanied because of the use of synthetic pesticides[4]. Further, there is a worldwide attention toward pest-management products based on natural products, having salient features such as low or no toxicity towards non-target organisms, availability from low-cost and renewable sources, and biodegradable as compared with their chemical synthetic counterparts. Many bacteria and fungi have been identified to display bioactivity against plant pests and diseases, mediated via secondary metabolites secreted in the environment.[5] These metabolites can affect several traits of insect biology, such as toxicity, mortality, growth inhibition, feeding activity, suppression of reproductive behavior, and reduction of fecundity and fertility.[6] The antifeedant activity is considered important from the perspective of pest management as the compounds inhibit the feeding capacity of the insects, thereby causing their starvation and death. Very few microbial metabolites are reported to exhibit antifeedant and insecticidal activities toward insect pests.[7–9] The natural ability of certain microorganisms to produce secondary metabolites that may be toxic toward insect pests prompted us to explore the bioactivity of these compounds against insect pests.

Bioprospecting is a term coined to refer to the search for novel products or microorganisms of economic importance from the world's biota. Although many early searches for bioactive compounds focused on soil microorganisms, the rate of discovery of interesting novel compounds from soil has diminished. In the recent years, research explorations on the isolation of microorganisms has focused on novel unexplored niches and/or extreme environments such as the tropical rain forests,[10] the marine sponges,[11] the mangroves,[12] Antarctica,[13] and Ladakh[14] regions for finding novel bioactive compounds exhibiting diverse biological activities. It is presumed that intensive sampling of unique habitats in a defined area will aide in the discovery of the less or undescribed fungi.[15] The Kaziranga National Park (26° 30'–26° 45'N and 93° 05'–93° 40'E), Assam, India is declared as a World Heritage site by *United Nations Educational, Scientific and Cultural Organization* (UNESCO) in 1985, which is located on the Southern bank of the River Brahmaputra at the foot of the Mikir—Kirbi Anglang Hills. This park is one of the last areas in Eastern India which is almost undisturbed by human with a unique natural riverine landscape of sheer forest, tall elephant grass, rugged reeds, marshes and shallow pools inhabited by the world's largest population of one-horned rhinoceroses and diverse range of other wild animals (http://whc.unesco.org/archive/periodicreporting/APA/cycle01/section2/337-summary.pdf). We conducted an intensive screening program on the isolation of fungi from different soil and dung samples collected from this hitherto unexplored biosphere and further tested the fungal culture extracts for antimicrobial activities.[16] The present study constitutes the purification, characterization, and structural elucidation of bioactive extrolites produced by a newly isolated *Aspergillus turcosus* strain Kaziranga-131 (KZR131) from the forest soil sample collected from this park and further assessed for their antimicrobial, antifeedant, and insecticidal activities.

16.2 MATERIALS AND METHODS

16.2.1 ISOLATION, IDENTIFICATION AND FERMENTATION CONDITIONS OF THE FUNGAL STRAIN KZR131

The fungal strain KZR131 was isolated from a forest soil sample collected from the Kaziranga National Park, Assam, India, as part of a bioactives screening program.[16] It was identified based on the analysis of the DNA sequences of the ITS1–5.8S–ITS2, ITS regions of the rRNA gene. Genomic DNA was extracted from the fungal mycelium cultured on potato dextrose agar (PDA) using ZR fungal/bacterial DNA identification kit (Zymo Research, Irvine, CA, USA). The primers ITS1 (TCC GTA GGT GAA CCT GCG G) and ITS4 (TCCTCCGCTTATTGATATGC) were used to amplify the ITS regions from the DNA extract. The PCR reaction was performed on an Eppendorf mastercycler epgradient S (Eppendorf AG, Hamburg, Germany) with the following cycles: (1) 94 °C for 5 min; (2) 45 cycles of 94 °C for 30 sec, 56 °C for 30 sec, and 72 °C for 30 sec, and (3) 72 °C for 7 min. The PCR product, spanning approximately 500–600 bp was checked on 1% agarose electrophoresis gel. It was then purified using Quick spin column and buffers (washing buffer and elution buffer) according to the manufacturer's protocol of GenElute gel extraction kit (Sigma-Aldrich, St. Louis, MO, USA). DNA sequencing was done using ABI 3130 Genetic Analyzer system (Applied Biosystems, CA, USA). The pure culture was maintained in the in-house culture repository of the institute with the accession number ICTF-028. The strain KZR131 was precultured aerobically in 4 × 1000 ml Erlenmeyer flasks containing 250 ml of potato dextrose broth (PDB) medium and agitated at 150 rev min^{-1} in an orbital shaker at 30 °C for 96 h. The fermented medium was later filtered through a muslin cheese cloth to remove the fungal biomass and then the infiltrate was centrifuged at 2000 rpm to obtain a cell-free supernatant.

16.2.2 EXTRACTION AND PURIFICATION OF EXTROLITES

The extrolite was extracted from the cell-free supernatant by absorption onto Diaion HP-20 (3%, Supelco, Bellafonte, PA, USA) resin. The resin was washed with water followed by extraction with methanol to obtain the crude extract fractions. The fractions were analyzed by thin layer chromatography (TLC) on silica gel 60 plates (F_{254}, Merck) and developed in ethyl acetate-hexane solvent mixture (50:50, v/v). The plates were visualized under the ultraviolet (UV) light at 254 nm which revealed the presence of a single compound. Further, the crude extract fractions containing the extrolite were pooled, concentrated under reduced pressure on a rotary vacuum evapo-rator (Rotavapor R-205, Büchi, Switzerland), and further profiled on silica gel (60–120 mesh) column (3 × 60 cm). The extrolite was eluted with a linear gradient of ethyl acetate-hexane solvent mixture (20:80, v/v) and after drying, the extrolite appeared as white crystals. This compound was further fractionated using a preparative RP HPLC (Gilson GX-281, Gilson, Inc., Middleton, WI, USA) equipped with a 155 UV/Vis detector, autosampler, and fraction collector, and interfaced with a Trilution LC software v2.1 on a Fortis C_{18} column (Waters) using acetonitrile and water (6:4, v/v) to yield two compounds.

16.2.3 STRUCTURAL CHARACTERIZATION

The HPLC purified extrolites were further subjected to ^1H and ^{13}C NMR, FT-IR, and electrospray ionization mass spectrometry (ESI-MS) studies for structural elucidation. The UV and visible spectra were measured by dissolving the sample in spectroscopic acetonitrile and recorded at 30 °C on a UV-visible double beam spectrophotometer (Lambda 25, Perkin-Elmer, Shelton, CT). The ^1H and ^{13}C NMR spectra were recorded on a Bruker Avance 300 MHz NMR spectrometer (Bruker, Switzerland) in deuter-ated chloroform at room temperature. The chemical shifts were given on a δ (ppm) scale with the solvent signal as reference and tetramethylsilane (TMS) as an internal standard. The Fourier transform infrared spectrum (FT-IR) was taken on a Thermo-Nicolet Nexus 670 FT-IR spectrophotom-eter (Thermo Fisher Scientific Inc., Madison, WI, USA) using KBr pellets and the spectrum was collected at a resolution of 4 cm^{-1} in the wave number region of 400–4,000 cm^{-1}. The high resolution mass spectrum (HR-MS) was recorded on a QSTAR XL Hybrid ESI-Q TOF mass spectrometer (Applied Biosystems Inc., Fosters City, CA, USA). An Electrothermal Digital 9000 Series melting point apparatus (Model IA9200, Barnstead, UK) was used for determining the melting point of the purified compounds. The puri-fied sample (1 mg) was placed in a glass capillary tube and this tube was placed in an aluminum heating block which was heated at a ramp rate of

10 °C min^{-1} from a temperature range of 25 to 300 °C for melting point measurement.

16.2.4 X-RAY CRYSTALLOGRAPHIC STUDIES

The purified extrolites were crystallized from the solvent mixture of hexane and ethyl acetate (60:40). Fine needles were grown on solvent evaporation after 3–4 days. A crystal suitable for the single crystal X-ray diffraction was selected and the X-ray crystal data for gliotoxin and acetylgliotoxin were collected at room temperature using a Bruker Smart Apex charge coupled device (CCD) diffractometer with graphite monochromated MoKα radiation (λ = 0.71073Å) using the ω-scan method.[17] Preliminary lattice parameters and orientation matrices were obtained from four sets of frames. The unit cell dimensions were determined using 8106 reflections for gliotoxin and acetylgliotoxin. Integration and scaling of the intensity data was accomplished using the SAINT program.[17] The structures were solved by Direct Methods using SHELXS97[17] and the refinement was carried out by full-matrix least-squares technique using SHELXL97.[18] Anisotropic displacement parameters were included for all non-hydrogen atoms. All H atoms were positioned geometrically and treated as riding on their parent C atoms, with the C-H distances of 0.93–0.97 Å, and with U_{iso}(H) = 1.2U_{eq} (C) or 1.5U_{eq} for methyl atoms. The O7-H7O bond distance was restrained to a target value of 0.90 Å.

16.2.5 ANTIMICROBIAL ACTIVITY

The antimicrobial activity was determined using the broth dilution method as described previously.[19] The target strains used for screening antimicrobial activity were procured from the Microbial Type Culture Collection (MTCC) and Gene Bank (CSIR-Institute of Microbial Technology, Chandigarh, India). These included *Micrococcus luteus* MTCC 2470, *Staphylococcus aureus* MTCC 96, *Staphylococcus aureus* MLS16 MTCC 2940, *Bacillus subtilis* MTCC 121, *Escherichia coli* MTCC 739, *Pseudomonas aeruginosa* MTCC 2453, *Klebsiella planticola* MTCC 530, and different *Candida* strains such as *Candida albicans* MTCC 183, *C. albicans* MTCC 227, *C. albicans* MTCC 854, *C. albicans* MTCC 1637, *C. albicans* MTCC 3017, *C. albicans* MTCC 3018, *C. albicans* MTCC 3958, *C. albicans* MTCC 4748, *C. albicans* MTCC 7315, *Candida parapsilosis* MTCC 1744, and *Issatchenkia orientalis* MTCC 3020. Other bacterial strains included *Staphylococcus aureus* ATCC 29213 and *Escherichia coli* ATCC 25922, and clinical strains of methicillin-resistant *Staphylococcus aureus* (MRSA) strain 15187 and vancomycin-resistant *Enterococci* (VRE) strain SP-346.

16.2.6 ANTIFEEDANT ACTIVITY

The antifeedant activity was tested against the larvae of two major agricultural pests, namely castor semilooper pest, *Achaea janata* L. (Lepidoptera: Noctuidae), and tobacco cutworm, *Spodoptera litura* L. (Lepidoptera: Noctuidae). The insects were maintained on the fresh leaves of castor, *Ricinus comminis* L., in the insectaries of the Biology Division, CSIR-Indian Institute of Chemical Technology, Hyderabad, India. Castor plants grown in the laboratory fields were used for rearing as well as for the experiments. The vapor toxicity of the test compounds was evaluated against three species of stored-product pest insects, the rice weevil *Sitophilus oryzae* L. (Coleoptera: Curculionidae), the pulse beetle *Callosobruchus chinensis* L. (Coleoptera: Bruchidae), and the red flour beetle *Tribolium castaneum* Herbst (Coleoptera: Tenebrionidae). The cultures of the pest insects were maintained in our laboratory for over 8 years without exposure to insecticides. The culture media for *S. oryzae* was wheat (*Triticum aestivum* L.), for *C. chinensis*—green gram (*Phaseolus mungo* L.), and for *T. castaneum*—broken rice (*Oryza sativa* L.). All the grains used for rearing were sterilized to avoid secondary infection by fungi. The cultures were maintained in the dark and in a growth chamber set at 27 ± 1 °C and $65 \pm 5\%$, and a relative humidity (RH). Adult insects, 1–7 days old, were used for fumigant toxicity tests. All the experiments were conducted in the laboratory with colonies under the same environmental conditions. Antifeedant activities of gliotoxin and acetylgliotoxin produced by *Aspergillus turcosus* strain KZR131 were tested by leaf disc bioassay[20] against the third instar larvae of two major agricultural pests, *Achaea janata* L. and *Spodoptera litura* (Fab). Stock solutions were prepared in methanol and a series of concentrations of gliotoxin and acetylgliotoxin were also prepared. The late second instar larvae of insect cultures were fed on castor leaves overnight. The newly molted third instar larvae were selected and starved for 3 h before the experiments to ensure hunger status of the larvae. Fresh castor leaves collected from the laboratory fields were washed with distilled water and the leaf discs (12.0 cm² area) were cut after drying and placed in clean glass petri dishes (15 cm diameter) containing moist filter paper discs, which provided humidity inside and aided in retaining the freshness of the leaf discs. Different doses (10, 25, 50, 75, and 100 µg cm^{-2} of the leaf) of the purified gliotoxin and acetylgliotoxin dissolved in methanol were sprayed uniformly on both sides of the leaf discs and were air-dried at room temperature. The control leaf discs were treated with the same quantity of solvent alone. In each petri dish, a pre-starved healthy third instar larva of *A. janata* or *S. litura* was introduced singly to assess the antifeedant activity. The experimental setups were kept in an illuminated growth chamber and the temperature was maintained at 28 ± 2 °C, 65 ± 5% relative humidity, and a photoperiod regime of 16:8 (light:dark). The food consumption by each larva was recorded by measuring the leaf area consumed after 24 h of both the control and treated leaf discs using an AM-300 portable leaf area meter (ADC BioScientific Ltd., Herts, England). The antifeedant index (AFI) was calculated from the formula: AFI (%) = (C − T)/(C + T) × 100, where C is the leaf area consumption of the control discs and T is the consumption of the treated discs.[21] For all the treatments, there were 10 replicates, and all the treatments and controls were replicated thrice.

16.2.7 INSECTICIDAL ACTIVITY (FUMIGANT TOXICITY BIOASSAY)

The vapor toxicity of the purified extrolites, gliotoxin, and acetylgliotoxin, was evaluated against the adults of the three stored-product pests, *S. oryzae*, *C. chinensis*, and *T. castaneum* by fumigant toxicity assay.[22] In brief, small plastic airtight containers (5 cm height × 5.8 cm diameter) (100 ml capacity) were used as fumigation chambers and filled with 20 g of the respective rearing diet of the tested insects. Gliotoxin and acetylgliotoxin were dissolved in methanol (20 µg mL^{-1}) and then applied to strips (2 × 0.5 cm) of filter paper (Whatman No. 1). The solvent was allowed to evaporate for 5 min and then the strips were stuck to the underside of the lid of a plastic container using cello tape. Twenty unsexed adults of each species 7–10 days old were released into the chamber and the containers were sealed. All the tests were carried out at 28 ± 2 °C temperature and 65 ± 5% relative humidity. The mortality was evaluated visually and by probing the body of the insect with a slender paintbrush. Dead insects were counted every 24 h for a total period of 96 h post-treatment. There were five replicates per treatment while the tests were repeated thrice on a different date each time to avoid any day-to-day variation. The LC$_{50}$ and LC$_{95}$ values were calculated by probit analysis for each species and treatment combination.[23]

16.3   RESULTS AND DISCUSSION

16.3.1   IDENTIFICATION OF THE FUNGAL STRAIN

The colony of the fungal strain KZR131 turned reddish on potato dextrose agar after 10 days. The aerial mycelium was velvety in texture, while the reverse of the colony was deep yellowish-orange. The conidiophores were smooth walled bearing flask-shaped phialides over two-thirds of the surface. The conidia were sub-globose, ovoid and smooth, and borne at the termini of the phialides. On the basis of the morphological characteristics and microscopic observation, the fungus was identified as *Aspergillus turcosus*. On the basis of the 18S rRNA gene sequence homology analysis with Genbank sequence database, the fungal isolate showed 99% homology with different strains of *Aspergillus turcosus* (Genbank accession numbers HF545008, HF545009 and HF545010), and was thus confirmed as *Aspergillus turcosus*. The 18S rRNA gene sequence has been deposited in the National Centre for Biotechnology Information (NCBI) with a Genbank accession number of KJ477085.

16.3.2   PURIFICATION AND STRUCTURAL CHARACTERIZATION

The culture filtrate of strain KZR131 revealed the presence of a single major spot in TLC with an *Rf* value of 0.5 and was developed in a solvent mixture of ethyl acetate-hexane (50:50, v/v). The extrolites present in the crude extract were fractionated by silica gel column chromatography and further purified by preparative RP HPLC to yield two compounds. The purified extrolites on the TLC plates were UV-active when visualized under the UV light at 254 nm. Further, these plates were developed by spraying with phosphomolybdic acid reagent followed by heating at 100 °C for 2–3 min which appeared as blue spots. Compound 1 was obtained as white crystalline solid with a yield of 25 mg L^{-1}; mp 221 °C, UV λ_{max} (in acetonitrile): 223.6, 250.2, and 273.4; infrared (IR) (KBr) (υ_{max}): 3433, 2926, 2856, 1664, 1415, 1371, 1238, 1061, 1013, 816, 720, 650, 611 cm^{-1}; ESI-MS: *m/z* 326.04. The molecular formula was determined as $C_{13}H_{14}N_2O_4S_2$ from the HR-MS (ESI) peak at *m/z* 327.0450 [M]$^+$ (calculated mass: *m/z* 327.04 [M+ H]$^+$). Compound 2 was obtained as a yellow crystalline solid with a yield of 25 mg L^{-1}; mp 162–163 °C, UV λ_{max} (in acetonitrile): 230.6, 250.2, and 268.4; IR (KBr) (υ_{max}): 3433, 2926, 2856, 1664, 1415, 1371, 1238, 1061, 1013, 816, 720, 650, 611 cm^{-1}; ESI-MS: *m/z* 368. The molecular formula was determined as

$C_{15}H_{16}N_2O_5S_2$ from the HR-MS (ESI) peak at m/z 391.04 [M+Na]$^+$ (calculated mass: m/z 368.05 [M+H]$^+$). On the basis of ^{1}H NMR, ^{13}C NMR, FT-IR, and ESI-MS spectral data (Figures 16-1–16-7 and Tables 16-1–16-3), the purified extrolites were identified as gliotoxin and acetylgliotoxin and their chemical structures are shown in Figure 16-8. The gliotoxin and acetylgliotoxin structures were in agreement with the earlier published literature.[24–26] Difficulties were encountered during the identification of gliotoxin by ^{1}H and ^{13}C NMR which was mainly because of their uncommon couplings that are temperature and solvent dependent as reported earlier.[25]

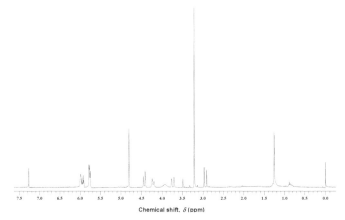

FIGURE 16-1 ^{1}H NMR spectrum of gliotoxin profiled from *Aspergillus turcosus* strain KZR131.

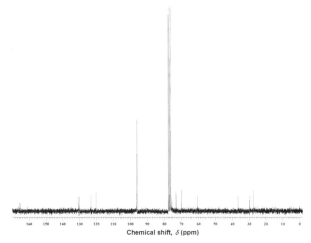

FIGURE 16-2 ^{13}C NMR spectrum of gliotoxin profiled from *Aspergillus turcosus* strain KZR131.

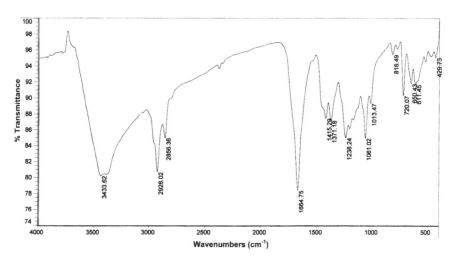

FIGURE 16-3 Fourier transform infrared spectrum (FT-IR) spectrum of gliotoxin profiled from *Aspergillus turcosus* strain KZR131.

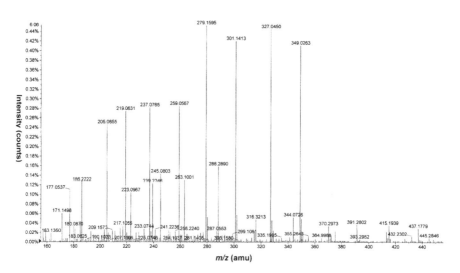

FIGURE 16-4 ESI-MS spectrum of gliotoxin profiled from *Aspergillus turcosus* strain KZR131.

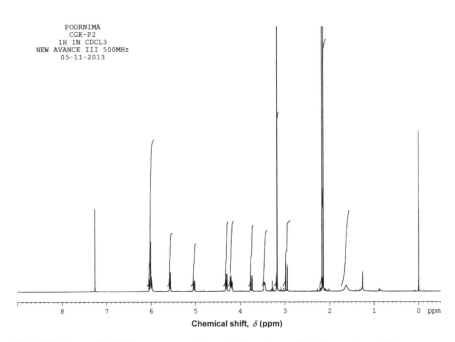

FIGURE 16-5 ¹H NMR spectrum of acetylgliotoxin profiled from *Aspergillus turcosus* strain KZR131.

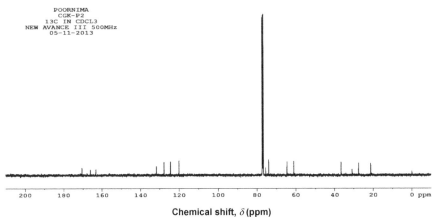

FIGURE 16-6 ¹³C NMR spectrum of acetylgliotoxin profiled from *Aspergillus turcosus* strain KZR131.

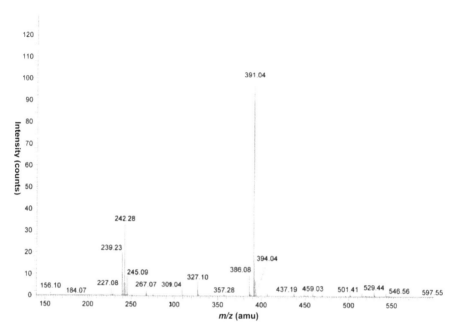

FIGURE 16-7   ESI-MS spectrum of acetylgliotoxin profiled from *Aspergillus turcosus* strain KZR131.

Gliotoxin   **Acetylgliotoxin**

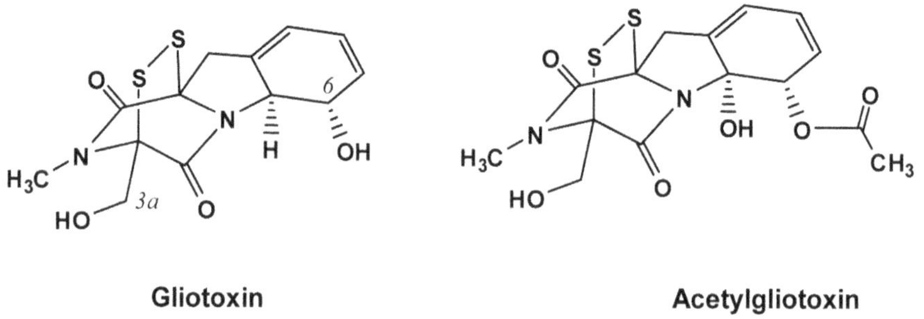

FIGURE 16-8 Chemical structures of gliotoxin and acetylgliotoxin profiled from *Aspergillus turcosus* strain KZR131.

TABLE 16-1 ¹H NMR Data Assignment for the Bioactive Compound 1 in CDCl₃ (deuterated chloroform) Profiled from *Aspergillus turcosus* Strain KZR131

Proton Assignment	Compound 1[a]	Cole and Cox (1981)	Kaouadji et al. (1990)
Me-2	3.20 s	3.22	3.21
H-3a$_A$	4.44 dd, J = 12.7 and 6.1 Hz	4.80	4.43
H-3a$_B$	4.2 dd, J = 12.7 and 9.5 Hz	4.80	4.26
3a-OH	3.5 dd, J = 9.5 and 6.1 Hz	–	3.55
H-5a	4.81 s	2.9	4.8
H-6	4.81 s	3.72	4.8
6-OH	5.78 s	–	5.8
H-7	5.96 br d, J = 5.1 Hz	5.9	5.99
H-8	5.96 br d, J = 10.1 and 5.1 Hz	5.9	5.99
H-9	5.96 br d, J = 10.1 Hz	5.9	5.95
H-10$_A$	3.71 br d, J = 18 Hz	4.44	3.76
H-10$_B$	2.97 br d, J = 18 Hz	4.18	2.96

[a] Recorded at 300 MHz, referenced to the residual solvent at δ_H 7.26 ppm.

TABLE 16-2 ¹³C NMR Chemical Shifts Assignment for the Bioactive Compound 1 in CDCl₃ Profiled from *Aspergillus turcosus* Strain KZR131

Carbon Assignment	Compound 1[a]	Cole and Cox (1981)	Kaouadji et al. (1990)
C-1	165.90	165.9	166.0
C-3	75.61	77.3	75.6
C-3a	60.5	60.5	60.7
C-4	165.1	165.1	165.3
C-5a	69.8	69.7	69.7
C-6	73.1	73.1	73.1
C-7	130.0	129.8	130.1
C-8	123.2	120.1	123.3
C-9	120.17	123.3	120.2
C-9a	130.0	130.7	130.8
C-10	36.6	36.6	36.6
C-10a	77.2 under CDCl₃ peak	77.3	–
2-Me	27.5	27.5	27.5

[a] Recorded at 300 MHz, referenced to residual solvent at δ_C 77.2 ppm.

TABLE 16-3 ^{1}H NMR and ^{13}C NMR Chemical Shifts Assignment for the Bioactive Compound 2 in CDCl$_3$ Profiled from *Aspergillus turcosus* Strain KZR131

Position	δ_C		Type	δ_H	
	Compound 2	Liang et al. (2014)		Compound 2	Liang et al. (2014)
1	166.1	168.4	C		
2			N		
3	77.2	78.2	C		
4	163.2	165.9	C		
5			N		
5a	64.5	64.9	CH	5.1	5.36, d (14.1)
6	74.0	74.3	CH	5.6	5.82 d (14.1)
7	127.7	128.6	CH	6.0	5.93, m
8	124.5	124.9	CH	6.0	5.93, m
9	120.04	120.7	CH	5.59	5.60, d (7.5)
9a	131.7	131.7	C		
10	36.7	41.4	CH$_2$	3.25	3.25, d (15.0)
				3.80	3.80, d (15.0)
10a	77.0	77.6	C		
11	27.5	29.0	CH$_3$	2.9	3.10, s
12	60.9	62.4	CH$_2$	4.3	4.32, d (12.3)
				4.2	4.00, d (12.3)
13	170.44	170.1	C̲OCH$_3$		
	21.3	21.4	COC̲H$_3$	2.1	2.17, s

^{a} Recorded at 300 MHz, referenced to residual solvent at δ_H 7.26 ppm.

The molecular structures of the purified compounds were further confirmed by X-ray crystallographic studies. The pure compound 1 was crystallized in the monoclinic chiral space group $P2_1$. The single crystal X-ray crystallographic data unambiguously confirmed the compound as gliotoxin as its crystal data matched with the reported gliotoxin crystal structure available in the Cambridge Structural Database, www.ccdc.cam.ac.uk (CSD reference code GLITOX10).[27–29] Further, the single crystal X-ray crystallographic data for acetylgliotoxin was established for the first time which is as follows: molecular formula $C_{15}H_{16}N_2O_5S_2$, $M = 368.42$, 0.43×0.38

× 0.32 mm³, orthorhombic, space group $P2_12_12_1$ (No. 19), $a = 10.240(2)$, $b = 10.243(2)$, $c = 30.716(6)$ Å, $V = 3221.8(11)$ Å,[30] $Z = 8$, $Z' = 2$, $D_c = 1.519$ g/cm³, $F_{000} = 1536$, MoKα radiation, $\lambda = 0.71073$ Å, $T = 293(2)$K, $2\theta_{max} = 50.0°$, 31,004 reflections collected, 5661 unique ($R_{int} = 0.0209$), final $GooF = 1.059$, $R1 = 0.0248$, $wR2 = 0.0655$, R indices based on 5625 reflections with $I > 2\sigma(I)$ (refinement on F^2), 446 parameters, 1 restraint, $\mu = 0.360$ mm⁻¹. Absolute structure Flack parameter = 0.01(4).[31] The single crystal X-ray data for acetylgliotoxin has the accession number CCDC 1054601, which also contains the supplementary crystallographic data of this paper. These data can be obtained free of charge from The Cambridge Crystallographic Data Center via www.ccdc.cam.ac.uk/data_request/cif. The ORTEP diagrams of gliotoxin and acetylgliotoxin with atom-numbering scheme are shown in Figure 16-9. Gliotoxin and its analogue, acetylgliotoxin, belong to the epidithiodioxopiperazine family having a diketopiperizine core with a disulfide bridge in an oxidized or reduced form.[32] Gliotoxin is a fungal toxin which was first reported from *Gliocladium fimbriatum*[33] and later from *Gliocladium flavofuscum*.[34] Moreover, *Penicillium terlikowskii*,[35] *Trichoderma virens*,[36] *Aspergillus fumigates*,[37] and marine *Aspergillus* sp.[38] were also reported to produce gliotoxin. Only two fungi, namely *Neosartorya pseudofischeri*[26] and *Dichotomomyces cejpii*[39] were reported to produce both gliotoxin and acetylgliotoxin.

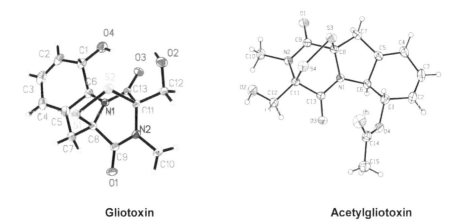

Gliotoxin   Acetylgliotoxin

FIGURE 16-9 ORTEP diagrams of gliotoxin and acetylgliotoxin profiled from *Aspergillus turcosus* strain KZR131 with the atom-numbering scheme. Displacement ellipsoids are drawn at the 30% probability level and the H atoms are shown as small spheres of arbitrary radius. There are two molecules in the asymmetric unit of the crystal structure ($Z' = 2$); however, only one molecule is shown here for clarity.

16.3.3 ANTIMICROBIAL SUSCEPTIBILITY TESTING

The antimicrobial susceptibility test results for gliotoxin and acetylglio-toxin profiled from *A. turcosus* strain KZR131 are shown in Table 16-4, which indicates that both these compounds exhibited significant antimicrobial activity against all the tested bacterial and *Candida* strains. The tested bacterial pathogens like *Bacillus subtilis* MTCC 121, *Micrococcus luteus* MTCC 2470, *Staphylococcus aureus* MTCC 96, *Staphylococcus aureus* MLS16 MTCC 2940, *Staphylococcus aureus* ATCC 29213, *Escherichia coli* ATCC 25922, *Escherichia coli* MTCC 739, *Klebsiella planticola* MTCC 530, and *Pseudomonas aeruginosa* MTCC 2453 showed greater susceptibility and the estimated minimum inhibitory concentration (MIC) values were 9.37, 0.5, 0.5, 37.5, 8, 64, 9.37, 37.5, and 2.3 $\mu g\ mL^{-1}$, respectively, while the different *Candida* strains showed an MIC value of 0.5 $\mu g\ mL^{-1}$. Gliotoxin also exhibited good inhibitory activity against MRSA and VRE strains which was comparable with standard ciprofloxacin. The extrolites (gliotoxin and acetylgliotoxin) possess a broad antibiotic spectrum against all the tested Gram-positive and Gram-negative bacteria, MRSA, VRE, and various *Candida* strains and were identified as potent and promising secondary metabolites. Gliotoxin was earlier reported to exhibit strong antimicrobial potential against various Gram-positive and Gram-negative bacteria, yeasts and filamentous fungi like *Rhizoctonia bataticola*, *Pythium debaryanum*, and *Sclerotium rolfsii* and inhibited bacterial multiplication and germination of fungal spores.[33,40] Gliotoxin produced by *A. fumigatus* was reported to be toxic to *Candida albicans* and *Cryptococcus neoformans*.[41] From the mode of action perspective, gliotoxin exhibits strong fungicidal effect by selectively binding to the thiol groups of the cytoplasmic membranes,[42] which induces reactive oxygen species (ROS) production where the reduced compound is oxidized to reform the disulfide bridge, producing hydrogen peroxide and superoxide radicals in the process. This in turn alters the structural and membrane integrity leading to cell damage and death.[43]

16.3.4 ANTIFEEDANT AND INSECTICIDAL ACTIVITIES

A renewed interest garnered on the development of environmentally safe pest control methods has inspired us to study the effect of potential secondary metabolites (extrolites) against destructive pests. There is paucity of information and studies being conducted on the application of fungal extrolites as

TABLE 16-4 Antimicrobial Activity of Purified Extrolites Profiled from *Aspergillus turcosus* Strain KZR131

Test Strains	Minimum Inhibitory Concentration[+] (MIC, µg mL⁻¹)				
	Acetyl Gliotoxin	Gliotoxin	Neomycin	Ciprofloxacin	Miconazole
Bacillus subtilis MTCC 121	9.37	9.37	18.75	0.5	—*
Micrococcus luteus MTCC 2470	0.5	0.5	18.75	0.5	—
Staphylococcus aureus MTCC 96	0.5	0.5	—	0.5	—
Staphylococcus aureus MLS16 MTCC 2940	37.5	37.5	18.75	0.5	—
Staphylococcus aureus ATCC 29213	8	8	ND[‡]	0.125	—
Methicillin-resistant *Staphylococcus aureus* strain 15187	ND	8	ND	8	—
Vancomycin-resistant Enterococci (VRE) strain SP-346	ND	16	ND	16	—
Escherichia coli ATCC 25922	64	64	ND	0.007	—
Escherichia coli MTCC 739	9.37	9.37	18.75	0.5	—
Klebsiella planticola MTCC 530	37.5	37.5	18.75	0.5	—
Pseudomonas aeruginosa MTCC 2453	2.3	2.3	18.75	0.5	—
Candida albicans MTCC 183	0.5	0.5	—	—	9.37
Candida albicans MTCC 227	0.5	0.5	—	—	9.37
Candida albicans MTCC 854	0.5	0.5	—	—	9.37
Candida albicans MTCC 1637	0.5	0.5	—	—	9.37
Candida albicans MTCC 3017	0.5	0.5	—	—	9.37
Candida albicans MTCC 3018	0.5	0.5	—	—	9.37
Candida albicans MTCC 3958	0.5	0.5	—	—	9.37

TABLE 16-4 *(Continued)*

Test Strains	Minimum Inhibitory Concentration[†] (MIC, μg mL⁻¹)				
	Acetyl Gliotoxin	Gliotoxin	Neomycin	Ciprofloxacin	Miconazole
Candida albicans MTCC 4748	0.5	0.5	–	–	9.37
Candida albicans MTCC 7315	0.5	0.5	–	–	9.37
Candida parapsilosis MTCC 1744	0.5	0.5	–	–	9.37
Issatchenkia orientalis MTCC 3020	0.5	0.5	–	–	9.37

[†]Minimum inhibitory concentration (MIC) is defined as the lowest concentration of the compound that inhibited 90% of the growth when compared with the growth of the control. The tests were repeated thrice and the average values are shown.

[‡]ND, Not determined.

* No activity.

potential biological control agents against agricultural pests like *Spodoptera litura* and *Achaea janata* and stored-product pests like *Sitophilus oryzae*, *Callosobruchus chinensis*, and *Tribolium castaneum* with limited information available only on entomopathogenic fungi. The antifeedant activity of glitoxin against *Spodoptera litura* and *Achaea janata* is shown in Table 16-5. The compound showed 100% antifeedant activity at a concentration of 40 μg cm^{-2} per leaf disc against both *S. litura* and *A. janata*. The feeding of the larvae with lower doses inhibited the larval growth in both the test insects. *Spodoptera litura* is an extremely dangerous polyphagous pest of many economically important crops,[44] especially in India, China, and many other Asian countries.[45] The caterpillars of *Achaea janata* L. feed on leaves of castor (*Ricinus communis* L.) and cause maximum damage between the months of August to November of the year. Owing to their high fecundity, both these pests occur in vast number and cause severe damage to the castor crop and reduce the yield. Several chemical insecticides are effective against these caterpillars; particularly the pyrethroids insecticides and are being widely used presently causing much pollution to the environment and are very cost-intensive. Most of the antifeedant compounds act as deterrents where they completely inhibit the insect from feeding in which the insect initially starts feeding and later stifles from further feeding activity, thus dying because of starvation. In our study, the purified gliotoxin at a dosage of 20 μg cm^{-2} per leaf disc seemed to possess good feeding deterrent property. The rejection of the treated diet was evident at the higher dose (40 μg cm^{-2} per leaf disc), and this could be because of the instantaneous suppression or rapid post-ingestive feedback.[46] Apart from this, the purified gliotoxin exhibited insecticidal

TABLE 16-5 Antifeedant Activity of Gliotoxin Profiled from *Aspergillus turcosus* Strain KZR131 Against the Third Instar Larvae of *Achaea janata* and *Spodoptera litura*

S. No.	Gliotoxin Dosage (μg cm^{-2} Leaf Area)	Mean Antifeedant Index ± SE[a]	
		Achaea janata	Spodoptera litura
1	10	66.4 ± 1.94	78.4 ± 1.69
2	25	72.6 ± 1.16	79.4 ± 1.63
3	50	72.4 ± 0.93	100 ± 0.0
4	75	81.8 ± 1.85	—[b]
5	100	90.4 ± 1.03	—

[a] Results of 30 replicates.

[b] No activity.

activities on the adults of *S. oryzae*, *C. chinensis*, and *T. castaneum*, the most widespread and very destructive primary pests on a variety of stored grains as well as stored products.[47] The LC_{50} values of gliotoxin on *S. oryzae, T. casteneum*, and *C. chinensis* were 5.52, 3.57, and 4.25 μg cm^{-1}, respectively, after 96 h (Table 16-6). As *C. chinensis* is a major pest on several kinds of pulses in India, the control of this pest with biological metabolites is highly advantageous. The results indicated that *C. chinensis* was more susceptible when compared with *S. oryzae* and *T. casteneum*. Gliotoxin caused 100% fumigant toxicity to *C. chinensis and S. oryzae* at 48 and 96 h, respectively; however, the compound produced a delayed mortality in *T. castaneum* taking about 96 h for killing 95% of the pest populations. Furthermore, it was observed that acetylgliotoxin did not exhibit antifeedant and insecticidal activities against these agricultural and stored-product pests.

16.4   CONCLUSION

In the present study, gliotoxin and acetylgliotoxin produced by a new *Aspergillus turcosus* strain KZR131 isolated from the forest soil collected from a biodiversity niche, Kaziranga National Park, Assam, India, were purified and characterized. The chemical structures of gliotoxin and acetyl-gliotoxin were elucidated based on ¹H and ¹³C NMR, FT-IR, and mass spectroscopic methods and further confirmed by X-ray crystallographic studies. They exhibited potent antimicrobial activity against various Gram-positive and Gram-negative bacteria, MRSA, VRE, and different *Candida* strains. It was further demonstrated for the first time that gliotoxin exhibited good antifeedant and insecticidal activities against several major and important pests, which may find use in pest management.

ACKNOWLEDGMENTS

The authors are thankful to Dr. T. C. Bora, Chief Scientist, CSIR-North Eastern Institute of Science and Technology, Jorhat, Assam, India, and Mr. Dharnidhar Boro, Forest Range Officer, Kaziranga National Park, Assam, for their kind help and providing the necessary logistics during the expedition of sample collection in the Kaziranga National Park, Assam. The financial support extended by the Council of Scientific and Industrial Research (CSIR), New Delhi, Government of India in the form of Network Project on *Exploitation of Microbial Wealth of India* is gratefully acknowledged.

TABLE 16-6 Insecticidal Activity of Gliotoxin Profiled from *Aspergillus turcosus* Strain KZR131 Against Three Stored Pests by Vapor Toxicity Method

Insects	Mean (%) Toxicity ± SD[b] (15 µg cc^{-1})				LC$_{50}$ (95% FL[a])	X^2 (df)	P-Level
	Day 1	Day 2	Day 3	Day 4			
Sitophilus oryzae	59 ± 2.8	76 ± 3.0	89 ± 2.0	100 ± 0.0	4.25 (3.93–4.57)	6.17 (4)	0.081
Tribolium castaneum	0.51 ± 2.7	63 ± 2.3	78 ± 1.8	95 ± 1.0	5.52 (5.12–5.92)	6.99 (4)	0.082
Callosobruchus chinensis	84 ± 5.4	100 ± 0.0	–[c]	–	3.57 (3.25–3.89)	4.11 (4)	0.249
Control	0.0 ± 5.0	0.0 ± 0.0	0.0 ± 0.0	0.0 ± 0.0	–	–	–

[a] Fumicidal limits, after 4 days of treatment.

[b] Each data is significantly different from another [one-way analysis of variance (ANOVA)]; Tukey test at $P < 0.05$; N-300.

[c] No activity.

The authors acknowledge the financial assistance provided in the form of a Senior Research Fellowship to Poornima Mongolla by CSIR, New Delhi. The authors are thankful to Dr. Inshad Ali Khan, Senior Scientist, Clinical Microbiology Laboratory, CSIR-Indian Institute of Integrative Medicine, Jammu Tawi, India, for his kind help to extend the antimicrobial activity determination of gliotoxin against MRSA and VRE strains.

KEYWORDS

- Acetylgliotoxin
- Antifeedant
- Antimicrobial
- Aspergillus turcosus
- Gliotoxin
- Insecticidal

REFERENCES

1. Ahmed, S.; Grainge, M.; Hylin, J. W.; Mitchel, W. C.; and Litsinger, J. A. Some promising plant species for use as pest control agents under traditional farming systems. In *Natural Pesticides from the Neem Tree and Other Tropical Plants*; Schmutterer, H., and Ascher, K. R. S., Eds.; GTZ: Eschborn, Germany, 1984; pp 565–580.

2. Schoonhoven, L. M.; Jermy, T.; and van Loon, J. J. A. *Insect-Plant Biology*. Chapman and Hall: Cambridge, UK, 1998; p 409.

3. Zeng, L. The advance of phosphine resistance in stored grain insects. *Nat. Enemies Insect.* **1996**, *18*, 37–42.

4. Tanaka, Y.; and Omura, S. Agroactive compounds of microbial origin. *Annu. Rev. Microbiol.* **1993**, *47*, 57–87.

5. Wood, R. K. S.; and May, M. J., Eds.; *Biological Control of Pests, Pathogens and Weeds, Development and Prospects*; The Royal Society: London, 1989.

6. Vey, A.; Hoagland, R.; and Butt, T. M. Toxic metabolites of fungal biocontrol agents. In *Fungi as Biocontrol Agents*; Butt, T. M., Jackson, C., and Magan, N., Eds.; CAB International: Wallingford, UK, 2001; pp 311–340.

7. Quesada-Moraga, E.; Carrasco-Diaz, J. A.; and Santiago-Alvarez, C. Insecticidal and antifeedant activities of protein secreted by entomopathogenic fungi against *Spodoptera litoralis* (Noctuidae, Lepidoptera). *J. Appl. Entomol.* **2006**, *130*, 442–452.

8. Ganassi, S.; De Cristofaro, A.; Grazioso, P.; Altomare, C.; Logrieco, A.; and Sabatini, M. A. Detection of fungal metabolites of various *Trichoderma* species by the aphid *Schizaphis graminum*. *Entomol. Exp. Appl.* **2007**, *122*, 77–86.

9. Valanarasu, M.; Kannan, P.; Ezhilvendan, S.; Ganesan, G.; Ignacimuthu, S.; and Agastian, P. Antifungal and antifeedant activities of extracellular product of *Streptomyces* spp. ERI-04 isolated from Western Ghats of Tamil Nadu. *J. Mycol. Medicale* **2010**, *20*, 290–297.

10. Strobel, G. A. Microbial gifts from rain forests. *Can. J. Plant Pathol.* **2000**, *24*, 14–20.

11. Holler, U.; Wright, A. D.; Matthee, G. F.; Konig, G. M.; Draeger, S.; Aust, H. J.; and Schulz, B. Fungi from marine sponges, diversity, biological activity and secondary metabolites. *Mycol. Res.* **2000**, *104*, 1354–1365.

12. Lin, Y.; Wu, X.; Feng, S.; Jiang, G.; Luo, J.; Zhou, S.; Vrijmoed, L. L. P.; Jones, E. B. G.; Krohn, K.; Steongröver, K.; and Zsila, F. Five unique compounds, xyloketales from mangrove fungus *Xylaria* sp. from the South China Sea coast. *J. Org. Chem.* **2001**, *66*, 6252–6256.

13. Moncheva, P.; Tishkov, S.; Dimitrova, N.; Chipeva, V.; Nikolova, S. A.; and Bogatzevska, N. Characteristics of soil actinomycetes from Antarctica. *J. Cult. Collect.* **2002**, *3*, 3–14.

14. Kamal, A.; Shaik, A. B.; Kumar, C. G.; Mongolla, P.; Rani, P. U.; Krishna, K. V. S. R.; Mamidyala, S. K.; and Joseph, J. Metabolic profiling and biological activities of bioactive compounds produced by *Pseudomonas* sp. strain ICTB-745 isolated from Ladakh, India. *J. Microbiol. Biotechnol.* **2012**, *22*, 73–83.

15. Hawksworth, D.; and Rossman, A. Y. Where are the undescribed fungi? *Phytopathology* **1987**, *87*, 888–891.

16. Kumar, C. G.; Mongolla, P.; Joseph, J.; Nageswar, Y. V. D.; and Kamal, A. Antimicrobial activity from the extracts of fungal isolates of soil and dung samples from Kaziranga National Park, Assam, India. *J. Mycol. Medicale* **2010**, *20*, 283–289.

17. Anonymous. *SMART and SAINT Software Reference Manuals, Versions 6.28a & 5.625*; Bruker Analytical X-ray Systems Inc.: Madison, Wisconsin, USA, 2001.

18. Sheldrick, G. M. *SHELXS97 and SHELXL97 Programs for Crystal Structure Solution and Refinement*; University of Gottingen: Germany, 1997.

19. Kumar, C. G.; and Mamidyala, S. K. Extracellular synthesis of silver nanoparticles using culture supernatant of *Pseudomonas aeruginosa*. *Coll. Surf. B Biointerfaces* **2011**, *84*, 462–466.

20. Devanand, P.; and Usha Rani, P. Insect growth regulatory activity of the crude and purified fractions from *Solanum melongena* L., *Lycopersicum esculentum* Mill. and *Capsicum annuum* L. *J. Biopest.* **2011**, *4*, 118–130.

21. Isman, B.; Koul, O.; Lucyzynski, A.; and Kaminski, J. Insecticidal and antifeedant bioactivities of neem oils and their relationship to azadirachtin content. *J. Agric. Food Chem.* **1990**, *38*, 1407–1411.

22. Usha Rani, P.; Venkateshwaramma, T.; and Devanand, P. Bioactivities of *Cocos nucifera* L. (Arecales, Arecaceae) and *Terminalia catappa* L. (Myrtales, Combretaceae) leaf extracts as post-harvest grain protectants against four major stored product pests. *J. Pest Sci.* **2011**, *84*, 235–247.

23. Finney, D. J. Ed.; *Probit Analysis,* 3rd eds; Cambridge University Press: London, 1971, p 318.

24. Cole, R. J.; and Cox, R. H., Eds.; *Handbook of Toxic Fungal Metabolites*; Academic Press: New York, 1981.

25. Kaouadji, M.; Steiman, R.; Seigle-Murandi, F.; Krivobok, S.; and Lucile, S. Gliotoxin, Uncommon ^{1}H couplings and revised ^{1}H- and ^{13}C-NMR assignments. *J. Nat. Prod.* **1990**, *53*, 717–719.

26. Liang, W. L.; Le, X.; Li, H. J.; Yang, X. L.; Chen, J. X.; Xu, J.; Liu, H. L.; Wang, L. Y.; Wang, K. T.; Hu, K. C.; Yang, D. P.; and Lan, W. J. Exploring the chemodiversity and biological activities of the secondary metabolites from the marine fungus *Neosartorya pseudofischeri. Mar. Drugs* **2014**, *12*, 5657–5676.

27. McCrone, W. C. Crystallographic data, Gliotoxin. *Anal. Chem.* **1954**, *26*, 1662–1663.

28. Fridrichsons, J.; and Mathieson, A. M. The crystal structure of gliotoxin. *Acta Cryst.* **1967**, *23*, 439–448.

29. Cambridge Structural Database. *ConQuest 1.14, ver. 5.33*, November 2011, May 2012 update, Cambridge Crystallographic Data Centre (CCDC), 2012.

30. Muller, P.; Herbst-Imer, R.; Spek, A. L.; Schneider, T. R.; and Sawaya, M. R. Crystal structure refinement. In *A Crystallographer's Guide to SHELXL*; Muller, P., Ed.; Oxford University Press: Oxford, New York, 2006, pp 57–91.

31. Flack, H. D. On enantiomorph-polarity estimation. *Acta Cryst.* **1983**, *A39*, 876–881.

32. Welch, T. R.; and Williams, R. M. Epidithiodioxopiperazines. Occurrence, synthesis and biogenesis. *Nat. Prod. Rep.* **2014**, *31*, 1376–1404.

33. Johnson, J. R.; Bruce, W. F.; and Dutcher, J. D. Gliotoxin, the antibiotic principle of *Gliocladium fimbriatum*. I. Production, physical and bioproperties. *J. Am. Chem. Soc.* **1943**, *65*, 2005–2009.

34. Avent, A. G.; Hanson, J. R.; and Truneh, A. Metabolites of *Gliocladium flavofuscum. Phytochemistry* **1993**, *1*, 197–198.

35. Waring, P.; Eichner, R. D.; Tiwari-Palni, U.; and Mullbacher, A. Gliotoxin-E, a new biologically active epipolythidioxopiperazine isolated from *Penicillium terlikowskii. Aust. J. Chem.* **1987**, *40*, 991–997.

36. Anitha, R.; and Murugesan, K. Production of gliotoxin on natural substrates by *Trichoderma virens. J. Basic Microbiol.* **2005**, *45*, 12–19.

37. Comera, C.; Andre, K.; Laffitte, J.; Collet, X.; Galtier, P.; and Isabelle, M. P. Gliotoxin from *Aspergillus fumigatus* affects phagocytosis and the organization of the actin cytoskeleton by distinct signalling pathways in human neutrophils. *Microbes Infect.* **2007**, *94*, 47–54.

38. Nguyen, V. T.; Lee, J. S.; Qian, Z. J.; Li, Y. X.; Kim, K. N.; Heo, S. J.; Jeon, Y. J.; Park, W. S.; Choi, I. W.; Je, J. Y.; and Jung, W. K. Gliotoxin isolated from marine fungus *Aspergillus* sp. induces apoptosis of human cervical cancer and chondrosarcoma cells. *Mar. Drugs* **2014**, *12*, 69–87.

39. Rodrigues, B. S. F.; Sahm, B. D. B.; Jimenez, P. C.; Pinto, F. C. L.; Mafezoli, J.; Mattos, M. C.; Rodrigues-Filho, E.; Pfenning, L. H.; Abreu, L. M.; Costa-Lotufo, L. V.; and Oliveira, M. C. F. Bioprospection of cytotoxic compounds in fungal strains recovered from sediments of the Brazilian coast. *Chem. Biodiversity* **2015**, *12*, 432–442.

40. Boutibonnes, P.; Auffray, Y.; Malherbe, C.; Kogho, W.; and Marais, C. Properties anti-bacteriennes et genotosciques de 33 mycotoxines. *Mycopathologia.* **1984**, *87*, 43–49.

41. Coleman, J. J.; Ghosh, S.; Okoli, I.; and Mylonakis, E. Antifungal activity of microbial secondary metabolites. *PLoS One* **2011**, *6*, e25321. DOI: 10.1371/journal.pone.0025321

42. Jones, R. W.; and Hancock, J. G. Mechanism of gliotoxin action and factors mediating gliotoxin sensitivity. *J. Gen. Microbiol.* **1988**, *134*, 2067–2075.

43. Lorito, M.; Leterbauer, C.; Hayes, C. K.; Woo, S. L.; and Harman, G. E. Synergistic combination of cell wall degrading enzymes and different antifungal compounds enhances inhibition of spore germination. *Microbiology* **1994**, *140*, 623–629.

44. Murali Krishna, T.; Devaki, K.; Raja Reddy, K.; and Venkateswarlu, U. Efficacy of certain new insecticide molecules against groundnut defoliator, *Spodoptera litura* (Fab.) (Noctuidae, Lepidoptera). *Curr. Biotica.* **2008**, *2*, 173–180.

45. Shivayogeshwar, A. B.; Mallikharjunaiah, H.; and Krishnaprasad, N. K. Integrated management of *Spodoptera litura* Fabricus (Noctuidae, Lepidoptera) in FCV tobacco crop. *Tobacco Res.* **1991**, *17*, 59–61.
46. Bernays, E. A.; Oppenheime, S.; Chapman, R. F.; Kwon, H.; and Gould, F. Taste sensitivity of insect herbivores to deterrents is greater in specialists than in generalists, A behavioural test of the hypothesis with two closely related caterpillars. *J. Chem. Ecol.* **2000**, *26*, 547–563.
47. Granousky, T. A. Stored product pests. In *Handbook of Pest Control*, 8th eds; Mallis, A., Hedges, S. A., and Moreland, D., Eds.; Mallis Handbook and Technical Training Co.: Cleveland, OH, 1997; pp 635–728.

CHAPTER 17

MORPHOLOGICAL, CULTURAL AND MOLECULAR DIVERSITY OF THE SALT-TOLERANT ALKALIPHILIC ACTINOMYCETES FROM SALINE HABITATS

S. D. GOHEL[1] and S. P. SINGH[1*]

[1]UGC- Centre of Advanced Studies, Department of Biosciences, Saurashtra University, Rajkot, Gujarat – 360005, India.

[*]Corresponding Author: S. P. Singh, PhD. E-mail: satyapsingh@yahoo. com

CONTENTS

Abstract 338
17.1 Introduction 338
17.2 Materials and Methods 339
17.3 Results and Discussion 342
17.4 Conclusion 349
Acknowledgments 349
Keywords 349
References 350

ABSTRACT

Actinomycetes are Gram-positive and filamentous soil bacteria that include many genera. Although mesophilic actinomycetes are widely distributed in nature and extensively explored, the salt tolerant alkalophilic actinomycetes from the saline habitats are scarcely studied. In the present report, therefore, we describe salt tolerant alkaliphilic actinomycetes of the saline habitats. The haloalkaliphilic actinomycetes are characterized based on their morphological and cultural properties and molecular traits. Cultural characteristics of the actinomycetes are studied by using various *International Streptomyces Project* media. The molecular traits employed for the diversity and phylogeny are usually based on the 16S rRNA gene homology and Denaturing Gradient Gel Electrophoresis (DGGE). Haloalkaliphilic actinomycetes have gained considerable attention because of their ability to produce a range of secondary metabolites and hydrolytic enzymes under different extremities. Most of the studies on actinomycetes so far have focused on antibiotic production and only a few have been explored for the enzymatic potential. The studies have suggested a wide occurrence of hydrolytic enzymes in haloalkaliphilic actinomycetes, where proteases occupy a prominent position.

17.1   INTRODUCTION

Actinomycetes are widely distributed in natural and manmade environments. However, they have largely been studied from soil. They are well known for their antibiotics and bioactive molecules, and thus are of considerable importance in the industries. Filamentous soil bacteria belonging to the genus *Streptomyces* is a rich source of bioactive and commercially significant compounds representing 70–80% of all isolated compounds.[1-4] These compounds were discovered by applying recent advances in understanding the genetics of the secondary metabolism of the actinomycetes and the development of new screening approaches.[5] *Streptomyces* are remarkable and focus special attention with regard to their morphological and metabolic differentiation phenomena manifested during development. *Streptomyces* species synthesize diverse natural secondary metabolites, the best known being the antibiotics used worldwide as pharmaceutical and agrochemical products.[6-8] While major studies focused on the diversity, phylogeny, and search for the bioactive molecules from the neutrophilic actinomycetes, only limited attention is evident on these organisms from extreme environment.

Although alkaliphilic bacteria are extensively studied, a similar account on the alkaliphilic actinomycetes is quite scarce. Culture-dependent and culture-independent methods have demonstrated diversity and novelty of the halotolerant and alkaliphilic actinomycetes of the saline and alkaline environments. Members of the genus *Nocardiopsis* are reported to predominate in saline or alkaline soils.[9] During the last few years, reports on the commercially significant enzymes of the salt tolerant, alkaliphilic *Nocardiopsis* sp. have appeared in the literature.[10,11] It is, therefore, necessary to focus attention on the extreme actinomycetes, as a possible way to discover novel taxa and consequently, new secondary metabolites and biocatalysts. Besides, their diversity and phylogeny would be of ecological significance.

Many extracellular proteinases have been reported from actinomycetes and characterized as serine-proteinases and metalloproteinases.[12–15] These enzymes are studied with respect to biochemical characterization, structure and functional relationship, and biotechnological relevance.[16] As protein constitutes an important constituent in nature, the utilization of agricultural by-products and other natural compounds by the proteolytic microorganisms may constitute an interesting alternative toward proteinases production at lower costs. Cheap fibrous substrates, like wheat bran and soybean are suitable for the production of protein hydrolyses by actinomycetes.[17,18] In the present study, occurrence and diversity of novel halophilic/halotolerant and alkaliphilic actinomycetes from saline habitats along Coastal Gujarat have been investigated.[1,2,10–14] The population heterogeneity in terms of extreme habitat, morphological features, growth patterns, and enzyme secretion were included as the parameters to judge the diversity, distribution, and strength of actinomycetes from saline habitats.

17.2 MATERIALS AND METHODS

17.2.1 ISOLATION OF ACTINOMYCETES

For the isolation of salt tolerant and alkaliphilic actinomycetes, soil and water samples were collected from diverse habitats of the saline habitats. The soil samples were differentiated on the basis of the color of soil and texture, such as stone/crystallized form, smooth, particulate, or powdery. The collected samples stored at 4–8 °C were used for the isolation of actinomycetes. The halotolerant and alkaliphilic actinomycetes were isolated using enrichment techniques as well as serial dilution and plating techniques.

17.2.2   MORPHOLOGICAL CHARACTERIZATION

The cell morphology of actinomycetes was studied by light and electron microscopic examinations. The cultures grown at 28 °C were examined periodically for the formation of aerial mycelium, spore-bearing hyphae, and spore chains by direct microscopic examination of the culture surface. The morphological characterization of actinomycetes was coupled with the Gram's reaction and colony morphology analysis. Colony characteristics such as size, shape, pigmentation, margins, texture, elevation, and opacity were followed to record the diversity and differentiation.

17.2.3   CHARACTERIZATION OF THE ISOLATES ON DIFFERENT MEDIA

The isolated actinomycetes were grown in different ISP media, actinomycetes isolation agar (AIA), starch casein agar (SCA), and starch agar (SA) and incubated for 10–14 days at 28 °C. The different media were supplemented with 5% NaCl (w/v) at pH 9.

17.2.4   MORPHOLOGICAL STUDIES USING SLIDE CULTURE TECHNIQUE

The morphological studies were based on macroscopic and microscopic methods. Microscopic characterization and cellular morphogenesis were followed using cover slip culture method. In this technique, a loopful of spores was inoculated in a thin film of 3% ISP medium on a sterile glass slide surface and then covered with a sterile cover slip to facilitate direct observation under the microscope. The cultures thus inoculated were incubated at 28 °C and examined periodically for the formation of aerial mycelium, spore-bearing hyphae, and spore chains by using direct microscopic examination of the culture surface.

17.2.5   SALT AND PH PROFILE

To study the salt and pH profile of actinomycetes, the cultures were grown in gelatin agar containing (g/L): gelatin, 30; peptone, 10; yeast extract, 10; NaCl, 0–200 g/L, and agar, 30. The pH of the medium was usually adjusted to 9 by adding separately autoclaved Na_2CO_3 (20% w/v). Similarly, for pH profiling, a constant salt concentration of the agar medium was maintained, varying the pH in the range of 7–11.

17.2.6 DENATURING GRADIENT GEL ELECTROPHORESIS (DGGE)

The detection of mutations by DGGE was based on the sequence-dependent electrophoretic mobility of the double stranded DNA fragments in a poly-acrylamide gel that contains a linear denaturing gradient. The DNA fragment was composed of at least two melting domains (blocks of sequence with a discrete melting temperature, or Tm). The DGGE involves electro-phoresis of double-stranded DNA fragments through a polyacrylamide gel containing a linear gradient of DNA-denaturing agents (e.g., a combination of formamide and urea) at a fixed temperature, usually 60 °C. Initially, the migration rate of the fragment depends on its molecular weight. However, at a specific point in the gel, the combination of denaturant concentration and temperature equals the Tm of the lowest melting domain, resulting in a partially single-stranded fragment. The mobility of these branched frag-ments in the polyacrylamide gel is abruptly retarded. The fact that the Tm for a given domain is determined by its sequence and base composition means that two DNA fragments that differ by a single base change (and thus in Tm) in the lowest melting domain will be separated from each other at the end of the run. In DNA, fragments of the same length but with different sequences can be separated. Separation is based on the decreased electrophoretic mobility of the single stranded DNA.[19–21] For DGGE analysis, we directly amplified 16S rRNA regions of the Actinomycetes sp. with two universal and three different species specific primer pairs. The DNA concentration of the samples was determined and 400 nanogram of DNA loaded onto 8% (w/v) polyacrylamide gel in TAE buffer.

17.2.7 EXTRACELLULAR ENZYME DETECTION

To study the effect of NaCl on the secretion of various extracellular enzymes, gelatin, starch, and Dubo's agar plates were prepared for protease, amylase, and cellulase, respectively, at NaCl concentrations, 0–20% w/v. The pH of each medium was adjusted to 9 by adding 20% Na_2CO_3. Similarly, the influ-ence of pH on the secretion of protease, amylase, and cellulose was studied on gelatin, starch, and Dubo's agar medium, respectively, at pH 7–11 and a constant NaCl of 5% w/v. Detection of alkaline protease was carried out by flooding the plates with Frazier's reagent, while alkaline amylase and cellulose were detected by pouring Gram's iodine into the plates.

17.3   RESULTS AND DISCUSSION

17.3.1   ACTINOMYCETES FROM THE EXTREME ENVIRONMENTS

Microorganisms from extreme environments have received great attention because of their special mechanisms of adaptation to the conditions of their habitats. Recently, some reports on the commercial significance of the novel metabolites and enzymes of such organisms have appeared in the literature.[10–14,22,23] Further, Tang and coworkers[24] investigated the biological characteristics of 43 actinomycete isolates from saline and alkaline soils in Xinjiang, Hebei, and Qinghai (China).

17.3.2   ISOLATION OF ACTINOMYCETES

The techniques used for the isolation of actinomycetes include enhancement of growth of desirable microorganisms in natural samples by enrichment culture technique or elimination of the undesirable bacterial population by pretreatment.[25] Many such attempts have led to the development of methods for the isolation of desirable actinomycetes that can produce useful secondary metabolites.[26–34] However, actinomycetes that can withstand harsh environmental conditions are rarely explored. In this direction, a research group of Prof. Satya P. Singh at the Biosciences Department of Saurashtra University has isolated large number of the actinomycetes from the saline habitats of Coastal Gujarat and investigated them for their diversity and enzymatic potential. Many enzymes from these organisms are characterized for structure and functional relationship.[10–14] Though many of these actinomycetes grow in the absence of salt, they can tolerate and grow in 10–15% NaCl (w/v) indicating that they are halotolerant rather than halophilic in nature. To explore the microbial diversity of the saline habitat, various dilution techniques and enrichment culture conditions were employed. Our studies revealed that the spores of most actinomycetes are resistant to desiccation or pretreatments of the natural habitat samples and thus, the isolation of rare actinomycetes was a reality. It has been reported that most actinomycetes spores show a higher resistance against wet or dry heat as compared with the corresponding vegetative hyphae.[35,36] Recently, Niyomvong et al.[37] revealed that the pretreatment of samples with moist and dry heat reduced the number of undesirable bacteria and enhanced the selective isolation of rare actinomycetes.

17.3.3 MORPHOLOGICAL CHARACTERIZATION

The actinomycete isolates are usually aerobic and Gram-positive with filamentous structures and well developed aerial and substrate mycelia develop when grown on YEME with NaCl 5% (w/v) and pH 9.0. The isolates were conformed to the group actinomycetes and distinguished on the basis of morphological characterization. The primary characterization of the actinomycetes was based on the colony and cell morphology, Gram reaction, and morphogenesis. Colony characteristics were recorded after growing the organisms on starch agar (5% w/v, NaCl; pH 9) or any other suitable medium until sporulation. The isolates were putatively assigned to the actinomycetes group on the basis of their tough, leathery colony, branched vegetative mycelia, and when aerial mycelia were present, spore formation was taken into consideration. The light microscopic examination revealed Gram stained, long filamentous structure as well as fragmented hyphae (Figures 17-1 and 17-2).

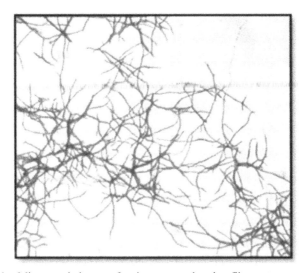

FIGURE 17-1   Microscopic image of actinomycete showing filamentous growth.

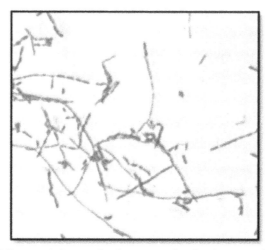

FIGURE 17-2   Microscopic image of actinomycete showing fragmentation of hyphae.

17.3.4   CULTURAL CHARACTERIZATION

Cultural characteristics of the actinomycetes were followed according to the protocol suggested in the *International Streptomyces project*.[38] The growth behavior of actinomycetes was studied on different media specific for actinomycetes and *Streptomyces* (Figure 17-3). Majority of the actinomycetes were able to grow well in most of the organic and chemically defined media.

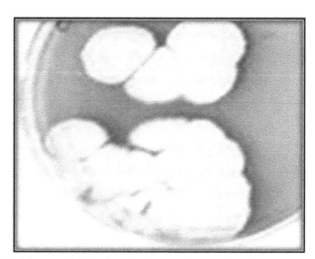

FIGURE 17-3   The growth behavior of actinomycete on oatmeal agar.

Typically, some of the colonies are elevated, spreading, and covered with gray to light pink and light blue colored aerial mycelia and spores. The spores were arranged in short to long chains, displaying smooth surfaces. Most of the pigmented colonies develop on ISP-2 and ISP-3 media followed by starch casein agar and starch agar.

17.3.5 IN SITU COLONY DEVELOPMENT STUDIES

In Situ colony development studies revealed that actinomycetes produce abundant aerial mycelium of white, gray, and yellow color, while substrate mycelium showing light yellow with soluble pigments of various colors. *Streptomycetes* sp. MARO1 and *Streptomyces violaceusniger* strain HAL64 have been reported to possess yellow to gray mycelia.[3,8] Further, dark gray to pale yellow colored aerial mycelium and gray, dark gray or black colored substrate mycelium of actinomycetes have been observed. However, besides gray and white colored mycelia, actinomycetes with reddish to pinkish mycelia were also observed with some of the actinomycetes described in literature.[39] The actinomycetes sporulated at different stages of the growth, as reflected in some reports.[40–42] Mature actinomycetes spores are shown in Figure 17-4.

FIGURE 17-4   Mature spores of actinomycetes.

17.3.6   SALT AND PH PROFILE

The isolation and characterization of halophilic and haloalkaliphilic actino-mycetes appears to be fairly recent. Majority of the actinomycetes studied from the saline habitats of Coastal Gujarat by our research group were capable to grow in the range of 0–10% and 0–15% salt and pH in the range of 7–11 and 8–11, displaying a scarce growth at pH 7.[10–12] The salt tolerance of the actinomycetes in our study was comparable with that reflected by *Nocardiopsis halotolerans* sp. Nov., a halo-tolerant actinomycetes which grows at salt concentrations of 0–10% w/v NaCl,[41] while the tolerance was at a reduced level when compared with *Streptimonospora salina* gen nov., sp. Nov. and *Streptomonospora alba* sp. Nov., the halophilic actinomy-cetes.[43,44] On the other hand, the pH requirement of the salt tolerant alka-liphlic actinomycetes was in agreement with *Nocardiopsis alkaliphila* in which optimum growth occurred at pH 10.[45] Recently, a novel alkaliphilic actinomycetes, *Streptomyces sodiphilus* sp. Nov., was reported to grow at an optimum pH of 9–10.[46] Quite recently, Li and coworkers[47] reported five novel species of the genus *Nocardiopsis* from hyper saline soil in China. Similarly, Li et al.[48] isolated a coccoid, non-motile novel actinomycete, *Kocuria aegyptia*, from a saline, desert-soil sample from Egypt having tremendous salt-tolerance. A considerably high salt and pH tolerance of these actinomycetes, along with their capacity to secrete commercially valuable primary and secondary metabolites and array of enzymes, can be considered as attractive features.

17.3.7   POLYMERASE CHAIN REACTION (PCR) AMPLIFICATION

The sequence analysis of the genes coding for the ribosomal subunits (16S, 23S, and 5S rRNA), in particular, the 16S rRNA gene, has become an impor-tant tool in bacterial identification, because it provides information about the phylogenetic relatedness of the species.[49] The DNA sequences of the ribosomal genes are highly conserved, but the genes also contain variable regions, which sometimes can be useful for species discrimination. However, the 16S rDNA sequence information alone may not be sufficient for species identification.[50] Further, repetitive intergenic DNA sequences (rep-PCR) and PCR-RFLP (Polymerase Chain Reaction-Restriction Fragment Length Polymorphism) of the 65-kDa heat shock protein gene have been used for the classification and identification of pathogenic and other clinically impor-tant *Streptomyces* species.[51] Antony-Babu et al.[52] reported that 24 isolates representing the color and rep-PCR groups grew well from pH 5 to 11, and optimally at pH 9. Similarly, phylogenetically close members of the *Streptomyces griseus* 16S rRNA gene clade were also reported. The representative alkaliphilic *Streptomycetes* form a heterogeneous but distinct clade in the *Streptomyces* 16S rRNA gene tree. An analysis of the complete 16S rRNA gene sequences demonstrated that the two strains analyzed in detail were most closely related to actinobacteria in the *Thermomonosporaceae* and the *Micromonosporaceae*.[53] Further, the analysis of 16S rDNA of a novel alkaliphilic actinomycetes strain TOA-producing Prp[Sc](Abnormal Prion Protein) degrading keratinase indicated that the strain belongs to the genus *Nocardiopsis*; however, it genetically differed from the other *Nocardiopsis* species.[54]

17.3.8   DENATURING GRADIENT GEL ELECTROPHORESIS

A wide variety of different methods to detect DNA sequence variations has been developed during the past few years. One of these methods, denaturing gradient gel electrophoresis (DGGE), has been shown to be sensitive and a method of choice to study mutations in large genes. The detection of mutations by DGGE is based on the sequence-dependent electrophoretic mobility of the double stranded DNA fragments in a polyacrylamide gel that contains a linear denaturing gradient. In DGGE and temperature gradient gel electrophoresis (TGGE), DNA fragments of the same length but with different sequences can be separated.[19,20,55–57] Separation on the gel is based on the decreased electrophoretic mobility of the single stranded DNA.[21] DGGE is an excellent, highly reproducible, comparative community analysis tool.[58,59] Further, PCR-DGGE is a flexible method that incorporates different approaches for accurate identification as functional genes in particular, bacterial populations or specific bacterial species are analyzed by hybridization with species-specific probes.[60] There is a need of the eminent methods for the rapid and accurate screening of total microbial populations in complex ecosystems. PCR-DGGE emerged as a powerful tool to assess the total gut microbial populations and was also used to detect previously unknown bacterial species in the GI tract of animals .[60–63] Understanding the relationship between the host and the disease-causing organisms will help in controlling efficient pathogens, an aspect of paramount importance in food safety and food processing, where quality control and assurance is highly significant. Although, the DGGE concept was originally developed for detecting single base changes and DNA polymorphisms, it has now become one of the most important molecular tools in microbial ecology. At present, there is considerable interest in the use of culture-independent methods for the characterization of the total microbial populations in different niches. In combination with the conventional culturing methods, molecular techniques such as DGGE have a significant role in molecular ecology.

17.3.9   EXTRA CELLULAR ENZYME SECRETION

During the last decade, there has been a dramatic increase in the need for bioactive compounds with novel activities and properties. Enzymes, after antibiotics are the most important biologically derived products with immense potential of commercial interest. During the last few years, extra-cellular enzymes from halophilic and alkaliphilic bacteria and actinomy-cetes have been studied.[10,64,65] However, it is evident that the exploration of the enzymatic potential of these microbes is just the beginning and till date only few enzymes are investigated in depth. Therefore, we screened our isolates for the production of extracellular alkaline protease, amylase, and cellulase. *Streptomycetes* are Gram-positive filamentous bacteria abundantly found in soils and widely cited in literature as great producers of antibiotics and important industrial enzymes. *Streptomycetes* produce a large number of extra cellular enzymes as part of their saprophytic mode of life. The ability to synthesize enzymes during the primary metabolism leads to the production of many proteins of industrial importance. The salt tolerant alkaliphilic actinomycetes from Coastal Gujarat have a varying degree of diversity with respect to their growth patterns and potential for extra cellular protease, amylase, and cellulase production with the majority of the isolates being able to produce all the three enzymes.[59,64] The isolates which tolerate higher salt (up to 10% w/v) are usually potent enzyme producers in comparison with those grown at lower NaCl concentrations. Recently, Bull et al.[48] reported marine actinomycetes capable of secreting extra cellular hydrolytic enzymes in the range of 0–10% NaCl. A fairly good salt and pH tolerance of these actinomycetes, along with their capacity to secrete commercially valuable enzymes can be biotechnologically attractive features. Studies on the production and purification of alkaline protease are reported from *Streptomyces clavuligerus*.[66] More recently, extracellular serine proteases secreted by alkaliphilic actinomycetes are reported and a data base on molecular diversity, phylogeny, and biocatalytic potential have been created.[67,68]

17.4 CONCLUSION

The salt-tolerant and alkaliphilic actinomycetes are explored to a very limited extent for their occurrence, growth characteristics, and secretion of enzymes. It is, therefore, important to pay attention to extremophilic actinomycetes from newer and unexplored habitats for their significance in ecological sustenance and search for novel secondary metabolites. The assessment of the aerial mycelium color appears as a strong phenotypic trait in Streptomycetes classification. The pigmentation of *Streptomyces* is distinct enough to allow ready delineation when combined with other fundamental features. Limited studies have demonstrated the wide occurrence of salt-tolerant and alkaliphilic actinomycetes from saline habitats. Polymerase chain reaction (PCR) combined with DGGE techniques distinguish different species belonging to the same genus. Moreover, DGGE is useful for the identification of large number of isolates and is a powerful fingerprinting method for revealing sequence heterogeneities in the 16S rRNA gene. Besides, the ability to secrete extra cellular enzymes, actinomycetes would provide additional criterion to assess their diversity.

ACKNOWLEDGMENTS

S. D. Gohel gratefully acknowledges the Research Fellowship in Science for Meritorious students (RFSMS) from the University Grant Commission (UGC), New Delhi, India. S. D. Gohel also acknowledges the CSIR (Council of Science and Industrial Research), New Delhi, India for Research Associateship. We acknowledge the support of UGC, New Delhi under various programs including UGC Centre of Advanced Studies. We are also grateful to Saurashtra University for its support and financial assistance.

KEYWORDS

- **DGGE**
- **Diversity**
- **Haloalkaliphilic Actinomycetes**
- **Hydrolytic Enzymes**
- **Phylogeny**

REFERENCES

1. Gohel, S. D.; Sharma, A. K.; Dangar, K. G.; Thakkar F. J.; and Singh S. P. Antimicrobial and biocatalytic potential of haloalkaliphilic actinobacteria. In *Exploitation of Halophilic Bacteria*; Maheshwari, D. K., Saraf, M., Eds.; Springer: Heidelberg, Germany, 2015.

2. Vasavada, S.; Thumar, J.; and Singh, S. P. Secretion of a potent antibiotic by salt-tolerant and alkaliphilic actinomycete *Streptomyces sannanensis* strain RJT-1. *Curr. Sci.* **2006,** *91*(10), 1393–1397.

3. El-Naggar, M. Y.; El-Aassar, S. A.; and Abdul-Gawad, S. M. Meroparamycin production by newly isolated local *Streptomyces* sp. Strain MAR01: Taxonomy, fermentation, purification and structural elucidation. *J. Microbiol.* **2006,** *44*, 432–438.

4. Dietera, A.; Hamm, A.; Fiedler, H. P.; Goodfellow, M.; Muller, W. E.; Brun, R.; and Bringmann, G. Pyrocoll, an antibiotic, antiparasitic and antitumor compound produced by a novel alkaliphilic *Streptomyces* strain. *J. Antibiot.* **2003,** *56*, 639–646.

5. Gullo, V. P.; McAlpine, J.; Lam, K. S.; Baker, D.; and Petersen, F. Drug discovery from natural products. *J. Ind. Microbiol. Biotechnol.* **2006,** *33* (7), 523–531.

6. Pamboukian, C. R. D.; and Facciotti, M. C. R. Production of the antitumoral retamycin during continuous fermentations of *Streptomyces olindensis*. *Process Biochem.* **2004,** *39*, 2249–2255.

7. Ben-Fguira, L. F.; Fosto, S.; Mehdi, R. B.; Mellouli, L.; and Laatsch, H. Purification and structure elucidation of antifungal and antibacterial activities of newly isolated *Streptomyces* sp. strain US80. *Microbiol. Res.* **2005,** *156*, 341–347.

8. El-Naggar, M. Y. Kosinostatin, a major secondary metabolite isolated from the culture filtrate of *Streptomyces violaceusniger* strain HAL64. *J. Microbiol.* **2007,** *45* (3), 262–267.

9. Tang, S. K.; Li, W. J.; Wang, D.; Zhang, Y. G.; Xu, L. H.; and Jiang, C. L. Studies of the biological characteristics of some halophilic and halotolerant actinomycetes isolated from saline and alkaline soils. *Actinomycetologica* **2003,** *17*, 6–10.

10. Gohel, S. D.; and Singh, S. P. Thermodynamics of a Ca²⁺ dependent thermostable alkaline protease from haloalkaliphilic actinomycetes. *Int. J. Bol. Macromol.* **2015,** *72*, 421–429.

11. Gohel, S. D.; and Singh, S. P. Characteristics and thermodynamics of a thermostable protease from a salt-tolerant alkaliphilic actinomycete. *Int. J. Bol. Macromol.* **2013,** *56*, 20–27.

12. Gohel, S. D.; and Singh S. P. Single step purification, characteristics and thermodynamic analysis of a highly thermostable alkaline protease from a salt-tolerant alkaliphilic actinomycete, *Nocardiopsis alba* OK-5. *J. Chromatogr. B.* **2012,** *889–890*, 61–68.

13. Gohel, S. D.; and Singh, S. P. Cloning and expression of alkaline protease genes from two salt-tolerant alkaliphilic actinomycetes in *E. coli. Int. J. Bol. Macromol.* **2012,** *50*, 664–671.

14. Thumar, J. T.; and Singh, S. P. Two - step purification of a highly thermostable alkaline protease from salt-tolerant alkaliphilic *Streptomyces clavuligerus* strain Mit-1. *J. Chromatogr. B.* **2007,** *854*, 198–203.

15. Lopes, A.; Coelho, R. R.; Meirelles, M. N. L.; Branquinha, M. H.; and Vermelho, A. B. Extracellular serine-proteinases isolated from *Streptomyces alboniger*: Partial characterization and effect of aprotinin on cellular structure. *Mem. Inst. Oswaldo. Cruz.* **1999,** *94* (6), 763–770.

16. Rao, M. B.; Tanksale, A. M.; Ghatge, M. S.; and Deshpande, V. V. Molecular and biotechnological aspects of microbial proteases. *Microbiol. Mol. Biol. Rev.* **1998**, *62*, 597–635.

17. Semedo, A. S.; Linhares, A. A.; Gomes, R. C.; Manfio, G. P.; Alviano, C. S.; Linhares, L. F.; and Coelho, R. R. Isolation and characterization of actinomycetes from Brazilian tropical soils. *Microbiol. Res.* **2001**, *155*, 259–266.

18. Nascimento, R. P.; Marques, S.; Alves, L.; Girio, F. M.; Bon, P. S.; Coelho, R. R.; and Amaral-Collaco, M. T. Production and partial characterisation of xylanases from *Streptomyces* sp. Strain AMT-3 isolated from Brazilian cerrado soil using agro-industrial by-products. *Enz. Microbial. Technol.* **2002**, *31*, 549–555.

19. Myers, R. M.; Maniatis, T.; and Lerman, L. S. Detection and localization of single base changes by denaturing gradient gel electrophoresis. *Methods Enzymol.* **1987**, *155*, 501–527.

20. Fischer, S. G.; and Lerman, L. S. Length independent separation of DNA restriction fragments in two dimensional gel electrophoresis. *Cell* **1979**, *16*, 191–200.

21. Muyzer, G.; and Smalla, K. Application of denaturing gradient gel electrophoresis (DGGE) and temperature gradient gel electrophoresis (TGGE) in microbial ecology. *Antonie Van Leeuwenhoek* **1998**, *73*, 127–141.

22. Imada, C. Enzyme inhibitors and other bioactive compounds from marine actinomycetes. *Antonie Van Leeuwenhoek* **2005**, *87* (1), 59–63.

23. Manam, R. R.; Teisa, S.; White, D. J.; Nicholson, B.; Neuteboom, S. T.; Lam, K. S.; Mosca, D. A.; Lloyd, G. K.; and Potts, B. C. Lajollamycin, a nitrotetraenespiro-beta-lactone-gamma-lactum antibiotic from the marine actinomycetes *Streptomyces nodosus*. *J. Nat. Prod.* **2005**, *68*, 240–243.

24. Tang, S. K.; Li, W. J.; Wang, D.; Zhang, Y. G.; Xu, L. H.; and Jiang, C. L. Studies of the biological characteristics of some halophilic and halotolerant actinomycetes isolated from saline and alkaline soils. *Actinomycetologica* **2003**, *17*, 6–10.

25. Tiwari, K.; and Gupta, R. K. Diversity and isolation of rare Actinomycetes: An overview. *Crit. Rev. Microbiol.* **2012**, *39*, 256–294.

26. Seong, C. N.; Choi, J. H.; and Baik, K. S. An improved selective isolation of rare Actinomycetes from forest soil. *J. Microbiol.* **2001**, *39*, 17–23.

27. Hamaki, T.; Suzuki, M.; Fudou, R.; Jojima, Y.; Kajiura, T.; Tabuchi, A.; Sen, K.; and Shibai, H. Isolation of novel bacteria and Actinomycetes using soil-extract agar medium. *J. Biosci. Bioeng.* **2005**, *99*, 485–492.

28. Tan, G. Y.; Ward, A. C.; and Goodfellow, M. Exploration of Amycolatopsis diversity in soil using genus-specific primers and novel selective media. *Syst. Appl. Microbiol.* **2006**, *29*, 557–569.

29. Qiu, D.; Ruan, J.; and Huang, Y. Selective isolation and rapid identification of members of the genus *Micromonospora*. *Appl. Environ. Microbiol.* **2008**, *74*, 5593–5597.

30. Qin, S.; Li, J.; Chen, H. H.; Zhao, G. Z.; Zhu, W. Y.; Jiang, C. L.; Xu, L. H.; and Li, W. J. Isolation, diversity, and antimicrobial activity of rare Actinobacteria from medicinal plants of tropical rain forests in Xishuangbanna, China. *Appl. Environ. Microbiol.* **2009**, *75*, 6176–6186.

31. Nakaew, N.; Pathom-aree, W.; and Lumyong, S. Generic diversity of rare Actinomycetes from Thai cave soils and their possible use as new bioactive compounds. *Actinomycetologica* **2009**, *23*, 21–26.

32. Baskaran, R.; Vijayakumar, R.; and Mohan, P. M. Enrichment method for the isolation of bioactive Actinomycetes from mangrove sediments of Andaman Islands, India. *Malaysian J. Microbiol.* **2011**, *7*, 26 32.

33. Istianto, Y.; Koesoemowidodo, R. S. A.; Saputra, H.; Watanabe, Y.; Pranamuda, H.; and Marwoto, B. Application of phenol pretreatment for the isolation of rare Actinomycetes from Indonesian soil. *Microbiology (Indonesia)* **2012**, *6*, 42–47.

34. Wang, D. S.; Xue, Q. H.; Zhu, W. J.; Zhao, J.; Duan, J. L.; and Shen, G. H. Microwave irradiation is a useful tool for improving isolation of Actinomycetes from soil. *Microbiology* **2013**, *82*, 102–110.

35. Seong, C. N.; Choi, J. H.; and Baik, K. S. An improved selective isolation of rare Actinomycetes from forest soil. *J. Microbiol.* **2001**, *39*, 17–23.

36. Kim, C. J.; Lee, K. H.; Shimazu, A.; Kwon, O. S.; and Park, D. J. Isolation of rare Actinomycetes in various types of soil. Korean *J. Appl. Microbiol. Biotechnol.* **1995**, *23*, 36–42.

37. Niyomvong, N.; Pathom-aree, W.; Thamchaipenet, A.; and Duangmal, K. Actinomycetes from tropical limestone caves. *Chiang Mai J. Sci.* **2012**, *39*, 373–388.

38. Shirling, E. B.; and Gottlieb, D. Methods for characterization of *Streptomyces* species. *Int. J. Syst. Bacteriol.* **1966**, *16*, 313–340.

39. Thangapandian, V.; Ponmurugan, P.; and Ponmurugan, K. Actinomycetes diversity in the rhizosphere soils of different medicinal plants in Kolly Hills-Tamilnadu, India, for secondary metabolite production. *Asian J. Plant Sci.* **2007**, *6* (1), 66–70.

40. Collins, M. D.; Lawson, P. A.; Labrenz, M.; Tindall, B. J.; Weiss, N.; and Hirsch, P. *Nesterenkonia lacusekhoensis* sp. Nov., isolated from hypersaline Ekho Lake, East Antarctica, and emended description of the genus *Nesterenkonia. Int. J. Syst. Evol. Microbiol.* **2002**, *52*, 1145–1150.

41. Al-Zarban, S. S.; Abbas, I.; Al-Musallam, A. A.; Steiner, U.; Stackebrandt, E.; and Kroppenstedt, R. M. *Nocardiopsis halotolerans* sp. nov. isolated from salt mars soil Kuwait. *Int. J. Syst. Evol. Microbiol.* **2002**, *52* (2), 525–529.

42. Kokare, C. R.; Mahadik, K. R.; Kadam, S. S.; and Chopade B. A. Isolation, characterization and antimicrobial activity of marine halophilic *Actinopolyspora* species AH1 from the west coast of India. *Curr. Sci.* **2004**, *15*, 81–88.

43. Li, W. J.; Xu, P.; Zhang, L. P.; Tang, S. K.; Cui, X. L.; Mao, P. H.; Xu, L. H.; Schumann, P.; Stackebrandt, E.; and Jiang, C. L. *Streptomonospora alba* sp. nov., a novel halophilic actinomycete, and emended description of the genus *Streptomonospora* Cui et al., 2001. *Int. J. Syst. Evol. MIcrobiol.* **2003**, *35*, 1421–1425.

44. Cui, X. L.; Mao, P. H.; Zeng, M.; Li, W. J.; Zhang, L. P.; Xu, L. H.; and Jiang, C. L. *Streptomonospora salina* gen. nov., sp. nov., a new member of the family Nocardiopsaceae. *Int. J. Syst. Evol. Microbiol.* **2004**, *50* (5), 1909–1913.

45. Hozzein, W. N.; Li, W. J.; Ibrahim, A. M.; Hammouda, O.; Mousa, A. S.; Xu, L. H.; and Jiang, C. L. *Nocardiopsis alkaliphila* sp. nov., a novel alkaliphilic actinomycete isolated from desert soil in Egypt. *Int. J. Syst. Evol. Microbiol.* **2004**, *54*, 247–252.

46. Li, W. J.; Zhang, Y. G.; Zhang, Y. Q.; Tang, S. K.; Xu, P.; Xu, L. H.; and Jiang, C. L. *Streptomyces sodiiphilus* sp. nov., a novel alkaliphilic actinomycete. *Int. J. Syst. Evol. Microbiol.* **2005**, *55*, 1329–1333.

47. Li, W. J.; Kroppenstedt, R. M.; Wang, D.; Tang, S. K.; Lee, C.; Park, J.; Kim, J.; Xu, H.; and Jiang, L. Five novel species of the genus *Nocardiopsis* isolated from hypersaline soils and emended description of *Nocardiopsis salina* Li et al. 2004. *Int. J. Syst. Evol. Microbiol.* **2006**, *56* (5), 1089–1096.

48. Li, J.; Zhang, Q.; Schumann, P.; Chen, H.; Hozzein, N.; Tian, P.; XU, H.; and Jiang, L. *Kocuria aegyptia* sp. nov., a novel actinobacteria isolated from a saline, alkaline desert soil in Egypt. *Int. J. Syst. Evol. Microbiol.* **2006**, *56* (4), 733–737.

49. Brenner, J.; Staley, T.; and Krieg, R. Classification of prokaryotic organisms and the concept of bacterial speciation. In *Bergey's Manual of Systematic Bacteriology*, 2nd eds; Boone, D. R, and Castenholz, R. W., Eds.; Springer-Verlag: New York, Berlin, Heidelberg, 2001, *1*, 27–48.

50. Rossello R.; and Amann, R. The species concept for prokaryotes. *FEMS Microbiol. Reviews* **2001**, *25*, 39–67.

51. Rintala, H.; Nevalainen, A.; and Suutari, M. Diversity of Streptomycetes in water-damaged building materials based on 16S rDNA sequences. *Lett. Appl. Microbiol.* **2002**, *34*, 439–443.

52. Antony-Babu, S.; and Goodfellow, M. Biosystematics of alkaliphilic streptomy-cetes isolated from seven locations across a beach and dune sand system. *Antonie Van Leeuwenhoek* **2008**, *94* (4), 581–591.

53. Valdes, M.; Perez, Nestor, O. P.; Santos, E.; Mellado, J. C.; Cabriales, J. P.; Normand, P.; and Hirsch, A. M. Non-*Frankia* actinomycetes isolated from surface sterilized roots of *Casuarina equisetifolia* fix nitrogen. *Appl. Env. Microbial.* **2005**, *71* (1), 460–466.

54. Mitsuiki, S.; Yasushi, M.; Masatoshi G.; Masaaki O.; Hiroaki K.; Kensuke F.; and Tatsuzo, O. Identification of an alkaliphilic actinomycetes producing PrpSc-degrading enzyme. *Mem. Fac. Agr. Kagoshima Univ.* **2007**, *42*, 11–16.

55. Heuer, H.; and Smalla, K. Application of denaturing gradient gel electrophoresis (DGGE) and temperature gradient gel electrophoresis (TGGE) for studying soil micro-bial communities. In *Modern Soil Microbiology*; Elsas, J. D., Wellington, E. M. H., and Trevors, J. T., Eds., Marcel Dekker: New York, 1997, pp 353–373.

56. Riesner, G.; Steger, R.; Zimmat, R.; Owens, A.; Wagenhofer, M.; Hillen, W.; Vollbach, S.; and Henco, K. Temperature-gradient gel electrophoresis of nucleic acids: Analysis of conformational transitions, sequence variations, and protein-nucleic acid interactions. *Electrophoresis* **1989**, *10*, 377–389.

57. Rosenbaum, V.; and Riesner, D. Temperature gradient gel electrophoresis; thermody-namic analysis of nucleic acids and proteins in purified form and in cellular extracts. *Biophys. Chem.* **1987**, *26*, 235–246.

58. Legatzki, A.; Ortiz, M.; Neilson, W.; Casavant, R.; Palmer, W.; Rasmussen, C.; Pryor, M.; Pierson, S.; and Maier, M. Factors influencing observed variations in the struc-ture of bacterial communities on calcite formations on Kartchner Caverns, AZ, USA. *Geomicrobiol. J.* **2012**, *29*, 422–434.

59. Singh, S. P.; Thumar, J. T.; Gohel, S. D.; and Purohit, M. K. Molecular diversity and enzymatic potential of salt-tolerant alkaliphilic actinomycetes. In *Current Research Technology and Education Topics in Applied Microbiology and Microbial Biotechnology*; Mendez-Vilas, A., Ed.; Formatex Research Center: Badajoz, Spain, 2010, pp 280–286.

60. Walter, J.; Tannock, W.; Tilsala, A.; Rodtong, S.; Loach, M.; Munro, K.; and Alatossava, T. Detection and identification of gastrointestinal *Lactobacillus* species by using dena-turing gradient gel electrophoresis and species-specific PCR primers. *Appl. Environ. Microbiol.* **2000**, *66* (1), 297–303.

61. McAuliffe, L.; Ellis, J.; Lawes, R.; Ayling, D.; and Nicholas, A. 16S rDNA PCR and denaturing gradient gel electrophoresis; a single generic test for detecting and differenti-ating *Mycoplasma* species. *J. Med. Microbiol.* **2005**, *54* (8), 731–739.

62. Al-Soud, W. A.; Bennedsen, M.; On, S. L.; Ouis, I. S.; Vandamme, P.; Nilsson, H. O.; Ljungh, A.; and Wadstrom, T. Assessment of PCR-DGGE for the identification of diverse *Helicobacter* species, and application to faecal samples from zoo animals to determine *Helicobacter* prevalence. *J. Med. Microbiol.* **2003**, *52* (9), 765–771.

63. Gong, J.; Forster, J.; Yu, H. Chambers R., Sabour, M., Wheatcroft, R., and Chen, S. Diversity and phylogenetic analysis of bacteria in the mucosa of chicken ceca and comparison with bacteria in the cecal lumen. *FEMS Microbiol. Lett.* **2002,** *208* (1), 1–7.
64. Singh, S. P.; Thumar, J. T.; Gohel, S. D.; Kikani, B.; Shukla, R.; Sharma, A.; and Dangar, K. Actinomycetes from marine habitats and their enzymatic potential. In *Marine Enzymes for Biocatalysis*; Tricone, A., Ed.; Woodhead Publishing Series in Biomedicine (Oxford) Ltd: Oxford, UK, 2013; pp 191–214.
65. Singh, S. P.; Raval, V. H.; Purohit, M. K.; Thumar, J. T.; Gohel S. D.; Pandey S.; Akbari V.; and Rawal C. Haloalkaliphilic bacteria and actinobacteria from the saline habitats: new opportunities for biocatalysis and bioremediation. In *Microbes in Environmental Management and Biotechnology*; Satyanarayan, T., Johri, B. N., and Prakash, A., Eds.; Springer: New York, 2012; pp 415–429.
66. Moreira, A.; Cavalcanti, T.; Duarte. S.; Tambourgi, B.; Magalhaes H.; Silva, L.; Porto, L.; and Filho, L. Partial characterization of proteases from *Streptomyces clavuligerus* using an inexpensive medium. *Braz. J. Microbiol.* **2001,** *32*, 623–629.
67. Sharma, K.; Gohel, S. D.; and Singh S. P. Actinobase: A database on molecular diversity, phylogeny and biocatalytic potential of salt tolerant alkaliphilic actinomycetes. *Bioinformation* **2012,** *8* (11), 535–538.
68. Mehta, V. J.; Thumar, J. T.; and Singh, S. P. Production of alkaline protease from an alkaliphilic actinomycete. *Bioresour. Technol.* **2006,** *97* (14), 1650–1654.

CHAPTER 18

MICROBIAL POPULATION DYNAMICS OF EASTERN GHATS OF ANDHRA PRADESH FOR XYLANASE PRODUCTION

G. RAMANJANEYULU*, A. RAMYA, and B. RAJASEKHAR REDDY

Department of Microbiology, Sri Krishnadevaraya University, Anantapuramu, Andhra Pradesh – 515003, India.

Corresponding author: G. Ramanjaneyulu.
E-mail: ramanj.003@gmail.com

CONTENTS

Abstract 356
18.1 Introduction 356
18.2 Materials and Methods 357
18.3 Results 361
18 4 Discussion 366
18.5 Conclusion 368
Acknowledgments 369
Keywords 369
References 369

ABSTRACT

The Eastern Ghats are a discontinuous hill range in peninsular India and stretch from the Mahanandi Basin in the north to Nilagiri Hills in the south, covering a distance of 1700 km and spreading over 75,000 km, with an average elevation of the mountain range of 600 m falling between 13°30'–19°07' N and 77°28'–84°45' E. The microbial populations are not explored in the Eastern Ghats for xylanases. In the present study, soil samples from the Eastern Ghats of Andhra Pradesh at different places with coverage of shrubs to large trees were collected and analyzed for soil properties and microbial populations on different media at two different temperatures 30 °C and 37 °C. The media used for the determination of the population of fungi, bacteria, and actinomycetes were Rose Bengal agar medium (RBA)/ Czapekdox agar medium (CZA), mineral salts agar medium (MSM), and starch-casein agar medium (SCAM), respectively. The carbon source of all the media was amended with 0.1% birch wood xylan or 10 g/L xylose, and the pH was adjusted to 5.0 for fungi and 7.0 for bacteria and actinomycetes. The microbial populations were expressed in CFU/g of dry weight soil. Xylose-utilizing bacterial populations ranged from 4×10^5 to 54×10^5, 8×10^5 to 52×10^5 in forest soils of the Eastern Ghats at 30 °C and 37 °C, whereas the xylan-utilizing bacterial population occurred at 2×10^4 to 45×10^4, 5×10^4 to 39×10^4 in the same forest soils at 30 °C and 37 °C, respectively. Xylose-utilizing actinomycetes populations at 2×10^4 to 14×10^4, 3×10^4 to 14×10^4 were recorded as against the xylan-utilizing actinomycetes population density of 2×10^3 to 11×10^3, 2×10^3 to 11×10^3 at 30 °C and 37 °C in the forest soil samples, respectively. Xylose-utilizing fungal populations at a density of 12×10^3 to 96×10^4, 4×10^3 to 35×10^4 were observed as against xylan-utilizing fungal populations at 12×10^3 to 94×10^4, 5×10^3 to 27×10^4 at 30 °C and 37 °C in the forest soils, respectively. Some of the fungal cultures isolated from forest soil samples exhibited xylanase activity within a range of 400 to 4000 U/ml and has the potential for harnessing several applications in paper, pulping, and bioethanol industry.

18.1   INTRODUCTION

Forest soils are considered to be a part of the natural ecosystem[1] and the ecosystem runs spontaneously because of the fewer disturbances.[2] Organic matter in the form of litter is continuously generated in forest soils because of the primary production of vegetation. Litter is mainly constituted with lignocellulosic mass that consists of cellulose, hemicellulose, lignin, and pectin. Soil microflora plays a pivotal role in the cycling of organic nutrients including lignocellulosic mass in the biogeochemical cycles.[3–7] Soil microorganisms, consisting largely of bacteria, fungi, and actinomycetes, are important in regulating the ecosystem processes such as decomposition, energy flow, carbon storage, and trace gas fuels.[8] Most of the nutrient requirements of plants are provided by soil microbial processes via the mineralization of soil organic nutrients, which greatly regulate the net primary production of natural ecosystems.[8]

Soil organic matter decomposition is one of the dynamic parameter that can be accomplished by bacteria, fungi, and actinomycetes. Xylan, a major constituent of hemicelluloses, occurs in considerable proportions in the organic matter (OM). Understanding of the population of xylan utilizing microflora in different ecosystems, forest in particular, is a basic step for the exploration of xylanolytic organisms. Extensive work has been reported on microbial ecology of different ecosystems, particularly, agriculture soils, grasslands, and barren fields with a focus on cellulolytic organisms, amylolytic organisms,[9–11] nitrifying microbes,[12–15] and proteolytic organisms.[16,17] As there are no reports on the studies related to population dynamics of xylan-utilizing microflora from the soils of the Eastern Ghats of Andhra Pradesh, India, the present study was aimed at focusing on the populations of xylan and xylose utilizing microflora in the forest soils of the Eastern Ghats of Andhra Pradesh.

18.2   MATERIALS AND METHODS

18.2.1   STUDY AREA

The population of xylan- and xylose-utilizing microflora was studied across the Eastern Ghats of Andhra Pradesh, India which includes Srisailam (M), Rudrakoduru (O), Mahanandi (G), Ahobilam (B), Gundla Brahmeswaram (A), Dornala (P), Atmakur (L) (Nallamala forest, Kurnool District), Lankamalla (N), Idupulapaya (J) (Kadapa District), Tirupati (E), Bakarapet (F), Talakona (C), and Horsley Hills (D) (Chittoor District). The Eastern Ghats are a discontinuous hill range in peninsular India stretching from the Mahanandi Basin in the north to the Nilagiri Hills in the south, covering a distance of 1700 km and spreading over 75,000 km. The average elevation of the mountain range is about 600 m and the highest peak is about 1700 m

Shevaroy hills which falls between 13°30'–19°07' N and 77°28'–84°45' E (Figure 18-1).

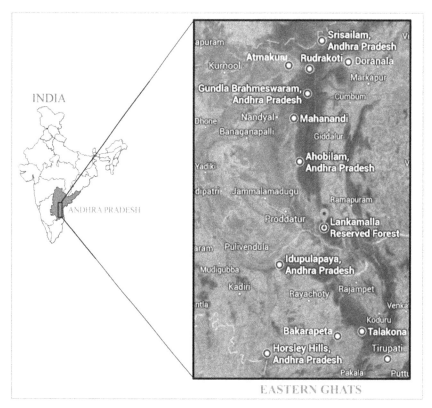

FIGURE 18-1   Sampling locations in the Eastern Ghats of Andhra Pradesh, India.

18.2.2   SOIL SAMPLE COLLECTION

The soil samples were collected from the rhizosphere of various plants by removing the superficial layer (approximately 0.5 mm) within a depth of 6–12 cm. At each site, 6–10 samples were collected individually within a range of 5–20 km distance and transported to the laboratory and stored in a refrigerator at 4 °C until further study.

18.2.3 PHYSICO-CHEMICAL PROPERTIES OF SOILS

The pH of the soils were measured with a pH meter (Elico) at 1:2.5 soil-water suspension and the salinity of the soils were determined in terms of electric conductivity (EC) by a conductivity meter (Elico).[18] Organic carbon was determined by the potassium dichromate method,[19] The available nitrogen was determined by the alkaline permanganate method,[2] phosphorous was estimated by the Olsen's method,[20] whereas potassium was determined by a flame photometer[21] with neutral normal ammonium acetate. Some of the essential micro nutrients were also quantified by Atomic Absorption Spectroscopy (AAS).

18.2.4 CULTURE MEDIA USED

The media used in this study for the culturing of fungi were Rose Bengal agar medium (RBA), mineral salts agar medium (MSM) for bacteria, and starch-casein agar medium (SCAM) for actinomycetes. The RBA was prepared by dissolving peptone: 5 g, KH_2PO_4: 1 g, $MgSO_4.7H_2O$: 1.0 g, streptomycin: 30 mg in 1000 ml. The MSM was prepared by dissolving NH_4NO_3: 1.5 g, KH_2PO_4: 2.5 g, NaCl: 1.0 g, $MgSO_4$: 1.5 g, $MnSO_4$: 0.01 g, $FeSO_4$: 0.005 g, $CaCl_2$: 0.05 g, in 1000 ml distilled water. Starch-casein agar medium included ingredients such as 0.1% casein (vitamin free) 0.3 g, KNO_3: 2.0 g, $MgSO_4.7H_2O$: 2.0 g, K_2HPO_4 : 2.0 g, $MnSO_4$: 0.05 g, $CaCO_3$: 0.02 g, $FeSO_4.7H_2O$: 0.01 g, agar : 20 g, nystatin: 50.0 mg, and penicillin 0.8 mg dissolved in 1000 ml of distilled water. All the above media were amended with xylose: 10.0 g/L or birch wood xylan 0.1%, and the pH was adjusted to 5.0 for fungi and 7.0 for bacteria and actinomycetes, and sterilized at 121 °C and 15 lbs of steam pressure for 20 min and poured into sterile petri dishes.

18.2.5 POPULATION STUDY

Each soil sample was dried at an ambient temperature in the laboratory under shade and sieved through 2 mm sieve and mixed to make composite soil sample for a particular location. Microbial population studies were carried out from the composite soil sample, 1 g of the composite soil sample was dissolved in 10 ml of sterile distilled water. This solution was diluted decimally (10^{-1} to 10^{-8}) and 0.1 ml of aliquots from appropriate dilutions for bacteria, fungi, and actinomycetes were plated on respective petri dishes containing 20–25 ml of different media and spread uniformly by a sterile glass spreader. After spreading, the plates were incubated at 30 °C and 37 °C in an incubator (Technico) and observed every day over a period of 7 days; the colony forming units (CFU) appeared in the plates were calculated by following formula:

CFU or viable cells of dry soil = mean plate count × dilution factor/dry weight of the soil taken.

18.2.6   SCREENING FOR XYLANASE-PRODUCING FUNGAL MICROFLORA

In view of the importance of xylanase in several industrial applications and its extracellular secretion nature, only fungal cultures derived from the present study were screened for xylanase production on plate assay followed by submerged state fermentation (SmF). Plate screening was performed on MSM amended with 0.1% xylan. The plates were inoculated with 5-day-old fungal cultures. An agar plug of 0.5 mm size of the 5-day-old culture was plucked with a cock borer and placed at the center of the MSM plate and incubated at 30 °C for five days. After five days, the plates were flooded with 10–15 ml of 0.01% Congo red[22] and de-stained with 1% NaCl and the clear zone formed around the fungal culture was measured in centimeters. On the basis of the diameter of the clear zone, some of the potential fungal cultures were selected and further checked for the production of xylanase in liquid medium in shake conditions (SmF). Thirty milliliters of liquid MSM in 250 ml Erlenmeyer flasks was inoculated with the addition of five 0.5 mm agar plugs of the 5-day-old culture and incubated at 30 °C and 150 rpm for 7 days and the xylanase activity was assayed according to the standard method[23] (Figure 18-2).

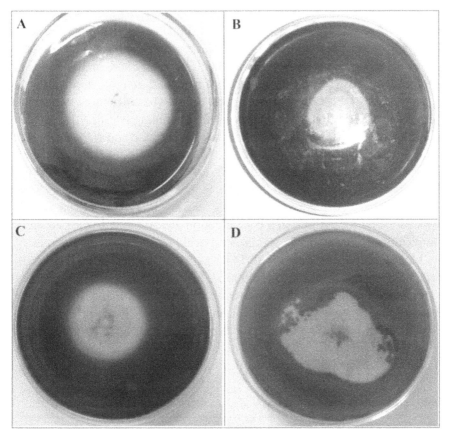

FIGURE 18-2 A–D plate assay for the screening of xylanase producing microflora on mineral salts agar medium (MSM) amended with 0.1% birch wood xylan.

18.3 RESULTS

18.3.1 PROPERTIES OF SOILS

The principle geological formations of soils were Charnockites and Khondalites in the northern part, and quartz and slate formation in the southern part. The average rainfall varies significantly between 600 mm in the southern Eastern Ghats to 1700 mm in the northern Eastern Ghats; maximum rainfall occurs between June and October. The forests in the Eastern Ghats are broadly classified into Tropical dry deciduous, Tropical moist deciduous,

Tropical semi evergreen, tropical thorn (scrubs), and Tropical dry evergreen types.[24, 25] The relative humidity is high at the Horsley Hills and Talakona followed by Srisailam and Ahobilam. The temperatures also vary seasonally in the study area between 12 and 31 °C in winter and 26 and 45 °C in summer. The soil samples in the study zone at different locations differed in the physico-chemical properties (Table 18-1). The soil samples collected from different locations broadly varied in pH ranging from 5.1–7.4 with minimum and maximum recorded for samples of the P and L locations. EC of the soil samples spanned from 0.1 to 0.58 for samples of L and P, respectively. Organic carbon ranged from 0.46–1.41 represented by samples of the P location on the lower side and samples of the L location on the higher side. Highest nitrogen concentration at 241 kg/ha was observed in samples of the B location and lowest nitrogen concentration at 150 kg/ha was recorded in samples of Q. Phosphorous was highest with 114.96 kg/ha in samples of the J location and was lowest with 54.61 kg/ha in samples of the L location. Potassium ranged between 1545.06 and –733.95 kg/ha in samples of J and L, respectively. All the tested micro nutrients Cu, Mn, Fe, and Zn were also present in below to above critical concentration levels in the present study. Zinc occurred in the soil samples within a range of 0.12 to 0.77 ppm; Cu, Fe, and Mn ranged from 0.945–2.1 ppm, 1.11–3.2 ppm, and 0.7–1.94 ppm in different soil samples, respectively.

TABLE 18-1   Physico-Chemical Properties of the Soil Samples

S. No.	Sample ID	pH	EC	N	P	K	Cu	Mn	Fe	Zn	OM
1	A	7.4	0.31	182	79.8	763.6	1.32	1.64	2.3	0.56	0.68
2	B	6.5	0.27	241	94.2	921.3	1.01	0.7	2.3	0.7	0.71
3	C	6.6	0.3	212	79.77	1072.11	1.96	0.89	2.46	0.67	0.99
4	D	6.0	0.4	218	82.2	892.1	2.1	1.94	3.2	0.6	0.63
5	E	6.4	0.32	202	81.4	998.4	1.2	0.84	2.3	0.54	0.73
6	F	6.2	0.28	198	74.1	1004.3	1.82	0.73	2.9	0.64	0.81
7	G	5.8	0.18	169	82.4	942.8	1.24	1.8	1.4	0.3	0.6
8	J	7	0.24	195	114.96	1545.06	1.68	1.64	1.96	0.74	0.77
9	L	6.6	0.1	184	54.61	733.95	1.45	0.89	2.46	0.64	1.41
10	M	5.4	0.32	178	85.26	1145.89	1.37	1.87	1.11	0.12	0.84
11	N	7.4	0.21	182	104.9	1148.0	1.39	1.42	1.7	0.77	0.94
12	O	6.1	0.14	172	68.4	894.1	1.23	1.4	2.8	0.6	0.74
13	P	5.1	0.31	174	98.4	1108.8	1.4	1.92	2.3	0.34	0.46

In the present study, the soil samples in the study zone at different locations (Table 18-1) in the Eastern Ghats have different pH, physical and chemical properties like organic matter content, nutrient concentration (macro and micro), and also various plant populations. Bacterial population counts were highest in the soil sample of P (Table 18-2) and lowest in the soil sample of A and L on xylose containing medium at both of the temperatures 30 °C and 37 °C. Highest density of the population of xylan-utilizing bacteria was also recorded in the soil samples of the P location followed by O and M. Lowest population of xylan-utilizing bacteria was present in locations - A and L, respectively at both the temperatures. Overall, the bacterial population occurred in the soil samples of different locations within a range of 2 to 54×10^5 cfu/g of dry weight of the soil at both xylose and xylan as the carbon source at 30 °C and 37 °C, respectively.

TABLE 18-2 Xylan and Xylose Utilizing Bacterial Population in Soil Samples of the Eastern Ghats

S. No.	Sample ID	Bacterial Population on MSM Medium Amended with			
		Xylose (Expressed as $\times 10^5$ cfu/g of soil)		Xylan (Expressed as $\times 10^4$ cfu/g of soil)	
		30 °C	37 °C	30 °C	37 °C
1	A	4	8	2	5
2	B	15	30	7	25
3	C	12	22	5	11
4	D	8	18	3	7
5	E	31	12	23	8
6	F	28	33	20	18
7	G	40	28	25	22
8	J	42	17	28	9
9	L	45	6	35	3
10	M	50	49	42	39
11	N	16	30	6	16
12	O	18	52	9	39
13	P	54	11	45	4

Soil samples of different locations in the Eastern Ghats harbored xylose-utilizing actinomycetes within a range of 2 to 16 × 10⁴ at both the temperatures as reflected by the colonies on the plates (Table 18-3), whereas the population of xylan-utilizing actinomycetes occurred in the soil samples to the extent of 2 to 11 × 10² at 30 and 37 °C (Table 18-3). The fungal populations were highest in the soil sample of P (Table 18-4) at both the temperatures on RBA followed by samples of L, C, J, and M, respectively.

TABLE 18-3 Xylan and Xylose Utilizing Actinomycetes Population in the Soil Samples of the Eastern Ghats

| S. No. | Sample ID | Actinomycetes Population on SCAM Medium Amended with | | | |
| | | Xylose (Expressed as ×10⁴ cfu/g of soil) | | Xylan (Expressed as ×10² cfu/g of soil) | |
		30 °C	37 °C	30 °C	37 °C
1	A	2	8	2	6
2	B	8	3	5	2
3	C	4	9	3	5
4	D	7	12	4	7
5	E	16	5	2	3
6	F	9	6	7	4
7	G	11	13	8	10
8	J	11	6	6	3
9	L	12	4	7	2
10	M	14	7	8	4
11	N	5	14	3	8
12	O	3	12	2	10
13	P	8	9	5	7

The total population of xylose-utilizing and xylan-utilizing microflora in the soil samples at different locations in the Eastern Ghats included bacterial microflora, fungal microflora, and actinomycetes and was computed by adding populations of the respective three categories at temperatures of 30 and 37 °C (Table 18-5). The total population of xylose-utilizing microflora in the soil samples of different locations in the Eastern Ghats at the two temperatures outnumbered the total population of xylan-utilizing microflora in same soil samples at the respective locations under identical conditions.

TABLE 18-4 Xylan and Xylose Utilizing Fungal Population in the Soil Samples of the Eastern Ghats

S. No.	Sample ID	Fungal Population (cfu/g of soil) on RBA Medium Amended with			
		Xylose		Xylan	
		30 °C	37 °C	30 °C	37 °C
1	A	29×10^5	18×10^5	15×10^5	26×10^4
2	B	38×10^2	24×10^2	11×10^6	47×10^3
3	C	12×10^3	18×10^5	11×10^3	33×10^3
4	D	33×10^3	19×10^3	12×10^4	86×10^3
5	E	58×10^4	46×10^4	76×10^4	58×10^3
6	F	42×10^4	39×10^3	12×10^4	7×10^4
7	G	67×10^5	48×10^4	76×10^4	87×10^4
8	J	23×10^4	17×10^3	24×10^4	29×10^4
9	L	26×10^3	9×10^3	19×10^3	16×10^3
10	M	46×10^3	37×10^3	61×10^3	52×10^3
11	N	32×10^6	18×10^2	24×10^2	27×10^2
12	O	89×10^2	55×10^2	48×10^2	68×10^2
13	P	36×10^2	32×10^2	62×10^2	47×10^2

The screening of the fungal cultures for xylanolytic activity in the plate method showed a clear zone after de-staining within a range of 0.5–11 cm (Figure 18-2A–D). The cultures showing the zone of clearance with a diameter greater than 5.0 cm were further tested for the production of xylanase in SmF. In SmF, all the selected cultures produced xylanase within a range of 400–4000 U/min.

TABLE 18-5   Xylan and Xylose Utilizing Total Microbial Population in the Soil Samples of the Eastern Ghats

S. No.	Sample ID	Total Microbial Population (Expressed as $\times 10^5$ cfu/g of soil)			
		Xylose		Xylan	
		30 °C	37 °C	30 °C	37 °C
1	A	34	27	15	3
2	B	16	30	111	3
3	C	13	23	0.6	1
4	D	9	19	2	2
5	E	38	17	10	14
6	F	33	34	3	26
7	G	108	34	10	11
8	J	45	17	5	39
9	L	46	6	35	0.5
10	M	52	50	5	4
11	N	336	31	0.6	2
12	O	18	53	1	40
13	P	55	12	46	0.5

18.4   DISCUSSION

The microbial populations in the forest soil were greatly affected by the physico-chemical properties of the soil and the cover of vegetation[26]. The pH, displayed by the soil samples at different locations in the Eastern Ghats in the present study, fall within a moderate range of 5–7.4, favorable for the growth of vegetation and microorganisms. Salinity is considered as a natural problem under arid and semi-arid conditions. However, the soil samples of the Eastern Ghats at different locations in the present study contained low amounts of salts as reflected by EC and corresponds to the non-saline category conducive for the establishment of vegetation and microbial growth under favorable conditions. The elements—Nitrogen, Phosphorous, and Potassium were available in considerable but varying proportions in the soil samples of the present study, for the growth of microorganisms and plants. Kanazawa et al. and Belay et al.[27,28] also reported a direct effect of NPK on the number of bacteria, actinomycetes, and fungi in a cultivated soil. On the other hand, the relationship between the population of fungi and actinomycetes and the soil pH was far weaker and organic carbon, total nitrogen, or the C:N ratio of the soil could not improve the soil microflora population over pH alone.[29]

Microelements such as Cu, Fe, Zn, and Mn were available in adequate amounts in the soil samples of the Eastern Ghats supporting the life of the microorganisms and plants. Soil organic matter plays a key role in the biological processes and also on the chemical and physical properties of soils. It provides energy for soil microbial community, increases cation exchange capacity, and ameliorates soil aggregate and structure.[30] Organic carbon was limited and did not exceed 1% in the soil samples of all the locations except for one (L).

Most of the information on microbial population in different ecosystems-agricultural fields, green pastures, degraded soils, and barren fields was obtained on the basis of culturing of culturable below ground life forms on energy rich medium. Microbial populations were further subcategorized on the basis of their ability to utilize the type of substrate into amylolytic microbes,[9–11] cellulolytic microbes,[10,31,32] ligninolytic microbes,[33–35] nitrifying microbes,[12–15] diazotroph microbes,[36–38] and proteolytic microbes.[16,17] Information on the population of these categories in different ecological niches is well documented. The population of xylan and xylose utilizing microorganisms in the soils of the Eastern Ghats of Andhra Pradesh was enumerated in the present study because of their importance in the utilization of xylan, which is a major constituent of hemicelluloses in the lignocellulosic biomass. The population of xylan—utilizing actinomycetes was lower than the population of xylan-utilizing fungi and xylan—utilizing bacteria in all the soil samples collected at different locations of the Eastern Ghats.

The present study on the enumeration of the population of xylan- and xylose-utilizing microbes in the forest soils appears to be the first report to the best of our knowledge. Xylan-utilizing fungal microflora was dominant in the soil samples at about 50% of the locations in the Eastern Ghats tested in the present study and constituted more than 50% of the total microbial population. In the forest soils, a greater diversity of the litter types can support a greater fungal diversity.[39,40] The distinct composition of the microbial communities in the different forest types is likely as a result of the content of labile and recalcitrant compounds released during litter decomposition. Similarly, the occurrence of xylan-utilizing and xylose-utilizing microflora in the forest soils of the Eastern Ghats in the present study could be attributed to the xylan and xylose released during the degradation of litter.

In forest soils, a greater diversity of litter types can support a greater fungal diversity.[39,40] The distinct composition of the microbial communities in different forest types is likely a result of the content of labile and recalcitrant compounds released during litter decomposition.

The population of microbial community depends on the interaction between plant species and soil.[41,42] Soil bacteria have been found to be dominant in microbial communities during the early successional stages, and fungi gradually became predominant with further succession.[43] The positive effects of plant diversity on soil microbial diversity may be caused by an increase in the quality and quantity of organic material and energy flow into the soils.[41,44–46] A higher litter diversity and decomposability resulting from a greater diversity of plant species[47,48] may result in a higher number of micro-niches available for the soil microorganisms. Some of the fungal cultures derived from this study exhibited xylanase activity as high as 4000 U/ml. Senthilkumar et al.,[49] Abdullah et al.,[50] Saha et al.,[51] Bekkarevich et al.,[52] Adhyaru et al.[53] and Pereira et al.[54] reported xylanase activity of fungal and bacterial cultures within a range of 182–3060 U and by following the same method employed in the present study.[23] Thus, forest soil samples are a massive treasure-house of xylan and xylose utilizing microflora for exploration.

18.5   CONCLUSION

The present study was carried out to examine soil samples at different locations in the Eastern Ghats of Andhra Pradesh for physico-chemical properties and xylan- and xylose utilizing microflora. Soil samples with moderate pH range and non-saline category, contained both macro and micro nutrients. Xylose utilizing bacteria as a group were a dominant component in the total microbial populations of the forest soils at majority of the locations of the Eastern Ghats at both the temperatures (30 and 37 °C). Xylan utilizing fungi and xylan utilizing bacteria shared equal honors by constituting more than 50% of the total microbial population in the soil samples at 50% of the locations each in the Eastern Ghats at the two temperatures. Xylan and xylose utilizing actinomycetes were a minor component of the total microbial populations in the soil samples at all the locations of the Eastern Ghats. The present study yielded fungal isolates with xylanase activity as high as 4000 U/ml. These isolates have a potential for the exploration of xylanase activities in the conversion of lignocellulosic wastes into value added products.

ACKNOWLEDGMENTS

We acknowledge the financial assistance provided by the University Grants Commission, New Delhi, in the form of fellowships to G. Ramanjaneyulu, A. Ramya, B. Shanti Kumari, and K. Dileep Kumar to carry out the above research.

KEYWORDS

- **Actinomycetes**
- **Bacteria**
- **Eastern Ghats**
- **Fungi**
- **Population Dynamics**

REFERENCES

1. Staddon, W. J.; Duchesne, L. C.; and Trevors, J. T. Conservation of forest soil microbial diversity, the impact of fire and research needs. *Environ. Rev.* **1996**, *4*, 267–275.
2. Sabi, G.; Bhuyan, M. K.; and Karmakar, R. M. Dynamics of microbial population in tea ecosystem. *J. Ind. Soc. Soil Sci.* **2003**, *51* (3), 252–257.
3. Wall, D. H.; and Virginia R. A. Controls on soil biodiversity: Insights from extreme environments. *Appl. Soil Ecol.* **1999**, *13*, 137–150.
4. Wardle, D. A. *Communities and Ecosystems: Linking the Aboveground and Belowground Components*; Princeton University Press: Princeton, NJ, 2002; pp 400.
5. Hackl, E.; Zechmeister-Boltenstern, S., Bodrossy, L., and Sessitsch, A. Comparison of diversities and compositions of bacterial populations inhabiting natural forest soils. *Appl. Environ. Microbiol.* **2004**, *70*, 5057–5065.
6. Kirk, J. L.; Beaudette, L. A.; Hart, M.; Moutoglis, P.; Klironomos, J. N.; Lee, H.; and Trevors, J. T. Methods of studying soil microbial diversity. *J. Microbiol. Methods* **2004**, *58*, 169–188.
7. Nandhini, B.; and Mary Josephine, R. A study on bacterial and fungal diversity in potted soil. *Int. J. Curr. Microbiol. Appl. Sci.* **2013**, *2* (2), 1–5.
8. Paul, E. A.; and Clark, F. E. *Soil Microbiology and Biochemistry*. Academic Press: San Diego, California, 1997; pp 340.
9. Sohail, M.; Ahmad, A.; Shahzad, S.; and Khan, S. A. A survey of amylolytic bacteria and fungi from native environmental samples. *Pak. J. Bot.* **2005**, *37* (1), 155–161.
10. Babu, G. V. A. K.; Viswanath, B.; Chandra, M. S.; Kumar, V. N.; and Reddy, B. R. Activities of cellulase and amylase in soils as influenced by insecticide interactions. *Ecotoxicol. Environ. Saf.* **2007**, *68*, 278–285.

11. Moradi, M.; Shariati, P.; Tabandeh, F.; Yakhchali, B.; and Khaniki, G. B. Screening and isolation of powerful amylolytic bacterial strains. *Int. J. Curr. Microbiol. App. Sci.* **2014**, *3* (2), 758–768.

12. Babu, G. V. A. K.; Kumar, V. N.; and Reddy, B. R. Effect of insecticides on nitrogen mineralization in a black vertisol soil. *Indian J. Microbiol.* **2006**, *46*, 129–134.

13. Gorlach-Lira, K.; and Coutinho, H. D. M. Population dynamics and extracellular enzymes activity of mesophilic and thermophilic bacteria isolated from semi-arid soil of northeastern Brazil. *Braz. J. Microbiol.* **2007**, *38*, 135–141.

14. Zeng, W.; Bai, X.; Zhang, L.; Wang, A.; and Peng, Y. Population dynamics of nitrifying bacteria for nitritation achieved in Johannesburg (JHB) process treating municipal wastewater. *Bioresour. Technol.* **2014**, *162*, 30–37.

15. Thomas, A. M.; Sanilkumar, M. G.; Vijayalakshmi, K. C.; Mohamed Hatha, A. A.; and Saramma, A. V. Dynamic changes in bacterial population and corresponding exoenzyme activity in response to a tropical phytoplankton bloom Chattonella marina. *J. Marine Biol.* **2014**, 1–6.

16. McSweeney, C. S.; Palmer, B.; Bunch R.; and Krause, D. O. Effect of the tropical forage calliandra on microbial protein synthesis and ecology in the rumen. *J. Appl. Microbiol.* **2001**, *90*, 78–88.

17. Wery, N.; Gerike, U.; Sharman, A.; Chaudhuri, J. B.; Hough, D. W.; and Danson, M. J. Use of a Packed-Column Bioreactor for Isolation of Diverse Protease-Producing Bacteria from Antarctic Soil, *Appl. Environ. Microbiol.* **2003**, *69* (3), 1457–1464.

18. Rayment, G. E.; and Higginson, F. R. *Australian Laboratory Handbook of Soil and Water Chemical Methods*; Inkata Press: Melbourne, 1992; pp 330.

19. Ghorbani-Nasrabadi, R.; Greiner, R.; Alikhani, H. A.; Hamedi, J.; and Yakhchali, B. Distribution of actinomycetes in different soil ecosystems and effect of media composition on extracellular phosphatase activity. *J. Soil Sci. Plant Nutr.* **2013**, *13* (1), 223–236.

20. Horta, M. C.; and Torrent, J. The Olsen P method as an agronomic and environmental test for predicting phosphate release from acid soils. *Nutr. Cycl. Agroecosys.* **2007**, *77*, 283–292.

21. Jackson, M. L. *Soil Chemical Analysis*; Prentice Hall of India Pvt. Ltd.: New Delhi, 1973; pp 498.

22. Teather, R. M.; and Wood, P. J. Use of Congo red-polysaccharide interactions in enumeration and characterization of cellulolytic bacteria from the bovine rumen. *Appl. Environ. Microbiol.* **1982**, *43* (4), 777–780.

23. Bailey, M. J.; Biely, P.; and Poutanen, K. Interlaboratory testing of methods for assay of xylanase activity. *J. Biotechnol.* **1992**, *23* (3), 257–270.

24. Champion, H. G.; and Seth, S. K. *The Revised Forest Types of India*; Manager of Publications: New Delhi, 1968; pp 404.

25. APSFR. *Andhra Pradesh State of Forest Report*; Andhra Pradesh Forest Department, Government of Andhra Pradesh: Hyderabad, 2013.

26. Mishra, R. R.; and Webster, R. K. Influence of soil environment and surface vegetation on soil microflora. *Proc. Nat. Acad. Sci.* (*India*). **1966**, *36* (11), 117–123.

27. Kanazawa, S.; Asakawa, S.; and Takai, Y. Effect of fertilizer and manure application on microbial numbers, biomass and enzyme activities in volcanic ash soils. *Soil Sci. Plant Nutr.* **1988**, *34*, 429–439.

28. Belay, A.; Claassens, A. S.; and Wehner, F. C. Effect of direct nitrogen and potassium and residual phosphorus fertilizers on soil chemical properties, microbial components and maize yield under long-term crop rotation. *Biol. Fert. Soils* **2002**, *35*, 420–427.

29. Johannes, R.; Erland, B.; Philip, C. B.; Christian, L. L.; Catherine, L.; Gregory, C. J.; Rob, K.; and Noah, F. Soil bacterial and fungal communities across a pH gradient in an arable soil. *ISME J.* **2010,** *4,* 1340–1351.

30. Wolf, D. C.; and Wagner, G. H. Carbon transformations and soil organic matter formation. In *Principles and Applications of Soil Microbiology*; Sylvia, D. M., Fuhrmann, J. J., Hartel, P. G., Zuberer D. A., Eds.; Pearson: Prentice Hall, USA, 2010, pp 285–332.

31. Russell, J. B.; Muck, R. E.; and Paul, J. W. Quantitative analysis of cellulose degradation and growth of cellulolytic bacteria in the rumen. *FEMS Microbiol. Ecol.* **2008,** *67,* 183–197.

32. Muhammad, I.; Asma, S.; Quratulain, S.; and Muhammad, N. Isolation and screening of cellulolytic bacteria from soil and optimization of cellulase production and activity. *Turk. J. Biochem.* **2012,** *37* (3), 287–293.

33. Sahoo, D. K.; and Gupta, R. Evaluation of ligninolytic microorganisms for efficient decolorization of a small pulp and paper mill effluent. *Proc. Biochem.* **2005,** *40* (5), 1573–1578.

34. Abd-Elsalam, H. E.; and El-Hanafy, A. A. Lignin biodegradation with ligninolytic bacterial strain and comparison of *Bacillus subtilis* and *Bacillus* sp. isolated from Egyptian soil. *Am-Euras. J. Agric. Environ. Sci.* **2009,** *5* (1), 39–44.

35. Luaine, B.; Nick, J. P. W.; Johannes, H. D. W.; and Harald, J. R. Isolation and characterization of novel bacterial strains exhibiting ligninolytic potential. *BMC Biotechnol.* **2011,** *11,* 94.

36. Karthikeyan, B.; Abdul Jaleel, C.; Lakshmanan, G. M. A.; and Deiveekasundaram, M. Studies on rhizosphere microbial diversity of some commercially important medicinal plants. *Colloids Surf. B. Biointer.* **2008,** *62* (1), 143–145.

37. Furczak, J.; and Joniec, J. Microbial populations and activity of biochemical processes related to carbon and nitrogen transformations in podzolic soil under willow culture in fifth year from treatment with sewage sludge. *Polish J. Environ. Stud.* **2009,** *18* (5), 801–810.

38. Subhajit, D.; Dipnarayan, G.; Tushar, K. M.; Abhishek, M.; Tapan, K. J.; and Tarun, K. D. A depth wise diversity of free living N$_2$ fixing and nitrifying bacteria and its seasonal variation with nitrogen containing nutrients in the mangrove sediments of Sundarban, WB, India. *Open J. Mar. Sci.* **2013,** *3,* 112–119.

39. Widden, P. Functional relationships between Quebec forest soil microfungi and their environment. *Can. J. Bot.* **1986,** *64,* 1424–1432.

40. Myers, R. T.; Zak, D. R.; White, D. C.; and Peacock, A. Landscape level patterns of microbial community composition and substrate use in upland forest ecosystems. *Soil Sci. Soc. Am. J.* **2001,** *65,* 359–367.

41. Susan, J. G.; Shenquiang, W.; Colin, D. C.; and Anthony, C. E. Selective influence of plant species on microbial diversity in the rhizosphere. *Soil Biol. Biochem.* **1998,** *30* (3), 369–378.

42. Vieira, F. C. S.; and Nahas, E. Comparison of microbial numbers in soils by using various culture media and temperatures. *Microbiol. Res.* **2005,** *160,* 197–202.

43. Wanze, Z. H. U.; Xiaohu, C. A. I.; Xingliang, L. I. U.; Jinxi, W.; Song, C.; Xiuyan, Z.; Dengyu, L.; and Maihe, L. Soil microbial population dynamics along a chronosequence of moist evergreen broad-leaved forest succession in Southwestern China. *J. Mt. Sci.* **2010,** *7,* 327–338.

44. Klein, D. A.; Frederick, B. A.; Biodini, M.; and Trlica, M. J. Rhizosphere microorganisms effects on soluble amino acids, sugars and organic acids in the root zone of *Agropyron cristatum, A. smithii* and *Bouteloua gracilis*. *Plant Soil* **1988,** *110,* 19–25.

45. Merbach, W.; Mirus, E.; Knof, G.; Remus, R.; Ruppel, S.; Russow, R.; Gransee, A.; and Schulze, J. Release of carbon and nitrogen compounds by plant roots and their possible ecological importance. *J. Plant Nutr. Soil Sci.* **1999**, *162*, 373–383.

46. Stephan, A.; Meyer, A.; and Schmid, B. Plant diversity affects culturable soil bacteria in experimental grassland communities. *J. Ecol.* **2000**, *88*, 988–998.

47. Sulkava, P.; and Huhta, V. Habitat patchiness affects decomposition and faunal diversity, A microcosm experiment on forest floor. *Oecologia* **1998**, *116*, 390–396.

48. Hansen, R. A. Effects of habitat complexity and composition on a diverse litter microarthropod assemblage. *Ecology* **2000**, *81*, 1120–1132.

49. Senthilkumar, S. R.; Ashokkumar, B.; Chandra Raj, K.; and Gunasekaran, P. Optimization of medium composition for alkali-stable xylanase production by *Aspergillus fischeri* Fxn 1in solid-state fermentation using central composite rotary design. *Bioresour. Technol.* **2005**, *96*, 1380–1386.

50. Roheena, A.; Kinza, N.; Aafia, A.; Mehwish, I.; and Shagufta, N. Enhanced production of xylanase from locally isolated fungal strain using agro-industrial residues under solid-state fermentation. *Nat. Prod. Res..* **2015**, *29* (1), 1006–1011.

51. Shyama, P. S.; and Shilpi, G. Optimization of xylanase production by *Penicillium citrinum* xym2 and application in saccharification of agro-residues. *Biocatal. Agr. Biotechnol.* **2014**, *3*, 188–196.

52. Bekkarevich, A. O.; Nemashkalov, V. A.; Koshelev, A. V.; Goryachev, D. A.; Bubnova, T. V.; Matys, V. Y.; Osipov, D. O.; Kondrat'eva, E. G.; Okunev, O. N.; and Sinitsyn, A. P. Cultivation of a novel cellulase/xylanase producer, *Trichoderma longibrachiatum* Mutant TW-1-59-27:13 Production of the enzyme preparation and the study of its properties. *Appl. Biochem. Microbial.* **2015**, *5*, (2) 229–235.

53. Dharmesh, N. A.; Nikhil, S. B.; and Hasmukh, A. M. Optimization of upstream and downstream process parameters for cellulase-poor-thermo-solvent-stable xylanase production and extraction by *Aspergillus tubingensis* FDHN1. *Bioresour. Bioprocess.* **2015**, *2* (3), 1–14.

54. de Cassia Pereira, J.; Marques, P. N.; Rodrigues, A.; Brito de Oliveira, T.; Boscolo, M.; da Silva, R.; Gomes, E.; and Bocchini Martins, D. A. Thermophilic fungi as new sources for production of cellulases and xylanases with potential use in sugarcane bagasse saccharification. *J. Appl. Microbiol.* **2015**, *118*, 928–939.

CHAPTER 19

PHYLOGENETIC AND PHENOGRAM BASED DIVERSITY OF HALOALKALIPHILIC BACTERIA FROM THE SALINE DESERT

HITARTH B. BHATT[1] and SATYA P. SINGH[1*]

[1]UGC-CAS Department of Biosciences, Saurashtra University, Rajkot, Gujarat – 360005, India.

*Corresponding author: Satya P. Singh. E-mail: satyapsingh@yahoo. com

CONTENTS

Abstract 374
19.1 Introduction 374
19.2 Materials and Methods 376
19.3 Results and Discussion 378
19.4 Conclusion 381
Acknowledgments 381
Keywords 381
References 381

ABSTRACT

While deserts account for 20% of the earth's terrestrial surface, their microbial diversity is least explored. The haloalkaliphilic bacteria have attracted a great deal of attention in the past couple of decades. The dual extremities of the halophiles and alkaliphiles make these microorganisms quite interesting to study the adaptation, metabolism, and biotechnological avenues. The bacterial diversity has been explored from the saline Kutch Desert. Microbial community analysis has been carried out in other deserts of the world, such as Antacama Desert (Chile), Monegros Desert (Spain), Negev Desert (Israel), Gobi Desert (Mongolia), Taklamaken Desert (China), Sahara Desert (Africa), and Sonoran Desert (America). Majority of the haloalkaliphilic bacteria from the saline deserts produce extracellular enzymes and some display a high resistance against several antibiotics. Phenotypic characters of the bacteria can be used for the cluster analysis to group them into phenons using Jaccard similarity coefficient and Unweighted Pair Group Mean Averages (UPGMA) algorithm. The bacterial diversity is well judged by employing biphasic approaches based on the genotypic and phenotypic characteristics

19.1   INTRODUCTION

Arid regions represent nearly 30% of the Earth's terrestrial surface. Deserts fall under the category of arid regions that receive <250 mm average annual precipitation or where evaporation and transpiration exceed precipitation. So far, a large number of environments have been studied for the evaluation of microbial community which provides intensive output about the microbial population and their interaction. Despite the large land area covered by arid regions and the habitat being most vulnerable to the climate changes, only limited is known about the arid/semi arid soil microbial diversity and the potential contributions of these communities to biogeochemical cycling and to the sustainability of vegetation in the desert area.

The desert conditions are a challenge to the microorganisms, as there is limited water and nutrient availability, a large range of temperature, and high exposure to ultraviolet (UV) irradiation. The study of desert microorganisms may help prevent the spread of deserts and to expand vegetation cover. The study may also offer to discover novel biological macromolecules such as new thermostable or haloalkaline-stable enzymes. Moreover, research on shrub islands has shown that because of the low levels of organic matter in semi arid and arid soils, plants become more dependent on the nitrogen and carbon fixation capabilities of the desert microbes. Besides, microbes can directly affect the solubilization of the primary minerals, releasing nutrients to be available to other living organisms. Much of the researches have so far focused on the vegetated area while only marginal information is available on the microbial diversity, metabolic potential, and community dynamics of the vast spaces of the sparsely green desert regions.

The saline desert of Kutch is unique as the north head of the Gulf of Kutch adjoins the Desert of Kutch with the regular flow of saline water during tides or through the water drifted by the south-west winds. Rainfall is fairly low, so that as water recedes and evaporates, there appears a crust of halite and gypsum crystals which grow in the clay and sands.[1] Thus, the study of microbes from this unique habitat may aid to understand their probable role in biogeochemical cycles and ecological significance. Further, the microbes from this unique habitat would provide opportunity for novel products such as extremozymes and other bioactive substances.

Extremophiles are a group of microorganisms that can grow and thrive in extreme environments of high or low temperatures, acidic or alkaline pH, high salinity, high metal concentrations, low nutrient content, low water activity, high radiation, high pressure, and low oxygen tension, usually considered as hostile to support life.[2] Extremophiles are a source of enzymes (extremozymes) having catalytic ability and stability under extreme conditions. Such enzymes have numerous applications in detergents, foods, pharmaceutical, leather, diagnostics, waste management, and silver recovery.[3] The outstanding stability of the extremophilic enzymes will reduce the gap between chemical and biological processes.[4] As in various extremophilic groups, the haloalkaliphilic microorganisms have developed different structural and physiological strategies to maintain their cell structure and function under the extreme conditions of salt and pH.[5]

Halophilic microorganisms have been isolated from salterns,[6] salt mines,[7] salt lakes,[8–10] and sea water.[11] These organisms are widespread in natural saline environments.[12,13] The halophilic organisms are distributed in all the three domains Bacteria, Archaea, and Eukarya,[14] displaying diversity with respect to salt tolerance, temperature, pH, and redox conditions. The haloalkaliphilic microorganisms are important to understand the mechanisms of adaptation to multiple stress conditions. Efforts have been focused on the discovery of the enzymes capable to function under harsh conditions.[15–19] Some of these enzymes, such as proteases, lipases, amylases, and DNases are important candidates for applications in food, detergent, chemical, pharmaceutical, paper, and pulp industries.

Nowadays, culture-independent methods are increasingly used to address the problems relating to the cultivability of the organisms.[20] The analysis of the microbial diversity has therefore, shifted from cultivation dependent approaches to culture independent metagenomic approaches. The metagenomic methods have a potential for the discovery of many novel microbial taxa and genes. However, because Polymerase Chain Reaction (PCR) based approaches have biases which can misrepresent community composition,[21,22] the cultivation of organisms from a given habitat still remains important for an accurate status of the physiology and complex ecological interactions.

To explore newer habitats for novel metabolites and bioactive molecules, there is a constant need to explore unexplored or unusual habitats. In this context, the arid or semi arid regions hold significance as they harbor an array of unique organisms with unexplored metabolic potential.

19.2 MATERIALS AND METHODS

19.2.1 SAMPLE COLLECTION AND PHYSICOCHEMICAL ANALYSIS

For the isolation of the haloalkaliphilic bacteria, soil samples were collected from the saline desert or any other arid habitat. The collected samples were kept in sterile plastic bags and stored at 4 °C. The samples were then subjected to physical and chemical analyses, such as pH, temperature, salinity, and conductivity.

19.2.2 ENRICHMENT AND ISOLATION OF THE ORGANISMS

Enrichment culture techniques were carried out in complex medium broth (CMB) consisting, (g/L): glucose, 10; peptone, 5; yeast extract, 5; KH_2PO_4, 5; with a varying concentration of NaCl (5% and 10%, w/v) at a constant pH of 9. The pH of the medium was adjusted by adding separately auto-claved Na_2CO_3 (20%, w/v). After 72 h of incubation, the cultures were serially diluted and an appropriate dilution was spread over a CMB agar (agar, 3%, w/v) plate and incubated at 37 °C. The colonies were selected based on colony morphology and appearance.

19.2.3 CELL MORPHOLOGY AND BIOCHEMICAL PROPERTIES

Gram's staining is the first criteria to differentiate bacteria, beside cell morphology, cell size, and cell arrangement. For further differentiation, biochemical properties of the isolates are followed. The biochemical tests include: production of catalase, oxidase, H_2S, indole, hydrolysis of urea, reduction of nitrate, and sugar fermentation. The biochemical media and test reagents were prepared as described earlier.[23] The production of amylase,[24] protease,[25] CMCase, and cellulose[26] by all the haloalakliphilic isolates at different NaCl concentrations had been carried out earlier. The antibiotic sensitivity/resistance of the bacterial isolates was assessed by Bauer-Kirby test using Dodeca discs specific for the Gram-negative and Gram-positive bacteria. The antibiotic sensitivity of the organisms was detected by measuring the zone of inhibition around the individual antibiotic disc.[27]

19.2.4 MOLECULAR IDENTIFICATION AND PHYLOGENETIC ANALYSIS

Identification of 16S rRNA is widely used for the identification of bacterial isolates up to genus and sometimes species level. The 16S rRNA gene sequences of the bacterial reference strains, type strains, and closest phylogenetic relatives are selected from GenBank by subjecting the nucleotide sequences of the bacterial isolates to similarity searches using BLASTn (http:// www.ncbi.nlm.nih.gov/blast). Multiple sequence alignment achieved with MUSCLE - a multiple sequence alignment method (v3.7)[28] configured for highest accuracy. After alignment, the ambiguous regions (containing gaps and/or poorly aligned) can be removed with Gblocks (v0.91b).[29] The phylogenetic tree is reconstructed using the maximum likelihood method implemented in the PhyML program - a web server for fast maximum likelihood-based phylogenetic inference (v3.0 aLRT).[30] The gamma shape parameter is estimated directly from the data (gamma = 0.486). Reliability for the internal branch is assessed using the aLRT test (Approximate Likelihood Ratio Test) (SH-Like). Graphical representation and edition of the phylogenetic tree can be performed with TreeDyn (v198.3).[31]

19.2.5 NUMERICAL ANALYSIS

For cluster analysis, phenotypic data is converted into a binary matrix, where the digit 1 represents the presence of a phenotypic character, and the digit 0 represents its absence. The Jaccard similarity measure is used to build a tree with the Unweighted Pair Group Mean Averages (UPGMA) algorithm. Analysis of the phenotypic data is performed by the software PAST (PAlaeontological STatistics).[32]

19.3 RESULTS AND DISCUSSION

The Desert is the example of an extreme environment, as it lacks free water, primarily required for the growth and survival of plants, animals, and microorganisms. The study of the extremophilic microbes facilitates to understand the interaction of the living organisms with the environment, and may eventually lead to the discovery of new organisms with distinctive survival strategies, physiological processes, or secondary metabolites.

It is of great significance to explore the unexplored world of organisms, as our knowledge is restricted to less than 1–5% of the total microbial population in nature. The majority of the halophiles and haloalkaliphiles so far have been isolated from the Soda Lakes, Dead Sea, solar salterns, and sea water.[9,33–36] The deserts are relatively less explored for these microbes. As the Desert of Kutch, in Gujarat (India) is an unexplored area for the halophiles and haloalkaliphiles, its study will be of great ecological significance. It is a typical ecosystem with the saline desert climate having unique floral and faunal diversity.

19.3.1 PHENOTYPIC CHARACTERIZATION

The haloalkaliphilic bacteria are usually Gram-positive.[27] However, the moderate halophiles with Gram-negative characteristics are also reported in literature.[37] Some halophiles, mainly *Halobacillus* are Gram variable in nature, which consistently stained Gram-negative despite repeated assessments at different salinities.[13,38] This strange observation has been previously reported in many other bacteria, including the halophilic bacillus *Filobacillus milosensis* which contains an unusual type of cell wall polymer.[39]

Despite the increasing emphasis on the molecular tools and chronometers, the metabolic and physiological features are important to diversify and differentiate the microorganisms. The biochemical characteristics of the organisms are controlled by the vital molecules of the cell. Most of the haloalkaliphilic bacterial isolates are positive for catalase and oxidase indicating their aerobic nature.[39,40] Many reports have highlighted the presence of oxidase and catalase in these bacteria.[39,41–43] Further, a majority of the moderate halophilic and alkaliphilic bacteria do not produce indole or utilize urea.[42,44,45]

Many halophiles and haloalkaliphiles produce extracellular enzymes, such as protease, amylase, cellulase, chitinase, and lipase.[17,46–50] However, the production and characterization of the extracellular enzymes in different groups of the moderately halophilic bacteria are still restricted.[50,51] Nevertheless, there are some reports on the extracellular hydrolytic enzymes, particularly proteases, from the moderately halophilic and alkaliphilic bacteria.[15,52–58]

Among the various approaches to characterize bacterial isolates, antibiotic profiling has emerged as a new tool to judge microbial diversity.[43,59–61] The antibiotic sensitivity and resistance profiling can be quite important to diversify the halophilic bacteria as they display natural resistance against structurally and functionally diverse compounds.[37] The antibiogram of some haloalkaliphilic bacteria is reported in literature.[34,43]

19.3.2 PHYLOGENETIC ANALYSIS

It was originally believed that archaea dominated the total microbial community of the hypersaline habitat over bacteria.[62] However, recent reports have suggested huge diversity of the members of the bacteria with a high occurrence of Gram-negative Betaproteobacteria, Gammaproteobacteria, and Deltaproteobacteria groups and a lower representation of the Gram-positive bacteria in saline soils.[63–65] The moderate halophiles with Gram-negative characteristics have been studied in detail, while information regarding the Gram-positive bacteria is rather limited.[37] In the deep-sea hyper saline lake anoxic sediments, Gram-positive bacteria are shown to dominate with a majority of the strains being members of *Bacillales*, including the genera of *Halobacillus*, *Virgibacillus*, and *Pontibacillus*.[66] Haloalkaliphilic bacteria of the Arid Natural Saline Systems of Southern Tunisian Sahara showed a high occurrence of Gram-positive bacteria with the dominance of the genera *Halobacillus*, *Virgibacillus*, *Bacillus*, *Oceanobacillus*, *Gracilibacillus*, *Peanibacillus*, and *Pontibacillus*.[67]

It is well known that Firmicutes are dominant among the members of the cultivable bacterial community in a variety of hyper saline habitats.[65,68–70] However, a number of other studies have also reported members of *Actinobacteria*,[71–73] Bacteroidetes,[13,74] and the Alphaproteobacteria[74] from different hyper saline ecosystems.

Subrahmanyam et al.[75] reported the dominance of *Bacillaceae* and *Staphylococcaceae* in the saline desert of Kutch (Great Rann). Similar results are reported for arid saline systems, such as Golea Salt Lake in Algeria Sahara,[76] Chott Djerid,[68] and Tunisian Multipond Solar Saltern,[65] Gobi Desert (Mongolia), and Taklamaken desert (China).[76]

19.3.3   COMPARISON OF PHENOGRAM AND PHYLOGRAM

Genotypic changes constantly occur in organisms irrespective of the phenotypic variations. In other words, the evolution is quasi-independent of the phenotypic characters. However, phenotypic characteristics of the microorganisms cannot be ignored. It is relevant to consider that microorganisms can modify their physiological characteristics on account of the availability of nutrients. Further, the regulation of gene expression can reduce the energy expenses. Thus, the physiological changes in bacteria can be explained on the basis of the availability of nutrients in the environment leading to changes in the phenotypic expression or acquisition of the inherited adaptations by horizontal gene transfer or selective pressure.[77]

We compared the phenotypic diversity with genotypic diversity by phenogram and phylogram, respectively, of the bacterial isolates from the saline desert.[27] While the phenogram supported the clustering and diversity patterns displayed by the phylogram, in certain cases, distinct phenotypic patterns were evident. Certain bacteria belonging to the same species on the basis of 16S rRNA gene sequences, displayed distinct phenotypic characteristics.[13,61,78–80] Similar pattern of results were also reported by Hedi et al.,[68] where halophilic bacteria from the salt lake in Tunisia were characterized using API (Analytical Profile Index) strips. The phenotypic characterization and 16S rDNA studies of the culturable and non-obligate halophilic bacterial communities from a hypersaline lake, La Sal del Rey, in extreme South Texas, USA have been compared, where several isolates had unique phenotypic profiles.[78] *Halomonas* spp. and its other phylogenetically related species were reported to have different phenotypic characteristics.[13,61,80]

19.4   CONCLUSION

The trends indicate that the haloalkaliphilic bacteria constitute an important part of the microbiota that inhabits arid and saline deserts. Huge phenotypic and phylogenetic diversities occur among these bacteria. Further, a combination of the phenotypic and genotypic diversity would truly reflect the bacterial diversity in these habitats.

ACKNOWLEDGMENTS

The authors gratefully acknowledge the financial and infrastructural supports by the University Grant Commission (UGC), New Delhi under various programs including CAS. They also acknowledge the Saurashtra University, Rajkot for its support. Hitarth B. Bhatt gratefully acknowledges a Research Fellowship in Sciences for Meritorious Students by UGC, New Delhi, India.

KEYWORDS

- **Coastal saline desert**
- **Haloalkaliphilic bacteria**
- **Microbial Diversity**
- **Phenogram**
- **16S rDNA**

REFERENCES

1. Gupta, V.; and Ansari, A. A. Geomorphic portrait of the Little Rann of Kutch. *Arab. J. Geosci.* **2014,** *7* (2), 527–536.
2. Gomes, J.; and Steiner, W. The biocatalytic potential of extremophiles and extremozymes. *Food Technol. Biotechnol.* **2004,** *42* (4), 223–235.
3. Karbalaei-Heidari, H. R.; Ziaee, A. A.; Amoozegar, M. A.; Cheburkin, Y.; and Budisa, N. Molecular cloning and sequence analysis of a novel zinc-metalloprotease gene from the *Salinivibrio sp.* strain AF-2004 and its extracellular expression in *E. coli. Gene* **2008,** *408,* 1–2.
4. Madigan, M. T.; and Marrs, B. L. Extremophiles. *Sci. Am.* **1997,** 66–71.
5. Horikoshi, K. Alkaliphiles: Some applications of their products for biotechnology. *Microbiol. Mol. Biol. Rev.* **1999,** *63* (4), 735–750.

6. Bardavid, R. E.; Ionescu, D.; Oren, A.; Rainey, F. A.; Hollen, B. J.; Bagaley, D. R.; Small, A. M.; and McKay, C. Selective enrichment, isolation and molecular detection of *Salinibacter* and related extremely halophilic Bacteria from hypersaline environments. *Hydrobiologia.* **2007,** *576,* 3–13.

7. Chen, Y. G.; Cui, X. L.; Pukall, R.; Li, H. M.; Yang, Y. L.; Xu, L. H.; Wen, M. L.; Peng, Q.; and Jiang, C. L. *Salinicoccus kunmingensis* sp. nov., a moderately halophilic bacterium isolated from a salt mine in Yunnan, south-west China. *Int. J. Syst. Evol. Microbiol.* **2007,** *57,* 2327–2332.

8. Swan, B. K.; Ehrhardt, C. J.; Reifel, K. M.; Moreno, L. I.; and Valentine, D. L. Archaeal and bacterial communities respond differently to environmental gradients in anoxic sediments of a California hypersaline lake, the Salton Sea. *Appl. Environ. Microbiol.* **2010,** *76,* 757–768.

9. Duckworth, A. W.; Grant, W. D.; Jones, B. E.; and Steenbergen, R. V. Phylogenetic diversity of soda lake alkaliphiles. *FEMS Microbiol. Ecol.* **1996,** *19,* 181–191.

10. Joshi, A. A.; Kanekar, P. P.; Kelkar, A. S.; Shouche, Y. S.; Vani, A. A.; Borgave, S. B.; and Sarnaik, S. S. Cultivable bacterial diversity of alkaline Lonar lake, India. *Microb. Ecol.* **2008,** *55,* 163–172.

11. Singh S. P.; Purohit, M. K.; Raval, V. H.; Pandey, S.; Akbari, V. G.; and Rawal, C. M. Capturing the potential of haloalkaliphilic bacteria from the saline habitats through culture dependent and metagenomic approaches. In *Current Research Technology and Education Topics in Applied Microbiology and Microbial Biotechnology*; Mendez-Vilas, A., Ed; Formatex Publishers: Spain, 2010, *1*, pp 81–87.

12. Sorokin, D. Y.; Tourova, T. P.; Galinski, E. A.; Bellach, C.; and Tindall, B. Extremely halophilic denitrifying bacteria from hypersaline inland lakes, *Halovibrio denitrificans* sp. nov. and *Halospina denitrificans* gen. nov., sp. nov., and evidence that the genus name *Halovibrio Fendich* 1989 with the type species *Halovibrio variabilis* should be associated with DSM 3050. *Int. J. Syst. Evol. Microbiol.* **2006,** *56,* 379–388.

13. Caton, T. M.; Witte, L. R.; Ngyuen, H. D.; Buchheim, J. A.; Buchheim, M. A.; and Schneegurt, M. A. Halotolerant aerobic heterotrophic bacteria from the Great Salt Plains of Oklahoma. *Microb. Ecol.* **2004,** *48,* 449–462.

14. Oren, A. Microbial life at high salt concentrations: Phylogenetic and metabolic diversity. *Saline Systems* **2008,** *4* (2), 13.

15. Raval, V. H.; Pillai, S.; Rawal, C. M.; and Singh, S. P. Biochemical and structural characterization of a detergent-stable serine alkaline protease from sea water haloalkaliphilic bacteria. *Process Biochem.* **2014a,** *49,* 955–962.

16. Pandey, S.; Rakholiya, K. D.; Raval, V. H.; and Singh, S. P. Catalysis and stability of an alkaline protease from a haloalkaliphilic bacterium under non-aqueous conditions as a function of pH, salt and temperature. *J. Biosci. Bioengg.* **2012,** *114* (3), 251–256.

17. Sahay, H.; Mahfooz, S.; Singh, A. K.; Singh, S.; Kaushik, R.; Saxena, A. K.; and Arora, D. K. Exploration and characterization of agriculturally and industrially important haloalkaliphilic bacteria from environmental samples of hypersaline Sambhar Lake, India. *World J. Microbiol. Biotechnol.* **2012,** *28* (11), 3207–3217.

18. Karan, R.; Capes, M. D.; and DasSarma, S. Function and biotechnology of extremophilic enzymes in low water activity. *Aquatic Biosystems* **2012,** *8* (1), 4.

19. Li, X.; Yu, H. Y. Purification and characterization of novel organic solvent-tolerant beta-amylase and serine protease from a newly isolated *Salimicrobium halophilum* strain LY20. *FEMS Microbiol. Lett.* **2012,** *329,* 204–211.

20. Spiegelman, D.; Whissell, G.; and Greer, C. W. A survey of the methods for the characterization of microbial consortia and communities. *Can. J. Microbiol.* **2005**, *51*, 355–386.
21. Martin-Laurent, F.; Philippot, L.; Hallet, S.; Chaussod, R.; Germon, J. C.; Soulas, G.; and Catroux, G. DNA extraction from soils: Old bias for new microbial diversity analysis methods. *Appl. Environ. Microbiol.* **2001**, *67*, 2354–2359.
22. Hur, I.; and Chun, J. A method for comparing multiple bacterial community structures from 16S rDNA clone library sequences. *J. Microbiol.* **2004**, *42*, 9–13.
23. Cappuccino, J. G.; and Sherman, N. *Microbiology, a Laboratory Manual*, 6th ed.; Pearson education: San Francisco, 2004.
24. Ramesh, B.; Reddy, P. R. M.; Seenayya, G.; and Reddy, G. Effect of various flours on the production of thermostable b-amylase and pullulanase by *Clostridium thermosulfurogenes* SV2. *Bioresour Technol.* **2001**, *76*, 169–171.
25. Manachini, P. L.; Fortina, M. G.; and Parini, C. Alkaline protease produced by *Bacillus thermoruber* a new species of *Bacillus*. *Appl. Microbiol. Biotechnol.* **1988**, *28*, 409–413.
26. Zvereva, E. A.; Fedorova, T. V.; Kevbrin, V. V.; Zhilina, T. N.; and Rabinovich, M. L. Cellulase activity of a haloalkaliphilic anaerobic bacterium, strain Z-7026. *Extremophiles* **2006**, *10*, 53–60.
27. Bhatt, H. B. Diversity and enzymatic potential of haloalkaliphilic bacteria from the Desert of Kutch in Gujarat. M.Phil Thesis, Saurashtra University, Rajkot, 2013.
28. Edgar, R. C. MUSCLE: Multiple sequence alignment with high accuracy and high throughput. *Nucleic Acids Res.* **2004**, *32*, 1792–1797.
29. Castresana, J. Selection of conserved blocks from multiple alignments for their use in phylogenetic analysis. *Mol. Biol. Evol.* **2000**, *17*, 540–552.
30. Guindon, S.; and Gascuel, O. A simple, fast, and accurate algorithm to estimate large phylogenies by maximum likelihood. *Syst. Biol.* **2003**, *52*, 696–704.
31. Chevenet, F.; Brun, C.; Banuls, A. L.; Jacq, B.; and Chisten, R. TreeDyn: Towards dynamic graphics and annotations for analyses of trees. *BMC Bioinform.* **2006**, *10*, 439.
32. Harper, D. A. T. *Numerical Palaeobiology*; John Wiley and Sons: New York, 1999.
33. Jones, B. E.; Grant, W. D.; Duckworth, A. W; and Owenson, G. G. Microbial diversity of soda lakes. *Extremophiles* **1998**, *2*, 191–200.
34. Pikuta, E.; Hoover, R. B.; Bej, A. K.; Marsic, D.; Detkova, E. N.; Whitman, W. B.; and Krader, P. *Tindallia californiensis* sp. nov., a new anaerobic, haloalkaliphilic, spore-forming acetogen isolated from Mono Lake in California. *Extremophiles* **2003**, *7*, 327–334.
35. Rees, H. C.; Grant, W. D.; Jones, B. E.; and Heaphy, S. Diversity of Kenyan Soda Lake alkaliphiles assessed by molecular methods. *Extremophiles* **2004**, *8* (1), 63–71.
36. Demergasso, C.; Casamayor, E. O.; Chong, G.; Galleguillos, P.; Escudero, L.; and Pedrós-Alió, C. Distribution of prokaryotic genetic diversity in athalassohaline lakes of the Atacama Desert, Northern Chile. *FEMS Microbiol. Ecol.* **2004**, *48* (1), 57–69.
37. Ventosa, A.; Nieto, J. J.; and Oren, A. Biology of aerobic moderately halophilic bacteria. *Microbiol. Mol. Biol. Rev.* **1998**, *62*, 504–544.
38. Schlesner, H.; Lawson, P. A.; Collins, M. D.; Weiss, N.; Wehmeyer, U.; Volker, H.; and Thomm, M. *Filobacillus milensis* gen. nov., sp. nov., a new halophilic spore-forming bacterium with Orn-D-Glutype peptidoglycan. *Int. J. Syst. Evol. Microbiol.* **2001**, *51*, 425–431.
39. Nowlan, B.; Dodia, M. S.; Singh, S. P.; and Patel, B. K. *Bacillus okhensis* sp. nov., a halotolerant and alkalitolerant bacterium from an Indian saltpan. *Int. J. Syst. Evol. Microbiol.* **2006**, *56*, 1073–1077.

40. Taprig, T.; Akaracharanya, A.; Sitdhipol, J.; Visessanguan, W.; and Tanasupawat, S. Screening and characterization of protease-producing *Virgibacillus*, *Halobacillus* and *Oceanobacillus* strains from Thai fermented fish. *J. App. Phar. Sci.* **2013**, *3*, 25–30.

41. Quesada, E.; Bejar, V.; Valderrama, M. J.; Ventosa, A.; and Ramos-Cormenzana, A. Isolation and characterization of moderately halophilic non motile rods from different saline habitats. *Microbiologia.* **1985**, *1*, 89–96.

42. Muntyan, M. S.; Tourova, T. P.; Lysenko, A. M.; Kolganova, T. V.; Fritze, D.; and Skulachev, V. P. Molecular identification of alkaliphilic and halotolerant strain *Bacillus sp.* FTU as *Bacillus pseudofirmus* FTU. *Extremophiles* **2002**, *6*, 195–199.

43. Zhang, Y. J.; Zhou, Y.; Ja, M.; Shi, R.; Chun-Yu, W. X.; Yang, L. L.; Tang, S. K.; and Li, W. J. *Virgibacillus albus* sp. nov., a novel moderately halophilic bacterium isolated from Lop Nur salt lake in Xinjiang province, China. *Antonie Van Leeuwenhoek* **2012**, *102* (4), 553–560.

44. Reddy, G. S.; Raghavan, P. U.; Sarita, N. B.; Prakash, J. S.; Nagesh, N.; Delille, D; and Shivaji, S. *Halomonas glaciei* sp. nov. isolated from fast ice of Adelie Land, Antarctica. *Extremophiles* **2003**, *7*, 55–61.

45. Dodia, M. S. Stability and folding of extracellular enzymes from haloalkaliphilic bacteria. Ph.D. thesis, Saurashtra University, Rajkot, Gujarat, India. 2005.

46. Hidri, D. E.; Guesmi, A.; Najjari, A.; Cherif, H.; Ettoumi, B.; Hamdi, C.; Boudabous, A.; and Cherif, A. Cultivation-dependant assessment, diversity, and ecology of haloalkaliphilic bacteria in arid saline systems of Southern Tunisia. *BioMed. Research Int.* **2013**, DOI: 10.1155/2013/648141.

47. Joshi, R. H.; Dodia, M. S.; and Singh, S. P. Production and optimization of a commercially viable alkaline protease from a haloalkaliphilic bacterium. *Biotechnol. Bioprocess. Eng.* **2008**, *13* (5), 552–559.

48. Purohit, M. K.; and Singh, S. P. Comparative analysis of enzymatic stability and amino acids sequences of thermostable alkaline proteases from two haloalkaliphilic bacteria isolated from coastal region of Gujarat, India. *Int. J. Biol. Macromol.* **2011**, *49* (1), 103–112.

49. Rohban, R.; Amoozegar, M. A.; and Ventosa, A. Screening and isolation of halophilic bacteria producing extracellular hydrolases from Howz Soltan Lake, Iran. *J. Ind. Microbiol. Biotechnol.* **2009**, *36*, 333–340.

50. Gohel, S. D.; and Singh, S. P. Purification strategies, characteristics and thermodynamic analysis of a highly thermostable alkaline protease from a salt tolerant alkaliphilic actinomycete, *Nocardiopsis alba* OK-5. *J. Chromatogr. B* **2012**, (889–890), 61–68.

51. Sanchez-porro, C.; Martin, S.; Mellado, E.; Bertoldo, C.; Antranikian, G.; and Ventosa, A. Diversity of moderately halophilic bacteria producing extracellular hydrolytic enzymes. *J. Appl. Microbiol.* **2003**, *94*, 295–300.

52. Coronado, M. J.; Vargas, C.; Hofemeister, J.; Ventosa, A.; and Nieto, J. J. Production and biochemical characterization of an α-amylase from the moderate halophile *Halomonas meridiana*. *FEMS Microbiol. Lett.* **2000**, *183*, 67–71.

53. Babavalian, H.; Amoozegar, M. A.; Pourbabaee, A. A.; Moghaddam, M. M.; and Shakeri, F. Isolation and identification of moderately halophilic bacteria producing hydrolytic enzymes from the largest hypersaline playa in Iran. *Microbiology* **2013**, *82*, 466–474.

54. Jogi, C.; Joshi, R. H.; Dodia; M. S.; and Singh, S. P. Extracellular alkaline protease from haloalkaliphilic bacteria isolated from sea water along coastal Gujarat. *J. Cell Tiss. Res.* **2005**, *5* (2), 439–444.

55. Mehta, V. J.; Thumar, J. T.; and Singh, S. P. Production of alkaline protease from an alkaliphilic actinomycete. *Bioresource Technol.* **2006,** *97* (14), 1650–1654.

56. Patel, R. K.; Dodia, M. S.; Joshi, R. H.; and Singh, S. P. Production of extracellular halo-alkaline protease from a newly isolated *Haloalkaliphilic Bacillus* sp. isolated from seawater in Western India. *World J. Microbiol. Biotechnol.* **2006a,** *22* (4), 375–382.

57. Dodia, M. S.; Joshi, R. H.; Patel, R. K.; and Singh, S. P. Characterization and stability of extracellular alkaline proteases from moderately halophilic and alkaliphilic bacteria isolated from saline habitat of coastal Gujarat, India. *Braz. J. Microbiol.* **2006,** *37,* 244–252.

58. Tikhonova, T. V.; Slutsky, A.; Antipov, A. N.; Boyko, K. M.; Polyakov, K. M.; Sorokin, D. Y.; Zvyagilskaya, R. A.; and Popov, V. O. Molecular and catalytic properties of a novel cytochrome c nitrite reductase from nitrate-reducing haloalkaliphilic sulfur-oxidizing bacterium *Thioalkalivibrio nitratireducens. Biochimica Biophysica Acta* (BBA) - *Prot. Proteomics* **2006,** *1764* (4), 715–723.

59. Gorshkova, N. M.; and Ivanova, E. P. Antibiotic susceptibility as a taxonomic characteristic of Proteobacteria of the genera *Alteromonas, Pseudoalteromonas, Marinomonas* and *Marinobacter. Russian J. Marine Biol.* **2001,** *27* (2), 116–120.

60. Mincer, T. J.; Fenical, W.; and Jensen, P. R. Culture-dependent and culture-independent diversity within the obligate marine actinomycete genus *Salinispora. Appl. Environ. Microbiol.* **2005,** *71,* 7019–7028.

61. Litzner, B. R.; Caton, T. M.; and Schneegurt, M. A. Carbon substrate utilization, antibiotic sensitivity, and numerical taxonomy of bacterial isolates from the Great Salt Plains of Oklahoma. *Arch. Microbiol.* **2006,** *185* (4), 286–296.

62. Hacine, H.; Rafa, F.; Chebhouni, N.; Boutaiba, S.; Bhatnagar, T.; Barratti, J. C.; and Ollivier, B. Biodiversity of prokaryotic microflora in El Golea Salt lake, Algerian Sahara. *J. Arid Environ.* **2004,** *58,* 273–284.

63. Anton, J.; Rosselló-Mora, R.; Rodriguez-Valera, F.; and Amann, R. Extremely halophilic bacteria in crystallizer ponds from solar salterns. *Appl. Environ. Microbiol.* **2000,** *66,* 3052–3057.

64. Azam, F.; and Malfatti, F. Microbial structuring of marine ecosystems. *Nature Rev. Microbiol.* **2007,** *5,* 782–791.

65. Baati, H.; Amdouni, R.; Gharsallah, N.; Sghir, A.; and Ammar, E. Isolation and characterisation of moderately halophilic bacteria from Tunisia solar saltern. *Curr. Microbiol.* **2010,** *60,* 157–161.

66. Sass, A. M.; McKew, B. A.; Sass, H.; Fichtel, J.; Timmis, K. N.; and Mcgenity, T. J. Diversity of Bacillus-like organisms isolated from deep-sea hypersaline anoxic sediments. *Saline Systems* **2008,** DOI: 10.1186/1746-1448-4-8.

67. Guesmi, A.; Ettoumi, B.; Hidri, D. E.; Essanaa, J.; Cherif, H.; Mapelli, F.; Marasco, R.; Rolli, E.; Boudabous, A.; and Cherif, A. Uneven distribution of *Halobacillus trueperi* species in arid natural saline systems of Southern Tunisian Sahara. *Microb. Ecol.* **2013,** *66,* 831–839.

68. Hedi, A.; Sadfi, N.; Fardeau, M. L.; Rebib, H.; Cayol, J. L.; Ollivier, B.; and Boudabous, A. Studies on the biodiversity of halophilic microorganisms isolated from El-Djerid Salt Lake (Tunisia) under aerobic conditions. *Int. J. Microbiol.* **2009,** DOI: 10.1155/2009/731786.

69. Xiang, W. L.; Guo, J. H.; Feng, W.; Huang, M.; Chen, H.; Zhao, J.; Zhang, J.; Yang, Z. R.; and Sun, Q. Community of extremely halophilic bacteria in historic Dagong brine well in southwestern China. *World J. Microbiol. Biotechnol.* **2008,** *24,* 2297–2305.

70. Yeon, S. H.; Jeong, W. J.; and Park, J. S. The diversity of culturable organotrophic bacteria from local solar salterns. *J. Microbiol.* **2005,** *43,* 1–10.

71. Jiang, H.; Dong, H.; Zhang, G.; Yu, B.; Chapman, L. R.; and Fields, M. W. Microbial diversity in water and sediment of Lake Chaka, an Athalassohaline Lake in Northwestern China. *Appl. Environ. Microbiol.* **2006,** *72* (6), 3832–3845.

72. Tsiamis, G.; Katsaveli, K.; Ntougias, S.; Kyrpides, N.; Andersen, G.; Piceno, Y.; and Bourtzis, K. Prokaryotic community profiles at different operational stages of a Greek solar saltern. *Res. Microbiol.* **2008,** *159,* 609–627.

73. Wu, Q. L.; Zwart, G.; Schauer, M.; Agterveld, M. P. K. V.; and Hahn, M. W. Bacterioplankton community composition along a salinity gradient of sixteen high-mountain lakes located on the Tibetan Plateau, China. *Appl. Environ. Microbiol.* **2006,** *72,* 5478–5485.

74. Benlloch, S.; Lopez-Lopez, A.; Casamayor, E. O.; Goddard, L. Ø. V.; Daae, F. L.; Smerdon, G.; Massana, R.; Joint, I.; Thingstad, F.; Pedrós-Alió, C.; and Rodríguez-Valera, F. Prokaryotic genetic diversity through-out the salinity gradient of a coastal solar saltern. *Environ. Microbiol.* **2002,** *4,* 349–360.

75. Subrahmanyam, G.; Khonde, N.; Maurya, D. M.; Chamyal, L. S.; and Archana, G. Microbial activity and culturable bacterial diversity in sediments of the Great Rann of Kachchh, Western India. *Pedosphere* **2014,** *24* (1), 45–55.

76. An, S.; Couteau, C.; Luo, F.; Neveu, J.; and DuBow, M. S. Bacterial diversity of surface sand samples from the Gobi and Taklamaken Deserts. *Microb. Ecol.* **2013,** *66,* 850–860.

77. Lima-Bittencourt, C. I.; Astolfi-Filho, S.; Chartone-Souza, E.; Santos, F. R.; and Nascimento, A. M. A. Analysis of *Chromobacterium sp.* natural isolates from different Brazilian ecosystems. *BMC Microbiology* **2007,** *7,* 58.

78. Phillips, K.; Zaidan, F.; Elizondo, O. R.; and Lowe, K. L. Phenotypic characterization and 16S rDNA identification of culturable non-obligate halophilic bacterial communities from a hypersaline lake, La Sal del Rey, in extreme South Texas (USA). *Aquatic Biosystems* **2012,** *8,* 5.

79. Bhatt, H. B.; and Singh, S. P. Phenotypic, metabolic and phylogenetic diversity of the haloalkaliphilic bacteria from the saline desert. *Science Excellence.* Gujarat University: Ahmedabad, 2014.

80. Wilson, C.; Caton, T. M.; Buchheim, J. A.; Buchheim, M. A.; Schneegurt, M. A.; and Miller, R. V. DNA-repair potential of *Halomonas* spp. from the salt plains microbial observatory of Oklahoma. *Microb. Ecol.* **2004,** *48,* 541–549.

CHAPTER 20

FUNGAL BIOTRANSFORMATION OF DRUGS: POTENTIAL APPLICATIONS IN PHARMA INDUSTRY

G. SHYAM PRASAD[1] and B. SASHIDHAR RAO[1*]

[1]Department of Biochemistry, University College of Science, Osmania University, Hyderabad, Telangana – 500007, India.

*Corresponding author: Prof. B. Sashidhar Rao. E-mail: sashi_rao@ yahoo.com

CONTENTS

Abstract 388
20.1   Introduction..388
20.2   Applications of Microbial Transformation Reactions 389
20.3   Selection of Microorganisms for Biotransformation 393
20.4   Maintenance of Microbial Cultures 394
20.5   Addition of Substrate 395
20.6   Factors Influencing Biotransformations 396
20.7   Incubation, Extraction, and Analysis 401
20.8   Scale Up 402
20.9   Utility of Microbial Transformations to the Pharma Industry 402
20.10   Conclusion 403
Acknowledgments...403
Keywords 404
References..404

ABSTRACT

Modification of organic compounds and synthetic drugs by simple chemically defined reaction catalyzed by isolated enzymes, whole cells is termed as biotransformation. These reactions are performed by plant cells, animal cells, and microbial cells, but microbial process is most efficient and economical because of rapid microbial growth and its high metabolic rate. Biotransformations are employed in enhancing the activity of drugs and organic molecules, in predicting mammalian drug metabolism, and in discovering novel compounds with enhanced biological activity.

This review highlights the methodology, applications, and factors influencing biotransformation of drugs in relation to enhanced bioactivity and production of biotransformed products with special reference to fungal systems.

20.1 INTRODUCTION

Microbial transformations of organic compounds have been employed as a powerful synthetic tool for introducing chemical functions into inaccessible sites of molecules, thus producing new analogs of natural and synthetic organic compounds which may be difficult to synthesize chemically. In microbial transformations, an intact microorganism usually in the form of growing or resting culture, using its enzyme systems, carries out an organic reaction on a selected substrate that has been brought into contact with its biomass. The term biotransformation is used for all those transformations which are brought about by pure enzymes, partially pure enzymes, microbial, plant, and animal cell culture. However, microbial process is far more efficient and economical because of the rapid growth and high metabolic rate. Further, these processes can be conducted in aqueous as well as in aqueous/organic environment; therefore polar organic compounds as well as water soluble compounds can be selectively and efficiently transformed with enzymes or active cells.[1] Microbial transformations encompass practically every type of possible chemical reaction. They are extremely useful, as these reactions can be carried out more economically. The interest in bioconversion is mainly because the product of the process is more useful or valuable than the precursor used. The spectrum of the type of reactions obtained with enzymatic system has now been widened to include not only oxidation, hydroxylation, reduction, hydrolysis, and condensation reactions, but also other reactions such as, dehydrogenation, β-oxidation, amination, deamination, hydration, dehydration, *N*- and *O*-dealkylation, decarboxylation, acylation, glycosylation, and isomerization. The general goals of biotransformations are specific modification of the substrate structure *via* selective transformation reactions. They include partial degradation of substrates into desirable metabolites by the means of controlled microbial reactions or reaction pathways, extension of the substrate structure by the use of biosynthetic reactions to create novel structural entities.

The main advantages of microbial- or enzyme-catalyzed reactions over chemical reactions are: (1) they are stereoselective and can be carried out at ambient temperature and atmospheric pressure. This minimizes problems of isomerization, racemization, epimerization, and rearrangement that generally occur during synthetic chemical processes[2] (2) Biocatalytic processes are generally carried out in aqueous solution, thus avoiding environmentally harmful chemicals that are employed in chemical processes, which need to dispose solvent waste.[1] (3) Microbial cells or their enzymes can be immobilized and reused for many cycles.

The physical state of biocatalysts that are used for biotransformation can be isolated enzyme systems or intact whole organisms depending on the factors such as the type of reaction, cofactors needed, and the scale in which the biotransformation has to be performed.

20.2 APPLICATIONS OF MICROBIAL TRANSFORMATION REACTIONS

Microbial transformation reactions can be employed

1. To enhance the activity of an organic compound.
2. In predicting mammalian drug metabolism.
3. In novel drug discovery.

20.2.1 TO ENHANCE THE ACTIVITY OF AN ORGANIC COMPOUND

Most of the drugs are marked as prodrugs whose metabolites are pharmacologically active. Chemical synthesis of active drug form is difficult, impossible, or is a costly affair. The best example is albendazole which is a benzimidazole carbamate with a broad antiparasitic spectrum.[3] Albendazole is practically insoluble in water and its oral bioavailability is very poor. In animals, after parenteral, oral, or intraruminal administration, albendazole is rapidly oxidized to sulfoxide,[3] and albendazole efficacy is attributed to this metabolite.[3] Chemical synthesis of direct albendazole sulfoxide is possible but not affordable, although there are reports in contemporary literature for the synthesis of albendazole sulfoxide and albendazole sulfone. However, these reactions are not easy to carry out, and have a high failure rate, because of the insolubility problems in albendazole, which often leads to mixtures of sulfoxides and sulfones that are difficult to separate.[4] To overcome these limitations, microbial transformation of albendazole leading to albendazole sulfoxide with enhanced activity and more solubulity, is a simple single step reaction with high yields employing the fungus *Rhizomucor pusillus* which has been achieved by Prasad et al.[5] as depicted below.

m/z=265
Albendazole
(Poor water solubility, less active)

m/z=282
Albendazole sulfoxide
(More water soluble, more active)

20.2.2   IN PREDICTING MAMMALIAN DRUG METABOLISM

In the process of drug discovery, the knowledge of drug metabolism and metabolite toxicity is very essential. Before the approval to use in humans, extensive studies to establish safety and efficacy of a drug are required which can be known by drug metabolism studies. Different animal models (rat, dog, etc.), microsomal preparations, or perfused organ systems, etc. are currently available and are associated with some disadvantages. The use of animal models is time consuming, expensive in terms of their care, mainte-nance, space, equipment and appropriate staff, and at times leads to suffering of the animals used.[6] Microsomes contain only phase I drug metabolism enzymes (DMEs) and uridine diphosphate (UDP)-glucuronosyl transfer-ases. The presence of a mixture of DMEs and wide variability in enzyme profiles between individual samples of liver microsomes, requirement of cofactor, and strict specific substrates and inhibitors for individual DMEs makes them difficult to use for establishing the role of a specific enzyme in the metabolism of a compound.[7] Requirement of healthy and fresh tissues, low viability of stored cells than that of freshly isolated hepatocytes, requirement of specific techniques and well established procedures, decrease of many DME levels rapidly during cultivation, no bile measurement, absence of organ specific cell to cell interaction, and no preserved anatomy are some of the disadvantages of hepatocytes.[8] Use of isolated organs in drug metabolism studies is also associated with some disadvantages like short-term viability (2–4 h), assessment of only few compounds with one organ, use of a high number of animals, and complexity of the setup.[8] Live slices also have some disadvantages like: difficulty in obtaining and need of fresh tissue, requirement of specific techniques and well established procedures, and problem in long-term storage.[7]

Microorganisms, especially fungi were found to possess cytochrome p450 enzyme system and oxidize organic compounds in the same way as mammalian hepatic cytochrome p450. Microbial transformation reactions mediated by fungi can be used for drug metabolism studies. Most of the microorganisms have the ability to modify a wide variety of drugs and other xenobiotic compounds similar to mammals, and the use of microbial systems as models for mimicking and predicting drug metabolism in humans and animals has received considerable attention.[9] Many mammalian phase I metabolic reactions (introduction of a functional group) and phase II metabolic reactions (conjugation with endogenous compounds), including hydroxylation, *O*- and *N*-dealkylation, dehydrogenation, and glucuronide and sulfate conjugation, also occur in microbial models.[10] Low cost, ease of handling, scale up capacity, and potential to reduce the use of animals[11] are some of the advantages of a microbial system as an *in vitro* model for producing drug metabolites. The use of mesophilic filamentous fungi of the genus *Cunninghamella*, as a model of mammalian biotransformation has been reported previously by Sun et al.[9] This mode of producing metabolites in large quantities using fungi is very convenient and a preparative method for the otherwise difficult to obtain metabolites, particularly when the structure is complex. It is pertinent to note that these biotransformed metabolites are difficult to isolate from a mammalian system, or they can be synthesized chemically. A comparative study of metabolism of Besipirdine in *C. elegans* and mammals[12] is shown in Figure 20-1 as an example. The use of microbial simulation of mammalian metabolism also gives an idea on the mechanism of action, toxicity, and pharmacological activity of the drugs and thus, helps in the discovery of new drug molecules. This approach has been successfully employed by other researchers in comparing the drug metabolism in fungal and mammalian systems.[9,13–17]

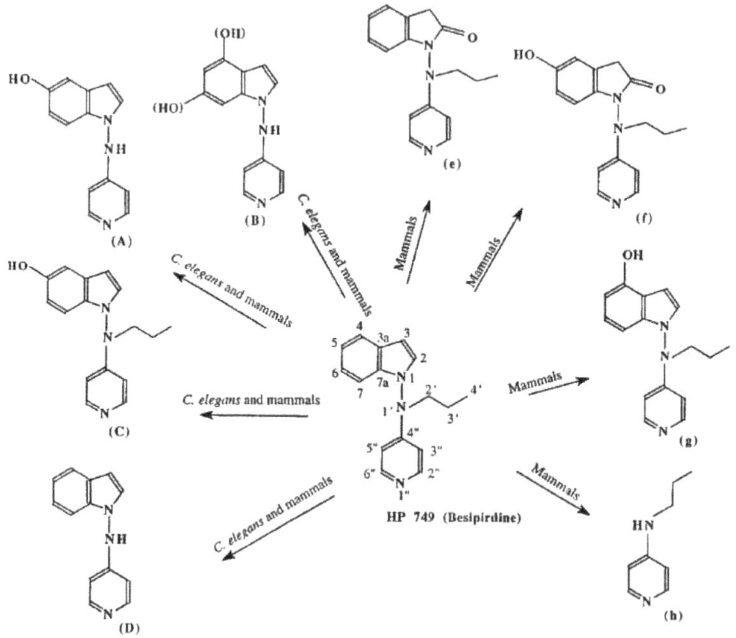

FIGURE 20-1 Microbial vs mammalian metabolism of HP 749 [*N*-(*n*-propyl) -*N*-(4-pyridinyl)-1*H*-indol-amine].

Microbial metabolites: (**A**) *N*-(4-pyridinyl)-*N*-(5-hydroxy-1*H*-indol)-1-amine, (**B**) *N*-(4-pyridinyl)-*H*-(4- or 6-hydroxy-1*H*-indol)-1-amine, (**C**) *N*-(*n*-propyl)- *N*-4-pyridinyl)-*N*-(5-hydroxy-1*H*-indol)-1-amine, and (**D**) *N*-(4-pyridinyl)-1*H*-indol-1-amine. *Mammalian metabolites:* (**e**) *N*-(propyl)-*N*-(4-pyridinyl)-1*H*-oxindolyl-1-amine, (**f**) *N*-(propyl)-*N*-(4-pyridinyl)-(5-hydroxy)-1*H*-oxindolyl-1-amine, (**g**) *N*-(propyl)-*N*-(4-pyridinyl)-*N*-(4-hydroxy)-1*H*-indol-1-amine, and (**h**) *N*-(4-pyridinyl)-(5-hydroxy)-1*H*-indol-1-amine.

20.2.3   IN NOVEL DRUG DISCOVERY

Drug development is a massive and challenging task from the laboratory scale to the market. The synthesized compounds or materials have to go through biological assays, *in vitro* and *in vivo* drug metabolism, pharmaco-kinetic studies, pharmacological, and toxicological studies. The compounds are to be ranked for clinical studies after performing all the studies and the compound that satisfies all the chemical and biological assays will be further tested. Great care is required to advance the compound from the preclinical stage onwards. The synthesis of compounds similar to that of the parent drug in the clinical trial will be a boon for the drug discovery programme.[18] Failure of drugs in the clinical trials will have to be supported by a group of compounds, whose structure or properties are similar to that of the drug under clinical trials. In the event of failure, the compounds with next best properties are to be tested.[18] Hence, the synthesis of compounds with structure and properties similar to that of the parent compound is necessary.

Biotransformation is one of the important experimental methods to identify compounds, whose structure and properties resemble the parent compound. Some of the examples of biosimilar compounds reported for diclofenac[19] and fenofibrate[20] are given below.

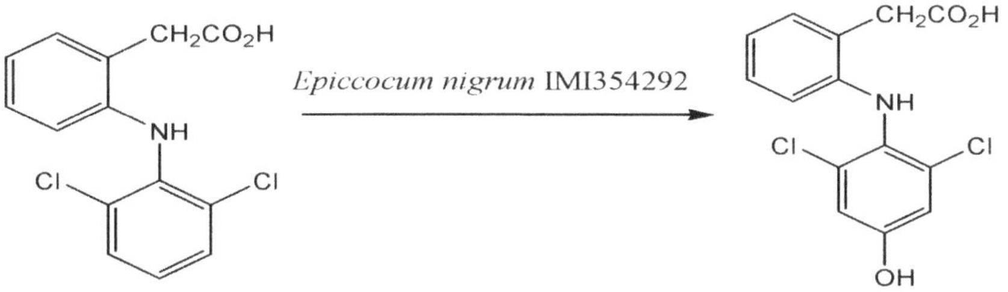

Biotransformation reaction of diclofenac by *Epiccocum nigrum* IMI354292:

CH_2CO_2H ... NH, Cl, Cl — m/z=296 — Diclofenac → CH_2CO_2H ... NH, Cl, Cl, OH — m/z=312 — 4-OH Diclofenac

Biotransformation reaction of fenofibrate by *Cunninghamella blakesleeana* NCIM 687:

Fenofibrate m/z=360 → Fenofibric acid m/z=318

20.3 SELECTION OF MICROORGANISMS FOR BIOTRANSFORMATION

20.3.1 BY ENRICHMENT METHOD

To perform a biotransformation reaction, it is desired to get a new strain of microorganism. Soil samples containing mixed microbial population are considered as a rich source for such organisms. Since these microorganisms are involved in the degradation of complex organic materials, some of them may have the potential to metabolize or modify the desired compound. Such microorganisms can be isolated by soil enrichment technique.[21] In this technique, soil sample or other samples where the organism is suspected is brought to the laboratory and serial dilutions are performed to reduce the load of the microorganisms and one ml of suitable dilution is added into the plates. Then a suitable nutrient agar medium is prepared and the carbon source of the medium is replaced by the compound to be transformed before adding into the plates and incubated at suitable temperature for bacteria and fungi. The organism which is enriched by the added compound will be grown which can be isolated, identified, and can be used in biotransformation studies. Through this method, it is possible to isolate many different types of microorganisms which can differentiate optically active compounds. Salokhe and Govindwar[22] isolated camphor degrading *Serratia marcescens* from soil by the soil enrichment technique. Uzura et al.[23] have isolated *Fusarium moniliforme* from soil by the enrichment technique which was capable of performing stereoselective oxidation of alkyl benzenes. Pentachlorophenol transforming *Penicillium veronii* was isolated by selective enrichment of the soil samples by Nam et al.[24] Isoeugenol transforming *Bacillus fusiformis* was isolated from soil by Zhao et al.[25] by enrichment technique. Dai et al.[26] have isolated the imidacloprid hydroxylating bacteria *Stenotrophomonas maltophilia* (China General Microbiological Culture Collection Center 1.1788) by enrichment technique. This process is laborious and time consuming, even though it is a time tested technique.

20.3.2   SCREENING OF SELECTED CULTURES

Biotransformation can also be performed by screening large number of bacteria and fungi identified from previous reports for their ability to transform the selected compound obtained in pure form from standard culture collection centers/culture banks. Because of their established identity and well documented information, it is a time saving process.

20.4   MAINTENANCE OF MICROBIAL CULTURES

The selected microbial cultures can be maintained in the laboratory by growing on various types of suitable agar slants, depending on the growth requirements of the specific class of microorganisms and can be stored at 2–8 °C. The cultures are transferred to fresh slants every 3 months to maintain viability and only fresh slants are used for the biotransformation studies.

20.5   ADDITION OF SUBSTRATE

An ideal substrate should be non-toxic to the organism, and should be readily soluble in an aqueous medium. The substrate is usually fed to a culture of microorganisms as sterile solution. Heat labile substrates can be filter sterilized before addition to the culture medium. The substrates which are not soluble in an aqueous medium are dissolved in a non toxic water miscible solvent before their addition to the culture. An emulsifying agent can be added for enhancing solubility. The optimal growth phase for feeding and optimal substrate concentration has to be optimized experimentally.

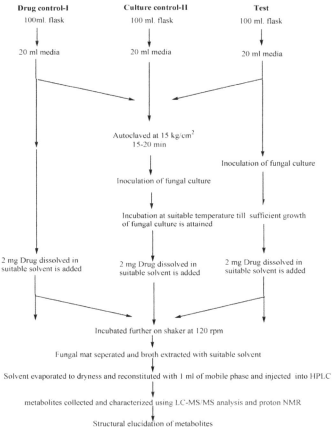

A typical Schematic Representation of Biotransformation Protocol

Maintenance of microbial cultures by subculturing
Preparation of media for growth of microbial cultures

Drug stock preparation

Drug control-I	**Culture control-II**	**Test**
100ml. flask	100 ml. flask	100 ml. flask
20 ml media	20 ml media	20 ml media

Autoclaved at 15 kg/cm^2
15-20 min

Inoculation of fungal culture

Inoculation of fungal culture

Incubation at suitable temperature till sufficient growth of fungal culture is attained

2 mg Drug dissolved in suitable solvent is added

2 mg Drug dissolved in suitable solvent is added

2 mg Drug dissolved in suitable solvent is added

Incubated further on shaker at 120 rpm

Fungal mat seperated and broth extracted with suitable solvent

Solvent evaporated to dryness and reconstituted with 1 ml of mobile phase and injected into HPLC

metabolites collected and characterized using LC-MS/MS analysis and proton NMR

Structural elucidation of metabolites

20.6   FACTORS INFLUENCING BIOTRANSFORMATIONS

Optimization of all the parameters is an important component to produce a maximal yield of the product. Optimization takes place once the feasibility of the production is demonstrated by the selected organism. This implies that working has been demonstrated based on theoretical viewpoint leading to an optimized process before proceeding to scale-up industrial process.

20.6.1   EFFECT OF MEDIA

Literature survey reveals that different synthetic media have been reported to influence biotransformation of organic compounds. Sutherland et al.[27] have reported maximum transformation of phenanthrene in malt extract glucose medium by employing *Phanerochaete chrysosporium*. Zhang et al.[28] have reported 95% of azatadine transformation by *Cunninghamella elegans* by using Sabouraud's dextrose broth. Similarly, Parshikov et al.[29] have reported the transformation of enrofloxacin in sucrose–peptone broth by employing *Mucor ramannianus*. A faster transformation of phenanthrene in Sabouraud's medium by *C.elegans* was reported by Lisowska and Dlugonski.[30] Zhang et al.[31] have also reported the hydroxylation of cinobufagin in potato dextrose medium by *Mucor spinosus*. Specific 12β-hydroxylation of cinobufagin in potato medium by using *Alternaria alternata* was reported by Ye et al.[32]

20.6.2   INCUBATION PERIOD

The rate and the amount of the product formed by the process of transformation were reported to be influenced by the incubation period. Miyazawa et al.[33] reported that the transformation of cedrol by *Glomerella cingulata* varied with the incubation period. Similarly, transformation of Limonene by *Pseudomonas putida* was maximal on Day 5 of the incubation period,[34] whereas 8 days of incubation period was optimum for the transformation of (L)-citronellal to (L)-citronellol by *Rhodotorula minuta*.[35]

20.6.3   EFFECT OF PH

Enzymes are amphoteric molecules containing a large number of acidic and basic groups, mainly situated on their surface. The charges on these groups will vary, according to their acid dissociation constants, with the pH of their environment. This will affect the total net charge of the enzymes and the distribution of charge on their exterior surfaces, in addition to the reactivity of the catalytically active groups. These effects are especially important in the neighborhood of the active sites. Taken together, the changes in charges with pH affect the activity, structural stability, and solubility of the enzyme. Even small changes in pH may cause significant changes in the ability of a given enzyme to catalyze certain reactions. In a similar manner to the effect on enzymes, the charge and charge distribution on the substrate (S), product, and co-enzymes will also be affected by pH.

All enzymes have a pH range in which they show maximum specific activity. The dependence of the activity on pH normally resembles a Gaussian distribution (bell shaped curve). That is, the activity is usually at a maximum in a relatively narrow pH range and decreases at both higher and lower pH. Although, for some enzymes, the optimum range of activity can actually be fairly wide. The pH at which the enzyme has maximum activity is called the optimum pH, where most of the product is formed. The pH of the medium was found to influence biotransformation reactions. Qazi et al.[36] have reported that a pH of 5.5 was optimum for the activity of glucose dehydrogenase enzyme and a pH of 3.5–4.0 for the activity of gluconate and keto gluconate forming dehydrogenases. While *Pseudomonas putida* preferred a pH of 5.0 for the transformation of limonene,[34] Silberbach et al.[37] have recorded a pH of 5.0 to be optimum for the biotransformation of glucose to 2, 5-diketogluconate by *G. oxydans*.

20.6.4   EFFECT OF TEMPERATURE

Temperature relationships of organisms have been one of the most active areas of study in comparative and environmental physiology.[38] Like most chemical reactions, the rate of an enzyme catalyzed reaction increases as the temperature is raised. A ten degree centigrade rise in temperature will increase the activity of most enzymes by 50–100%. Variation in the reaction temperature as small as 1 or 2 degrees may introduce changes of 10–20% in the results. In the case of enzymatic reactions, this is complicated by the fact that many enzymes are adversely affected by a high temperature. The reaction rate increases with temperature to a maximum level, then abruptly declines with a further increase of temperature. As the temperature increases, molecular motion increases resulting in more molecular collisions. If, however, the temperature rises above a certain point, the heat will denature the enzyme, causing to loose its three-dimensional functional shape by denaturing its hydrogen bonds. Cold temperature, on the other hand, slows down enzyme activity by decreasing molecular motion. Enzymes, however, are proteins and undergo essentially irreversible denaturation (i.e., conformational alteration entailing a loss of biological activity) at temperatures above those to which they are ordinarily exposed in their natural environment. These denaturing reactions have standard free energies of activation of about 200–300 KJ mol[-1] which means that, above a critical temperature, there is a rapid rate of loss of activity. The actual loss of activity is the product of this rate and the duration of incubation. It may be because of the covalent changes such as the deamination of asparagine residues or non-covalent changes such as the rearrangement of the protein chain. Inactivation by heat denaturation has a profound effect on the enzyme productivity.

Literature survey suggests that temperature has a profound influence on biotransformation. *A. niger* and *A. cellulosae* preferred 30 °C for the transformation of dehydro pinguisenol.[39] A temperature of 28 °C was found to be optimum for the bioconversion of diterpenes by *Cephalosporium aphidicola*.[40] Chattarjee and Bhattacharya[34] have reported that 30 °C was most favorable for the transformation of limonene by *Psudomonas putida*. Harshad and Mohan[35] have reported 27 °C to be the optimum temperature for the biotransformation of (L)–Citronellal to (L)–Citronellol by *Rhodotorula minuta*. Hydroxylation of 10-dexoartemisinin to 15-hydroxy-10-dexoxoartemisinin by *Aspergillus niger* was recorded at 28 °C by Parshikov et al.[41]

20.6.5 CARBON SOURCES

Carbon is the major structural and functional component of microbial cells and plays an important role in the nutrition of fungi. A wide variety of carbon sources including carbohydrates, organic acids, and amino acids, along with their derivatives and some polycyclic compounds and alkaloids are also used by fungi as sources of carbon. Many fungal species are able to thrive on different kinds of aliphatic hydrocarbons. Thus, fungi are capable of using a wide variety of carbon compounds but most of the fungi prefer simple sugars. Jones et al.[42] by using the simple sugar glucose as the carbon source have reported good transformation of 4-ethyl phenol by *Aspergillus fumigatus*. Diez et al.[43] in his studies with glucose as the carbon source have reported the hydroxylation of sclareal by *Rhizopus stolonifer*. Ye et al.[30] by employing glucose as carbon source have reported specific 12β-hydroxylation of cinobufagin by *Alternaria alternata*. Similarly, Herath et al.[44] by using dextrose as the carbon source have reported transformation of 3-hydroxy flavone by employing *Beauveria bassiana*. Fiaux de Medeiros et al.[45] by using sucrose as carbon source reported the transformation of 10-deoxoartemisinin by *Mucor ramannianus*. Jin and Li[46] by using fructose as the carbon source reported enantio selective hydrolysis of o-Nitrostyrene oxide by *A. niger*.

20.6.6   NITROGEN SOURCES

In nature, both organic and inorganic forms of nitrogen are encountered. The efficiency of the nitrogen source in inducing growth of the fungus depends upon the capacity of the organism to convert the complex nitrogen source into the assimilable form. Fungi are reported to exhibit great specificity for the nitrogen source present in the medium. Like the carbon source, nitrogen is also used both for functional and structural purposes by fungi. The source of nitrogen has a profound influence on the metabolism of microorganisms.

Similarly, Diez et al.[43] in their studies have shown that using ammonium nitrate as the nitrogen source resulted in maximal transformation of sclareol by employing *Rhizopus stolonifer*. Farooq et al.[47] preferred sodium nitrate as the nitrogen source and reported the transformation of (-)-β-pinene to a maximum extent by using *Botrytis cinerea*.

20.6.7   EFFECT OF VITAMINS

Vitamins are organic compounds which are necessary for the growth of microorganisms. They function as co-enzymes or constituent parts of co-enzymes which catalyze special reactions. Vitamins are necessary for proper utilization of carbohydrates, fats, and proteins. The capacity of different organisms to synthesize the vitamins varies. Fungi with regard to their vitamin requirement occupy a position in between the totally independent green plants and completely dependent animals. The topic has been the subject of many investigations and the importance of vitamins in fungal nutrition has been shown by many workers.

20.6.8   INFLUENCE OF SUBSTRATE CONCENTRATION

The amount of drug to be added into the fermentation broth for the biotransformation should be optimized, as higher concentrations of the drug may be toxic and influence the growth of the microorganisms. Literature survey also revealed the effect of drug concentration on its transformation. Agarwal et al.[48] reported the critical role of benzaldehyde concentration on L-phenyl acetyl carbinol production by *Saccharomyces cerevisiae*. Shukla et al.[49] studied the transformation of benzaldehyde at various concentrations by using *Torulospora delbruecki*. Chatterjee and Bhattacharya[34] have recorded maximum transformation of limonene by *Psuedomonas putida* at its 0.2% concentration. Similarly, transformation of 2,4-dichlorophenol and penta-chlorophenol by *Trametes versicolor* by their concentration in the basal medium was also reported by Sedarati et al.[50]

20.6.9   INFLUENCE OF SOLVENTS

Enzymes are highly specific and extremely enantio- and regio-selective catalysts. Because of these attributes, they are widely used in biotransformation for the production of fine chemicals and optically active compounds of industrial importance in processes that are effective and ecofriendly alternative to chemical synthesis.[1,51] Over the past two decades, biocatalyst in organic solvents has emerged as an area of intensive research, fueled mainly by chemical and pharmaceutical interest. Attention has been especially focused on enzymes as catalysts for asymmetric synthetic transformations.[52,53] However, enzymes are denatured or inactivated in the presence of organic solvents and specific catalytic activity of enzymes that are stable in non-aqueous environments are generally lower than those in the aqueous systems.[54,55] Protein solubility in organic solvents is variable and depends on many parameters including the nature of the solvent, the properties of proteins, and the physicochemical conditions at the protein solvent interface. Until some years ago, the vast majority of organic solvents were considered to be unsuitable for proteins because of their insolubility in these solvents with an exception of the classical protein dissolving organic solvents like dimethyl sulfoxide (DMSO), ethylene glycol, and formamide.[56] However, solvents such as DMSO and formamide tend to render enzymes inactive. Systems involving mixtures of water and organic solvents are moderately used to enable the bioconversion of substrates that are moderately soluble in water, to modify the enantioselection of enzymes or to decrease the water content in the reaction medium to favor synthesis hydrolysis.[57] Moreover, intimate contact between the organic solvent and the enzyme can lead to changes in the enzymatic reaction that is catalyzed. For example, when dimethyl formamide (DMF) was added to a reaction medium containing subtilisin Bacterial Protease Nagase (BPN), the aminolysis reaction was increased and the hydrolysis reaction was suppressed. This was associated with changes in the active site of the enzyme histidine residue caused by the presence of 50% DMF.[58] In addition, the enantioselectivity of subtilisin for ethyl-2-(4-substituted phenoxy) propionates was enhanced, owing to the deformation of the enzyme that occurred in water—DMSO mixture.[59]

20.6.10   EFFECT OF GLUCOSE CONCENTRATION

Monosaccharides usually are easily assimilable forms of carbohydrates, among which glucose has been reported to be the most favored carbon source for most of the fungi. The concentration of glucose was found to influence biotransformation. Kakimito et al.[60] reported that the transformation of L-arginine to L-citrulline by *P. putida* was influenced by the concentration of glucose.

20.6.11   INFLUENCE OF AGITATION

A proper agitation speed is important for appropriate air supply and proper mixing of the media components. The degree of agitation required for a fermentation study will be dependent on the organism and the composition of the fermentation medium.[12] Many researchers reported that agitation has an influence on the biotransformation reaction.[25,61,62]

20.7   INCUBATION, EXTRACTION, AND ANALYSIS

Incubation time may be carried out up to 7 days and biotransformation products contained in the reaction medium are recovered from the whole broth by using a suitable solvent after the removal of the biomass by filtration. Aqueous samples can be directly analyzed by high-performance liquid chromatography (HPLC) or repeatedly extracted with organic solvents for thin-layer chromatography (TLC) of gas chromatography (GC) analysis. The important parameters to be taken into consideration include stability of the drug under varying pH conditions, extraction of the drug and possible metabolites from aqueous media with organic solvents under varying conditions of pH, and sensitive and well resolved chromatographic systems capable of detecting as minute as 1% transformation.

Liquid chromatography-mass spectrometry/mass spectrometry (LC-MS/MS) analysis is used in structure elucidation and thereby in the identification of metabolites. The LC part of the system will separate the metabolites into individual peaks and the MS-MS part will identify them by giving their structures after reading their masses and helps in working out the structures based on the fragmentation pattern. Representative LC-MS/MS spectra[20] of fenofibric acid and its fragmentation pattern are shown in Figure 20-2.

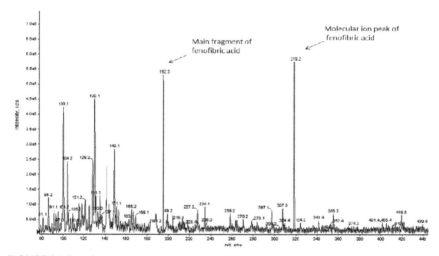

FIGURE 20-2   Showing LC-MS/MS spectra of fenofibric acid and its fragmentation pattern.

20.8   SCALE UP

The scale-up of such a process is usually the final step in any research and development program leading to the large-scale manufacture of such products by fermentation. For structural elucidation and biological activity studies, large quantities of metabolites can be obtained after optimizing different parameters either in large size flasks or in stirred fermenters.

20.9   UTILITY OF MICROBIAL TRANSFORMATIONS TO THE PHARMA INDUSTRY

Chiral drug intermediates can be prepared by asymmetric synthesis by either chemical or biocatalytic processes using microbial cells or enzymes.

Chirality is a key factor in the efficacy of many drugs and agrochemicals. Production of single enantiomers of the intermediates has become increasingly important in the pharmaceutical industry. Single enantiomers can be produced by chemical or chemo-enzymatic synthesis. The advantages of biocatalysis over chemical synthesis are that the enzyme-catalyzed reactions are often highly enantioselective and regioselective. They can be carried out at ambient temperature and atmospheric pressure, thus avoiding the use of more extreme conditions that could cause problems with isomerization, racemization, epimerization, and rearrangement. Microbial cells and enzymes derived from them can be immobilized and reused for many cycles. In addition, enzymes can be overexpressed to make biocatalytic processes economically efficient, and enzymes with modified activity can be tailor-made. The preparation of thermostable and pH-stable enzymes by random and site-directed mutagenesis has led to the production of novel biocatalysts for the biotransformation of drugs.

20.10 CONCLUSION

Biotransformation reactions mediated by microorganisms, especially fungi are more advantageous because of the versatility of their enzyme systems. Almost all chemically defined reactions are performed and metabolites can be produced in high yields most economically and in an ecofriendly method for performing pharmacological studies. Rare compounds not possible or difficult by synthetic methods are also made possible by microbes. The products of biotransformation are biosimilar and possess similar physical and chemical properties; thus, it is easy to identify novel compounds with an enhanced biological activity. Biotransformation reactions are stereoselective, regioselective, and highly advantageous in the pharma industry Interestingly, fungi produce drug metabolites that are similar to mammalian xenobiotics. Hence, fungal models are appropriate to study mammalian drug metabolism.

ACKNOWLEDGMENTS

Dr. G. Shyam Prasad is thankful to the University Grants Commission (UGC), New Delhi, India for providing Dr. D. S. Kotari fellowship at the Department of Biochemistry, Osmania University, Hyderabad, Telangana.

KEYWORDS

- **Biotransformation**
- **Pharma and drug discovery**

REFERENCES

1. Schmid, A.; Dordick. J. S.; Hauer, B., Kiener, A.; Wubbolts, M.; and Witholt, B. Industrial biocatalysis today and tomorrow. *Nature* **2001,** *409,* 258–268.
2. Prasad, G. S.; Girisham, S.; and Reddy, S. M. Microbial transformation of albendazole. *Indian J. Exp. Biol.* **2009,** *48,* 415–420.
3. Prasad, G. S.; Srisailam, K.; Girisham, S.; and Reddy, S. M. Biotransformation of albendazole by fungi. *World. J. Microb. Biot.* **2008,** *24,* 1565–1571.
4. Soria-Arteche, O.; Castillo, R.; Hernández-Campos, A.; Pena, M. H.; Navarrete-Vazquez, G.; Medina-Franco, J. L.; and Gómez-Flores, K. Studies on the selective S-oxidation of albendazole, fenbendazole, triclabendazole, and other benzimidazole sulfides. *J. Mex. Chem. Soc.* **2005,** *49,* 353–358.
5. Prasad, G. S.; Girisham, S.; and Reddy, S. M. Potential of thermophilic fungi *Rhizomucor pusillus* in biotransformation of anti-helmintic drug albendazole. *Appl. Biochem. Biotechnol.* **2011,** *165,* 1120–1128.
6. Arora, T.; Mehta, A. K.; Joshi, V.; Mehta, K. D.; Rathor, N.; Mediratta, P. K.; and Sharma, K. K. Substitute of animals in drug research, an approach towards fulfillment of 4Rs. *Indian J. Pharm. Sci.* **2011,** *73,* 1–6.
7. Wrighton, S. A.; Branden, M. V.; Stevens, J. C.; Shipley, L. A.; Ring, F. J.; Rettie, A. E.; and Cashman, J. R. *In vitro* methods for assessing human hepatic drug metabolism, their use in drug development. *Drug Metab. Rev.* **1993,** *25,* 453–484.
8. Groneberg, D. A.; Siestrup, C. G.; and Fischer, A. A. *In vitro* models to study hepatotoxicity. *Toxicol. Pathol.* **2002,** *30,* 394–399.
9. Sun, L.; Huang, H. H.; Liu, H.; and Zhong, D. F. Transformation of verapamil by *Cunninghamella blakesleeana. Appl. Environ. Microbiol.* **2004,** *70,* 2722–2727.
10. Abourashed, E. A.; Clark, A. M.; and Hufford, C. D. Microbial models of mammalian metabolism of *xenobiotics,* an updated review. *Curr. Med. Chem.* **1999,** *6,* 359–374.
11. Zhang, D.; Freeman, J. P.; Sutherland, J. B.; Walker, A. E.; Yang, Y.; and Cerniglia, C. E. Biotransformation of chlorpromazine and methdilazine by *Cunninghamella elegans. Appl. Environ. Microbiol.* **2006,** *62,* 798–803.
12. Rao, G. P.; and Davis, P. J. Microbial models of mammalian metabolism, biotransformation of HP 749 (besipiridine) using *Cunninghamella elegans. Drug. Metab. Dispos.* **1997,** *25,* 709–715.
13. Zhong, D. F.; Sun, L.; Liu, L.; and Huang, H. H. Microbial transformation of naproxen by *Cunninghamella* species. *Acta Pharmacol. Sin.* **2003,** *24,* 442–447.
14. Cha, C. J.; Doerge, D. R.; and Cerniglia, C. E. Biotransformation of malachite green by the fungus *Cunninghamella elegans. Appl. Environ. Microbiol.* **2001,** *67,* 4358–4360.

15. Moody, J. D.; Freeman, J. P.; and Cerniglia, C. E. Biotransformation of doxepin by *Cunninghamella elegans*. *Drug. Metab. Dispos.* **1999,** *27,* 1157–1164.

16. Moody, J. D.; Heinze, T. M.; Hansen, E. B.; and Cerniglia, C. E. Metabolism of the ethanol amine type antihistamine diphenhydramine (Benadryl) by the fungus *Cunninghamella elegans. Appl. Microbiol. Biotechnol.* **2000,** *53,* 310–315.

17. Hezari, M; and Devis, P. J. Microbial models of mammalian metabolism. Furosemide glucoside formation using the fungus *Cunninghamella elegans. Drug Metab. Dispos.* **1993,** *21,* 259–267.

18. Ravindran, S.; Basu, S.; Surve, P.; Lonsane, G.; and Sloka, N. Significance of biotransformation in drug discovery and development. *J. Biotechnol. Biomaterial.* **2012,** *S13,* 005.

19. Webster, R.; Pacey, M.; Winchester, T.; Johnson, P.; and Jezequel, S. Microbial oxidative metabolism of diclofenac, production of 4-OH diclofenacx using *Epicoccum nigrum* IMI354292. *Appl. Microbiol. Biotechnol.* **1998,** *49,* 371–376.

20. Prasad. G. S.; Govardhan, P.; Girisham, S.; and Reddy, S. M. Fungal mediated generation of mammalian metabolites of fenofibrate and enhanced pharmacological activity of the main metabolite fenofinric acid. *Drug Metab. Lett.* **2014,** *8,* 88–95.

21. Salokhe, M. D.; and Govindwar, S. P. Effect of carbon source on the biotransformation enzymes in *Serratia marcescens. World J. Microb. Biotechnol.* **1999,** *15,* 259–263.

22. Carglie, N. L.; and McChesney, J. D. Microbiological steroid conversions, utilization of selected mutants. *Appl. Microbiol.* **1974,** *27,* 991–994.

23. Uzura, A.; Katsuragi, T.; and Tani, Y. Stereoselective oxidation of alkyl benzenes by fungi. *J. Biosci. Bioeng.* **2001,** *91,* 217–221.

24. Nam, I. H.; Chang, Y. S.; Hong, H. B.; and Lee, Y. E. A novel catabolic activity of *Pseudomonas veronii* in biotransformation of pentachlorophenol. *Appl. Microbiol. Biotechnol.* **2003,** *62,* 284–290.

25. Zhao L. Q.; Sun, Z. H.; Zheng, P.; and Zhu, L. L. Biotransformation of isoeugenol to vanillin by a novel strain of *Bacillus fusiformis. Biotechnol. Lett.* **2005,** *19,* 1505–1509.

26. Dai, Y. J.; Yuan, S.; Ge, F.; Chen, T.; Xu, S. C.; and Ni, J. P. Microbial hydroxylation of imidacloprid for the synthesis of highly insectidal olefin imidacloprid. *Appl. Microb. Biotechnol.* **2006,** *71,* 927–934.

27. Sutherland, J. B.; Selby, A. L.; Freeman, J. P.; Evans, F. E.; and Cerniglia, C. E. Metabolism of phenanthrene by *Phanerochaete chrysosporium. Appl. Environ. Microbiol.* **1991,** *57,* 3310–3316.

28. Zhang, D.; Hansen, E. B. Jr.; Deck, J.; Heinze, T. M.; Sutherland, J. B.; and Cerniglia, C. E. Fungal biotransformation of the antihistamine azatadine by *Cunninghamella elegans. Appl. Environ. Microbiol.* **1996,** *62,* 3477–3479.

29. Parshikov, I. A.; Freeman, J. P.; Lay, J. O. Jr.; Beger, R. D.; Williams, A. J.; and Sutherland, J. B. Microbiological transformation of enrofloxacin by the fungus *Mucor ramannianus. Appl. Environ. Microbiol.* **2000,** *66,* 2664–2667.

30. Lisowska, K.; and Dlugonski, J. Concurrent corticosteroid and phenanthrene transformation by filamentous fungus *Cunninghamella elegans. J. Steroid Biochem. Mol. Biol.* **2003,** *85,* 63–69.

31. Zhang, W.; Ye, M.; Qu, G. Q.; Wu, W. Y.; Chen, Y. J.; and Guo, D. A. Microbial hydroxylation of cinobufagin by *Mucor spinosus. J. Asian. Nat. Prod. Res.* **2005,** *7,* 225–229.

32. Ye, M.; Qu, G.; Guo, H.; and Guo, D. Specific 12 beta-hydroxylation of cinobufagin by filamentous fungi. *Appl. Environ. Microbiol.* **2004,** *70,* 3521–3527.

33. Miyazawa, M.; Nankai, H.; and Kameoka, H. Biotransformation of (+)-cedrol by plant pathogenic fungus, *Glomerella cingulata. Phytochemistry* **1995**, *40*, 69–72.
34. Chatterjee, T.; and Bhattacharya, D. K. Biotransformation of limonene by *Pseudomonas putida. Appl. Microbiol. Biotechnol.* **2001**, *55*, 541–546.
35. Harshad, R. V.; and Mohan, R. H. Biotransformation of (L)-citronellal to (L)-citronellol by free and immobilized *Rhodotorula minuta. Electronic. J. Biotechnol.* **2003**, *6*, 1–10.
36. Qazi, G. N.; Parshad, R.; Verma, V.; and Chopra, C. L. Di Keto gluconate fermentation by *Gluconobacter oxydans. Enzyme Microb. Technol.* **1991**, *13*, 504–507.
37. Silberbach, M.; Maier, B.; Zimmermann, M.; and Buchs, J. Glucose oxidation by *Gluconobacter oxydans*, characterization in shaking-flasks, scale-up and optimization of the pH profile. *Appl. Microbiol. Biotechnol.* **2003**, *62*, 92–98.
38. Prosser, C. L.; and Heath, J. E. Temperature. In *Comparative Animal Physiology, Environmental and Metabolic Animal Physiology,* 4th ed; Prosser, C. L., Ed.; Wiley-Liss: New York, **1990**; pp 109–165.
39. Lahlou, E. H.; Noma, Y.; Hashimoto, T.; and Asakawa, Y. Microbial transformation of dehydropinguisenol by *Aspergillus* sp. *Phytochemistry* **2000**, *54*, 455–460.
40. Takahashi, J. A.; Barroso, H. A.; and Oliveira, A. B. Optimization of diterpenes Bioconversion process by the Fungus *Cephalosporium aphidicola. Braz. J. Microbiol.* **2000**, *31*, 83–86.
41. Parshikov, I. A.; Muralieedharan, K. M.; Miriyala, B.; Avery, M. A.; and Williamson, J. S. Hydroxylation of 10-deoxoartemisinin by *Cunninghamella elegans. J. Nat. Prod.* **2004**, *67*, 1595–1597.
42. Jones, K. H.; Trudgill, P. W.; and Hopper, D. J. Ethylphenol metabolism by *Aspergillus fumigatus. Appl. Microbiol. Biotechnol.* **1994**, *60*, 1978–1983.
43. Díez, D.; Sanchez, J. M.; Rodilla, J. M.; Rocha, P. M.; Mendes, R. S.; Paulino, C.; Marcos, I. S.; Basabe, P.; and Urones, J. G. Microbial hydroxylation of sclareol by *Rhizopus stolonifer. Molecules* **2005**, *10*, 1005–1009.
44. Herath, W.; Mikell, J. R.; Hale, A. L.; Ferreira, D.; and Khan, I. A. Microbial metabolism. Metabolites of 3- and 7-hydroxyflavones. *Chem. Pharm. Bull.* **2006**, *54*, 320–324.
45. Fiaux de Medeiros, S.; Avery, M. A.; Avery, B.; Leite, S. G. F.; Freitas, A. C. C.; and Williamson, J. S. Biotransformation of 10-deoxoartemisinin to its 7β-hydroxy derivative by *Mucor ramannianus. Biotechnol. Lett.* **2002**, *24*, 937–941.
46. Jin, H.; Li, Z. Y.; and Dong, X. W. Enantioselective hydrolysis of various substituted styreneoxides with *Aspergillus niger* CGMCC 0496. *Org. Bimol. Chem.* **2004**, *2*, 408–414.
47. Farooq, A.; Choudhary, M. I.; Tahara, S.; Rahman, A. U.; Baser, K. H.; F. and Demirci, F. The microbial oxidation of (-)-beta-pinene by *Botrytis cinerea. Z. Naturforsch C.* **2002**, *57*, 686–690.
48. Agarwal, S. C.; Basu, S. K.; Vora, V. C.; Mason, J. R.; and Pirt, S. J. Studies on the production of L-acetyl phenyl carbinol by yeast employing benzaldehyde as precursor. *Biotechnol. Bio. Eng.* **1987**, *29*, 783–785.
49. Shukla V. B.; and Kulkarni, P. R. Biotransformation of Benzaldehyde to L-Phenyl acetyl carbinol (L-PAC) by free cells of *Tosulaspora delbrueckii* in presence of Beta-cyclodextrim. *Braz. Arch. Biol. Technol.* **2002**, *45*, 265–268.
50. Sedarati, M. R.; Keshavarz, T.; Leontievsky, A. A.; and Evans, C. S. Transformation of high concentrations of chlorophenols by the white-rot basidiomycete *Trametes versicolor* immobilized on nylon mes. *Electronic. J. Biotechnol.* **2003**, *6*, 2226–2254.
51. Carrea, G.; and Riva, S. Properties and synthetic applications of enzymes in organic solvents. *Angew. Chem. Int. Ed. Engl.* **2000**, *39*, 2226–2254.

52. Jaeger, K. E.; Dijkstra, B. W.; and Reetz, M. T. Bacterial biocatalysts, molecular biology, three-dimensional structure and biotechnological applications of lipases. *Ann. Rev. Microbiol.* **1999,** *53,* 315–351.

53. Garcia, J. E.; Garcia, G. J. F.; Bastida, A.; and Mayoralas, F. A. Enzymes in the synthesis of bioactive compounds, the prodigious decades. *Bioorg. Med. Chem.* **2004,** *12,* 1817–1834.

54. Ogino, H.; and Ishikawa, H. Enzymes which are stable in the presence of organic solvents. *J. Biosci. Bioeng.* **2001,** *91,* 109–116.

55. Ru, M. T.; Dordick, J. S.; Reimer, J. A.; and Clark, D. S. Optimizing the salt induced activation of enzymes in organic solvents, effects of lyophilization time and water content. *Biotechnol. Bioeng.* **1999,** *63,* 233–241.

56. Singer, S. J. The properties of proteins in nonaqueous solvents. *Adv. Protein Chem.* **1962,** 17, 1–68.

57. Castro, G. R.; and Knubovets, T. Homogenous biocatalysis in organic solvents and water-organic mixtures. *Crit. Rev. Biotechnol.* **2003,** *23,* 195–231.

58. Kidd, R. D.; Sears, P.; Huang, D. H.; Witte, K.; Wong, C. H.; and Farber, G. K. Breaking the low barrier hydrogen bond in a serine protease. *Protein Sci.* **1999,** *8,* 410–417.

59. Watanabe, K.; and Ueji, S. Dimethyl sulfoxide induced high enantio-selectivity of subtilisin Carlsberg for hydrolysis of ethyl 2-(4-substituted phenoxy) propionates. *Biotechnol. Lett.* **2000,** *22,* 599–603.

60. Kakimoto, T.; Shibatani, T.; Nishimura, N.; and Chibata, I. Enzymatic production of L-citrulline by *Pseudomonas putida. Appl. Microb.* **1971,** *22,* 992–999.

61. Liras, P.; and Umbreit, W. W. Transformation of morphine by resting cells and cell-free systems of *Arthrobacter* species. *App. Microbiol.* **1975,** *30,* 262–266.

62. Bastida, J.; de Andres, C.; Cullere, J.; Busquets, M.; Manresa, A. Biotransformation of oleic acid into 10-hydroxy-8E-octadecenoic acid by *Pseudomonas* sp. 42A2. *Biotechnol. Lett.* **1999,** *12,* 1031–1035.

INDEX

α-helices, 5, 11
α-ketoamides, 40
α-ketoesters, 40
β-1,4 glucose units, 221
β-1,4-D-glycosidic bond, 221
β-avenohtionins, 6
β-D-glucosidase assay, 209
β-glucosidase, 202, 203, 209–216, 221, 251
β-glucuronidase, 7, 15
β-sheets, 5, 11
β-zeathionins, 15
γ-thionins, 7

A

Abiotic stress, 6
Acetylcholine deficiency, 65
Acetylcholine esterase (AChE), 65, 66
Acetylcholinesterase inhibitory activity, 83
Acetylgliotoxin, 332
Acetylthiocholine iodide (ATCI), 66
Achaea janata, 311, 316, 329
Actinomadura hibisca, 223, 232, 233
Actinomycetes, 145, 221, 234, 242,
    338–349, 356–360, 364–368
    amylases, 221
    cellulases, 221
    chitinases, 221
    hemicellulases, 221
    keratinases, 221
    pectinases, 221
    peptidases, 221
    proteases, 221
    xylanases, 221
Actinomycetes grown ISP media, 340
    actinomycetes isolation agar (AIA), 340
    starch agar (SA), 340
    starch casein agar (SCA), 340
Actinosynnema, 221
Activation energy, 253, 255, 256
Active site analysis, 81
Acute phase reactants, 95

C-reactive protein, 95
    lipopolysaccharide-binding protein
    (LBP), 97
    procalcitonin, 96
    serum amyloid A (SAA), 97
Acute phase proteins
    alpha 1 acid glycoprotein, 98
    markers, 98
    transthyretin, 99
Acute phase response (APR), 95
Adrenomedullin (ADM), 99, 103
Agro forestry system, 267
Agrobacterium rhizogenes, 5
Ailanthus altissima, 137
Alfalfa defensin, 16
Algal sources, 279, 282
Alkaliphilic
    actinomycetes, 338, 339, 346–349
    bacteria, 339, 348, 379
Allelochemicals, 135–139, 150, 151
Allelopathic
    compounds, 135–137
    inhibition, 138
    plants, 137, 138, 150
Allelopathy, 134, 135, 151
Alpha 1 acid glycoprotein, 98
Alpha-melanocyte-stimulating hormone
    (alpha-MSH), 99
Alternaria
    alternata, 396, 398
    longipes, 16, 19
    solani, 16, 19
Alzheimer's disease (AD), 65
Amaranthus caudatus, 10
Ammonium sulphate precipitation, 220,
    225, 238, 243, 244
Amphipathicity, 5, 11
Amylases, 221, 375
Anacardium, 65–67
Anaerobic
    bacteria, 278
    microorganisms, 299

Analysis of variance (ANOVA), 117, 160, 163, 189, 190
Analytical assays, 268
  laccase enzyme assay, 268
  lignin peroxidise assay, 268
  manganese peroxidase assay, 268
  protein estimation, 268
Anionic phospholipids, 5
Anisotropic displacement, 70, 315
Antiarthritisproperties, 66
Antibacterial, 5, 9, 13, 15, 17, 37, 66
Antibiotic sensitivity/resistance, 377
Anticancer, 4, 6, 18, 20, 81, 110, 111
  peptides (ACPs), 13
  properties, 6
Antifeedant, 316, 329, 332
  activity, 311, 312, 316, 317, 329
  compounds, 329
Antifungal, 5, 9–11, 15–17, 66, 158
Anti-inflammatory, 66, 93–95, 102
  activities, 99
Antimicrobial, 1, 3–7, 9–15, 17–20, 332
  activity, 5–12, 19, 33, 134, 311, 315, 326, 330, 332
    cyclotides, 10
    hevein/knottin-like peptides, 10
    lipid transfer proteins, 8
    plant defensins, 7
    plant peptides, 6
    snakins, 9
    thionins, 6
  compounds, 17
  peptides, 4–6, 10–20
  susceptibility testing, 326
  therapy, 91, 104
Antimycobacterials, 33
Antineoplastic
  agents, 111
  drug treatments, 18
Antioxidant, 66, 110, 126, 127, 187
  defense, 110, 126
Antiparasitic spectrum, 389
Antithrombin, 103
Antitubercular agents, 29–32, 34–40, 42, 43, 46–52, 54
  cyclic peptides, 18
  drugs, 29, 31, 32, 37, 45, 52, 53

high throughput screening (HTS) technique, 32
  molecules, 54
  research, 34
Antiviral activities, 4, 13
Apocynaceae, 10
Apoplasm, 177
Apoplast, 180
Apoptosis, 127
Arabinose, 40, 234, 237, 251
Arbuscular mycorrhizal (AM), 156
Arginine glycerol agar, 221, 226
Aromatic amines, 299
Arrhenius equation, 256
Artificial lipid membranes, 9
Aspergillus
  fumigatus, 398
  niger, 202–211, 213, 215, 216, 398, 399
  terreus, 78
  turcosus, 311, 312, 316–325, 329–332
Atomic absorption spectrophotometry, 110, 114

B

B. glumae, 15
Bacillaceae, 380
Bacillomycin, 17
Bacillus, 17, 203, 221, 299, 315, 326, 379, 394
  cereus, 203, 299
  pumiluss, 203
  subtilis, 17, 299, 315, 326
Bacterial
  infections, 96, 97
  lipopolysaccharides, 91
  membranes, 13
  microflora, 364
  protease nagase (BPN), 401
  sources, 280
  treatment, 172
Bathocuproine disulfonic acid disodium salt (BCS), 113
Benzimidazole carbamate, 389
Benzofurobenzopyran nucleus, 45
Benzoquinones, 136
Benzothiazole derivatives, 39
Bioaccumulation process, 299
Bioactive

compounds, 38, 65, 66, 78, 312, 348
    tricyclic systems, 46
Biochemical characterization, 339
Biodegradable, 202, 298, 312
Biofuel production, 250
Biogeochemical cycles, 267, 357, 375
Bioinformatics, 20
Bioisosterism, 34, 36
Biological
    activity, 4, 41, 43, 388, 398, 402, 403
    conversion, 279
    nitrogen fixing (BNF), 156
    properties, 45, 54
    systems, 7, 14, 135
Biomarkers, 90–93, 96, 100–104
    CD14 receptor markers/presepsin, 100
    micro-RNA (MI RNA), 103
    pro-ADM, 103
    soluble trem-1, 102
    soluble upar, 101
Bio-photolysis, 281
Bio-Rad protein assay, 114
Biotechnological applications, 279
Biotransformation, 387–404
BLAST, 278, 282, 283, 290
Botrytis cinerea, 16, 399
Bovine serum albumin (BSA), 224
Brassicaceae family, 7
Broad spectrum activity, 19
Burkholderia plantarii, 15

C

Cadmium, 172–177, 179, 181
    toxicity, 182
Callosobruchus chinensis, 316, 329
Cambridge Crystallographic Data Center
    (CCDC), 311
Cancer cells, 6, 13, 14, 18, 69
Candida albicans, 315, 326
Candida krusei, 298, 302–305
Candida strains, 311, 315, 326, 330
Carbohybrids, 39
Carboxymethyl cellulase method, 209
Carboxymethyl cellulose (CMC), 188, 206,
    220, 223
Cardiolipin (CL), 13
Caseinase activity, 234, 237
Cationic peptides, 5, 13

Cauliflower mosaic virus promoter
    (CaMV35S), 15
Cellobiohydrolase, 203, 212
Cell-penetrating peptides (CPPs), 19
Cellular morphogenesis, 340
Cellulase
    activity, 206, 220, 228, 229, 237, 239,
    242
    enzyme activity, 237–240
    production, 202, 203, 206, 207,
    211–214, 228, 348
Cellulases, 251
    enzymes types, 251
        cellobiohydrolases or exoglucanases,
        251
        endoglucanases, 251
        β-glucosidases, 251
Cellulolytic
    actinomycetes, 243
    enzymes, 207–212, 216
    organisms, 202, 203, 357
Cellulomonas biazotea, 203
Cellulosic biomass, 250, 280
Cephalosporium aphidicola, 398
Chaetomium arcuatum, 65–73, 75, 77–83
    strain, 65, 66, 78–83
Charge coupled device (CCD), 315
Chemical mutagenesis, 205, 211
Chemical oxygen demand (COD), 281
Chemical Technology, 27, 63, 68, 83, 133,
    277, 283, 290, 309, 316
Chemotherapeutic agents, 32
Chemotherapy, 30, 90, 91
Chitinases, 15, 221
Chlamydomonas reinhardtii, 278, 279, 282
Chlamydospores, 67, 159
Chlorella, 278, 282, 283, 289
Chlorella variabilis (CV), 278, 282, 289
Chlorpyrifos, 266–274
Cholinergic drugs, 65, 71
Chromatographic systems, 401
Cisplatin, 110–115, 117–121, 123, 125–127
Clavibacter michiganensis, 5
Clostridium
    acetobutylicum (CA), 278, 282, 289
    pasteurianum, 278, 282
    thermocellum, 280
Coastal saline desert, 381

Cocos nucifera, 18
Coinoculation, 172
Colony formation assays, 113, 118, 119
Colony forming units (CFU), 360
Community acquired pneumonia, 95
Copper-dependent metallothionein expression (*CUP1*), 125
Copper-responsive element (CuRE), 112
Corynebacterium, 7, 14
Council of Scientific and Industrial Research (CSIR), 32, 68, 83, 330
C-reactive protein (CRP), 90, 92, 95–99, 101–104
Crimean-Congo hemorrhagic fever (CCHF), 101
Crystallinity, 250, 251, 260
    indexes, 250
Crystallographic
    data, 79, 324, 325
    studies, 324, 330
Crystallography, 9, 65, 83
Curcubitaceae, 10
Curtobacterium flaccumfaciens, 5
Cyanobacteria, 279
Cycle progression, 110
Cyclic lipopeptides, 17
Cyclopeptides, 17
Cyclopsychotride, 5
Cyclotides, 4–6, 10, 18, 19
Cycloviolacin, 18
Cysteine amino acids, 9, 10
Cytoplasmic
    membrane, 17
    organelles, 8
Cytotoxic, 6, 7, 13, 14, 17, 18, 80, 111, 113
Czapek Dox
    agar medium (CZA), 210, 356
    agar plates, 205
    agar slants, 203

D

Decolorization, 303, 306
Degree of polymerization, 251
Deltaproteobacteria groups, 379
Dementia, 71
Denaturing gradient gel electrophoresis (DGGE), 338, 341, 347
Dendrophtora clavata, 6

Densitometric quantitation, 122, 123
Department of Biotechnology (DBT), 151
Department of Energy (DOE), 284
Depolymerization, 202
Desulfovibrio vulgaris, 278, 282
Dialysis, 220, 225, 238, 242–244
Dimethyl formamide (DMF), 400
Dimethyl sulfoxide (DMSO), 66, 400
Dinitrosalicyclic (DNS) method, 224
Dinitrosalicylic acid method, 209
Directly observed treatment short-course (DOTS), 30
Dithiothreitol (DTT), 140
Down's syndrome, 71
Drechslera teres, 7, 14
Drosomycin, 16
Drug
    concentration, 114, 400
    metabolism, 390, 391, 392
        enzymes, 390
    sensitivity, 111
Dual agar plate method, 182
Dubo's agar medium, 342
Dubo's agar plates, 341
Dye decolorization, 300–302, 305
Dye degradation, 301

E

Economical process, 250
Ecosystem processes, 357
Electric conductivity (EC), 359
Electron microscopy, 173
Electrophoresis, 68, 114, 116, 134, 138, 140, 141, 149, 220, 226, 313, 341, 347
    electricus, 66
Electrospray ionization mass spectrometry (ESI-MS), 314
Electrostatic interaction, 13
Embryonic fibroblasts, 126
Endoglucanase activity, 209
Endophyte, 83
Endophytic fungus, 65, 67, 78, 79, 82, 83
Energy consumption, 5, 280
Energy-dispersive X-ray microanalysis, 173
Entomopathogenic fungi, 329
Environmental
    conditions, 8, 136, 299, 316, 342
    pollution, 203, 267, 298

Enzymatic
    activity, 17, 122
    degradation, 221
    enzyme synergy, 251
        composition, 251
        origin, 251
    hydrolysis, 203, 250–252
        cellobiohydrolase, 203
        cellulase, 251
        composition, 251
        crystallinity, 251
        degree of polymerization, 251
        endoglucanase or carboxymethylcel-
        lulase, 203
        physical properties of substrate, 251
        β-glucosidases, 203
    intrinsic kinetics, 251
    mass transfer, 251
        bulk/pore diffusion, 251
        substrate adsorption, 251
Enzymatic metal transformation, 173
Enzyme
    active site, 259
    activity, 7, 136, 206, 207, 220, 224, 225,
    238–243, 258, 268, 269, 398
    assays, 208
    catalyzed reactions, 403
    commission (EC), 267
    extraction, 204
    kinetics, 257–260
    linked immunosorbent assay (ELISA),
    91
    synergy, 251
Epidithiodioxopiperazine, 325
Epimerization, 389, 403
Erlenmayer conical flasks, 207
Erwinia, 5, 7, 14, 16
Erwinia carotovora, 5, 16
Escherichia coli, 5, 315, 316, 326
Ethyl acetate-hexane mixture, 68, 69
Ethyl methane sulfonate (EMS), 202, 205,
    211
Ethylene diaminetetra acetic acid (EDTA),
    140
Eugenetin, 65, 76, 78, 80–83
Evan's blue, 176, 177
Evolutionary analyses, 72
Extra cellular

cellulase, 206, 214
enzymes, 212, 220, 221, 267, 341, 348,
374, 379
    detection, 341
    secretion, 348

F

Fenton's reaction, 112
Fermentation, 188, 202–204, 207, 211–216,
    250, 251, 279, 280, 281, 360, 377,
    399–402
Filobacillus milosensis, 378
Filter paper assay, 208
Filter paper unit (FPU), 209
Flavonoids, 136
Fourier transform infrared spectrum (FT-
    IR), 65, 311, 314, 320
Fructose, 234, 237, 399
Fumigant toxicity bioassay, 317
Fungal
    biomass, 68, 188, 268, 313
    decolorization, 299
    infections, 96
    microflora, 364, 367
    species, 78, 158, 398
Fungicidal effect, 326
Fusarium tricinctum, 78

G

Galactolipids, 9
Galanthamine hydrobromide, 66
Gas chromatography (GC), 401
Gas diffusion paths, 281–288
Gatifloxcin, 39
General therapy, 71
Genotypic changes, 380
Geographical areas, 104
Germination, 9, 15, 134–139, 142–145, 172,
    176, 179, 188, 326
Glaucoma, 71
Gliotoxin, 311, 317, 325, 326, 329–332
Glomus constrictum, 156–162, 164–168
Glomus constrictum Trappe, 156, 157
Glucanases, 15
Glycerol asparagine agar, 221, 226
Glycosylated mucins, 13
Gmelina arborea, 137

Gracilibacillus, 379
Gram-negative bacteria, 5, 13, 91, 311, 326, 330
Gram-positive bacteria, 5, 14, 91, 96, 377, 379
Graphic furnace system, 114
Graphite furnace, 114

H

Haddowia longipes, 79
Haematological toxicity, 33
Haemocele, 7, 14
Haemocytometer, 205, 208
Haloalkaliphilic
    actinomycetes, 338, 346
    bacteria, 374–381
    microorganisms, 375
Halobacillus, 378, 379
Halophiles/haloalkaliphiles, 379
    amylase, 379
    cellulase, 379
    chitinase, 379
    lipase, 379
    protease, 379
Heavy metal contamination, 172
Heliomicin, 16
Hematoxylin, 177
    root staining, 177
    staining, 172, 178
Hemicellulases, 221
Hemocyte, 13
Henderson–Hasselbalch equation, 254
Hepatotoxicity, 71
Heptapeptide cherimolacyclopeptide C, 18
Herpes simplex virus, 13
Heterobasidium annosum, 16
Heterocycle, 46, 51
Heterogeneity, 93, 339
Hevein and knottin-like peptides, 10
High Mobility Group Box Protein-1 (HMGB1), 94
High through-put screening (HTS) technique, 32, 33
High-performance liquid chromatography (HPLC), 401
High-resolution mass spectrometry (HR-MS), 69, 75, 78
Holocellulose, 258

Homeostasis, 95, 112
Homo sapiens, 82
Homoisoflavonoids, 45
Homology, 278, 282, 283, 288, 290
Honest significant difference, 191
Hordothionins, 8, 15, 16
Hormonal balance, 136
Human breast adenocarcinoma cells, 69, 70
Human chronic myelogenous leukemia cell line, 70
Human cytomegalovirus, 13
Human immunodeficiency virus (HIV), 13, 29
Human papilloma virus, 13
Hyacinthaceae, 45
Hybond-membranes, 114
Hybrid
    drugs, 37
    molecules, 29, 32, 37, 38, 52, 54
Hybridization, 29, 32, 34, 38, 40–47, 53, 54, 114, 115, 121, 347
    strategies, 32, 41, 54
Hydrogen bonding, 251
Hydrogen metabolism, 279
Hydrogenase, 278, 279, 281, 283, 285, 287–290
Hydrolysis rate, 259, 261
Hydrolytic enzymes, 202, 338, 348, 349, 379
Hydrophobicity, 6, 11
Hypoxanthine, 237

I

Immobilization, 272, 274
Immune and neuroendocrine systems, 99
Immunoluminometric assays, 91
Immunomodulatory effects, 7, 14
Immunosuppressive procedures, 90, 91
Information hyperlinked over protein (iHOP), 92
Inhibitor binding, 82
Innate immunity, 4
Inoculum density, 202
Insecticidal, 5, 11, 311–313, 329, 330
Intracellular oxidation, 123
Intraruminal administration, 390
Intrinsic kinetics, 251
Isoaminobutyric acid, 14

Isoelectric focusing (IEF), 141
Isolation of actinomycetes, 221, 226, 339, 342
Isomerization, 389, 403
Isoniazid (H), 30
Iturin, 17

K

Keratinases, 221
Kin recognition, 151
Kinetic behavior, 255, 257
Klebsiella oxytoca, 5
Klebsiella pneumoniae, 91
Kocuria aegyptia, 346
Korsakoff disease, 71

L

Lactoferrin, 13, 92
Leiurus quinquestriatus hebraeus, 8
Leptosphaeria maculans, 16
Lewy bodies, 71
Lignification, 175, 180, 181
Lignin degradation, 266
Lignin peroxidase (LiP), 266, 267, 269
Ligninase, 220
Lignin-degrading organisms, 267
Ligninolytic enzymes, 266–273
    laccase (LAC), 266
Lignocellulose materials, 266
Lignocellulosic
    biomass, 202, 221, 266, 267, 367
    substrates, 204
    wastes, 202, 368
Limulus polyphemus, 13
Lipid
    peroxidation, 110, 115, 123, 126
    transfer proteins, 6, 8
Lipinski's rule, 52
Lipoperoxidation, 123
Lipopolysaccharide, 104
    binding protein (LBP), 97
Lipopolysaccharides (LPS), 5, 13
Lipoteichoic acid, 91
Liquid chromatography-mass spectrometry/ mass spectrometry (LC-MS/ MS), 402
Liquid-liquid chromatography (LLC), 300
Logarithmic-phase cells, 119

Lycopersicon esculentum, 187
Lymphocytes, 101
Lymphocytic leukemia cells, 18

M

Macrophomina phaseolina, 156–159, 161–165, 168
Malignant tumor, 13
Mammalian
    cells, 6, 13, 15, 112
    drug metabolism, 388–390, 403
    systems, 391
Manganese peroxidase (MnP), 266, 268
Mannose-Binding Lectin (MBL), 90
Mass spectroscopic techniques, 65, 311
Matrix assisted laser desorption ionization time-of-flight, 142
Mean difference (MD), 65, 80, 191–194
Metabolic processes, 13
Metal
    biosorption, 173
    contaminations, 172
    precipitation, 173
    toxicity, 173, 180
Metallothionein (MT), 111, 126
Methicillin-resistant *Staphylococcus aureus* (MRSA), 311, 316
Michaelis Menten kinetics, 254, 259, 260
Michaelis pH function, 257–260
Microbial
    cells, 299, 388, 398, 402
    degradation, 137, 138, 299
    diversity, 381
    process, 388
    symbionts types, 156
        mycorrhizal fungi, 156
        nitrogen fixing bacteria, 156
    system, 20, 391
    transformation, 388–391
    type culture collection, 315
Microbialor enzyme-catalyzed reactions, 389
Micrococcus luteus, 5, 315, 326
Microkjeldahl method, 160
Microorganisms for biotransformation, 393
    addition of substrate, 395
    agitation influence, 401
    carbon sources, 398

enrichment method, 393
glucose concentration effect, 401
incubation period, 396
maintenance of microbial cultures, 394
media effect, 396
microbial transformations, 402
nitrogen sources, 399
pH effect, 396
scale up, 402
screening of selected cultures, 394
solvents influence, 400
substrate concentration influence, 399
temperature effect, 397
vitamins effect, 399
Microscopic characterization, 340
Microsomal preparations, 390
Mineral salts agar medium (MSM), 356, 359, 361
Minimum inhibitory concentration (MIC), 326
Modified Bristol Medium (MBM), 283
Molecular
    docking studies, 81
        active site analysis, 81
        inhibitor binding, 82
    hybridization, 29, 32, 37, 53, 54
Monocyte chemoattractant protein (mcp)-1, 94
Mononuclear cells (MNC), 98
Multicomponent cyclocondensation (MCC), 43
Multi-dimensional phenomenon, 135
Multidrug therapy program, 30
Multiple drug resistant (MDR), 18
Murine xenograft model, 18
Mutagenesis, 205
    chemical mutagenesis, 205
    UV irradiation, 205
Myasthenia gravis, 71
Mycobacterium smegmatis, 40
Mycobacterium tuberculosis, 29
Mycorrhizal
    fungi, 138, 156, 157, 163, 165
    inoculum, 159
    Rhizobium treatments, 164
    root colonization, 156, 157, 162
    root infection, 156

N

National Center for Agricultural Utilization Research, 228
National Centre for Biotechnology Information (NCBI), 72, 92, 142, 174, 223, 304, 318
Nectria galligena, 78
Neighbor-joining method, 233
Neurodegenerative disorder, 65
Neuropeptides, 99
Neurospora crassa, 14
Neurotoxicity, 33
Neutrophil Gelatinase Associated Lipocalin (NGAL), 102
Neutrophilic actinomycetes, 338
Neutrophils, 95, 101
Nocardiopsis, 221, 339, 346, 347
    aegyptia, 242
    alkaliphila, 346
Nociceptin, 90, 100
Non-motile novel actinomycete, 346
Northern Regional Research Laboratory (NRRL), 228
Nutrient uptake (NPK), 156, 157
Nutritional and acute phase indicator (NAPI), 98

O

Oceanobacillus, 379
Oligomerization, 11
Open source model for drug discovery (OSDD), 32
Organic agar gause 2, 221, 226
Organophosphorus pesticides (OP), 267
Orosomucoid, 98
Oryza sativa, 134, 139, 142, 144, 145, 150, 151, 316
Oxazolidinone ring, 41
Oxidation, 110, 123–127, 142, 269, 278, 279, 388, 394
Oxygen sensitivity, 290

P

Parkinson's dementia, 71
Pathogenic
    microbes, 266
    processes, 92

P-coumaryl alcohol, 266
Peanibacillus, 379
Penicillium
   chrysogenum, 210
   veronii, 394
Peptidases, 221
Pest-management, 311
Phanerochaete chrysosporium, 272, 396
Pharmaceutical
   agrochemical industries, 10
   preparations, 19
Pharmacophoric behavior, 54
Phaseolus vulgaris, 18
Phenogram, 375–381
Phenolic acids, 136
Phenolic compounds, 180
Phenotypic and taxonomic characterization,
   220, 224
Phenylpropanoid units, 266
   coniferyl, 266
   sinapyl, 266
   p-coumaryl alcohol, 266
Phloroglucinol, 172, 175, 180, 181
Phosphatidylcholine, 9, 13
Phosphatidylethanolamine, 9, 13
Phosphatidylglycerol (PG), 13
Phosphatidylinositol, 9
Phosphatidylserine (PS), 13
Phosphomolybdic acid reagent, 318
Photoproduction, 279, 284
Photosynthesis, 136, 279, 280, 284
Photosynthetic
   ferredoxin (PetF), 279
   organisms, 279
Phylogenetic tree analysis, 72, 232, 233,
   377
Physico-chemical properties, 359, 362, 366,
   368
Physiological mechanisms, 136
Phytolacca americana, 11
Phytopathogenic
   bacteria, 7, 14, 15
   fungi, 7, 14, 15, 187
Phytopathogens, 4, 8, 9, 14–16
Phytophthora parasitica, 16
Phytoremediation process, 173
Piperazine-thiosemicarbazone hybrids, 51
Plamodiophora brassicae, 15

Plant
   defensins, 7, 8
   hormones, 173
   insect, 135
   lignocellulosic biomass, 202
   microorganisms, 135
   peptides, 20
   soil-plant interactions, 135
   virus, 135
Plastoquinone (PQ), 284
P-nitrophenol, 209
Podosphaera fusca, 17
Polyacetylenes, 136
Polycyclic compounds, 398
Polymerase chain reaction (PCR), 67, 68,
   92, 223, 313, 346–349, 376
Polymerase Chain Reaction-Restriction
   Fragment Length Polymorphism, 346
Polyphasic approach, 220
Polyurethane cubes, 271–274
Pontibacillus, 379
Postural Tachycardia Syndrome (PTS), 71
Potassium dichromate method, 359
Potato dextrose agar (PDA), 67, 140, 146,
   158, 187–189, 313
Prealbumin (PAL), 99
Procalcitonin, 90, 92, 96
Procalcitonin (PCT), 92
Production of cellulase enzyme, 203, 222,
   242
Protein carbonylation, 123, 126
Protein Data Bank (PDB), 282
Proteome analysis, 134
Pseudomonas, 5–9, 14–16, 221, 299, 315,
   326, 396, 397
   aeruginosa, 5, 315, 326
   putida, 398, 400
Purothionins, 6–8, 14–16
Pyrazinamide (Z), 30
Pyricularia oryzae, 9, 15
Pyrularia pubera, 7
Pythium debaryanum, 326

Q

Quinoline-oxazolidinone hybrids, 41

R

Racemization, 389, 403

Radiotherapy, 90, 91
Ralstonia eutropha, 288
Rapid Grant for Young Investigators, 151
Reactive dyes, 298–302, 305, 306
Reactive oxygen species (ROS), 326
Receiver operative characteristic (ROC), 102
Regenerated cellulose (RC), 253
Relative humidity (RH), 316
Resazurin microtitre plate assay (REMA), 38
Rhizobia, 138, 157, 164
Rhizobium, 5, 156–168
Rhizobium meliloti, 5
Rhizoctonia bataticola, 326
Rhizopus stolonifer, 398, 399
Rhizospheres, 137
Rhodotorula minuta, 396, 398
Ribonucleotide redutase, 7, 14
Rifampicin (R), 30
Root Mean Square Deviation (RMSD), 283, 289
Rose Bengal agar medium (RBA), 356, 359
Rubia cordifolia, 18

S

Saccharomyces cerevisiae, 110–113, 299–306, 400
Saccharomyces species, 298
Sarcotoxin, 16
Scenedesmus obliquus (SO), 278, 282, 289
Sclerotium rolfsii, 326
S-deprived culture, 284
  aerobic phase, 284
  anaerobic phase, 284
  hydrogen production phase, 284
  oxygen consumption phase, 284
  termination phase, 284
Secalethionins, 6
Semecarpus anacardium, 65–67, 82, 83
Sephadex column, 243
Septic shock, 90–94, 102, 103
Sequential Organ Failure Assessment (SOFA), 93, 102
Serratia marcescens, 394
Serum amyloid A (SAA), 90, 97
Siderophores, 173, 195
Simple precision (SP), 71

Sitophilus oryzae, 316, 329
Slouable triggering receptor expressed on myeloid cells (sTREM), 90
Sodium dodecyl sulphate (SDS), 114
Soil enrichment technique, 393, 394
Soil sample collection, 358
Solid State Fermentation (SSF), 202
Soluable form of urokinase type plasminogen receptor (suPAR), 90, 101
Soluable tumor necrosis factor receptors (sTNFRs), 93
Spectrophotometer, 69, 115, 209, 252, 301, 314
Sphingomyelin (SM), 13
Spodoptera litura, 311, 316, 329
Spore chain morphology, 234, 237
Staphylococcaceae, 380
Staphylococcus aureus, 5, 10, 91, 315, 316, 326
Starch agar (SA), 340
Starch casein agar (SCA), 340
Starch-casein agar medium (SCAM), 356, 359
Starched casein agar, 221
Stereum ostrea, 266–268, 270–274
Sterilization, 67, 204–207
Streptimonospora salinagen, 346
Streptomyces, 220–224, 228–233, 242, 243, 338, 344–349
  alanosinicus, 242
  gancidicus, 242
  globosus, 242
  ruber, 242
Streptoverticillum morookaense, 242
Structure-activity relationship (SAR), 33, 80
Structure-property relationship (SPR), 33
Superoxide dismutase (SOD) activity, 110, 112, 116, 121–127
Sweet sorghum bagasse (SSB), 251, 253, 255, 257, 259, 261
Systemic inflammatory response syndrome (SIRs), 91, 92, 102

T

Temperature gradient gel electrophoresis (TGGE), 347
Tetrahydroquinoline nucleus, 46
Tetramethylsilane (TMS), 69, 314

Theoretical model, 256–260
Thielaviopsis paradoxa, 7, 14
Thinlayer chromatography (TLC), 68, 73, 314, 318, 401
Thyroxine-binding protein, 99
Toroidal model, 12
Transmembrane orientation, 11
Trans-phenylacrylamide, 38
Triazole based hybrids, 42
Tribolium castaneum, 316, 329
Trichoderma, 186–197, 203, 325
    cultures, 189
    harzianum, 186
    species, 186–190, 193, 195
    specific agar, 187
    viride, 186
Trifluoperazine (TPZ), 48
Trifluoroacetic acid (TFA), 142
Triticum aestivum L., 6, 316
Trizma hydrochloride, 66
Tuberculosis (TB), 29, 32, 54
Tumor necrosis factor (TNF-α), 90

U

Ultraviolet (UV), 202, 314, 374
    exposed spores, 210
    exposure, 205, 211
    irradiation, 203, 216
    light, 205, 318
    radiation, 205, 210, 215
    visible spectra, 69
United Nations Educational, Scientific and Cultural Organization (UNESCO), 312
University Grant Commission (UGC), 151, 181, 349, 381, 403
Unweighted Pair Group Mean Averages (UPGMA), 374, 378
Uridine diphosphate (UDP), 390
Urocortin, 99

V

Vacuum evaporator, 68, 314
Van der Waals forces, 251
Vancomycin-resistant Enterococci (VRE), 311, 316
Vascular dementia, 71
Vasoactive intestinal peptide (VIP), 99
Verticillium alboatrum, 16

Vesicular stomatitis virus (VSV), 13
Vigna radiata, 156, 158, 172, 173, 175, 177–182
Vigna sesquipedalis, 18
Viola odorata, 18
Violaceae, 10
Viomycin, 30
Viral infections, 96, 97
Virgibacillus, 379
Viscacea, 6
Viscotoxins, 6, 7, 14–18
Volatile fatty acids (VFAs), 281

W

Waste management, 375
Whatman filter, 208, 268
Whatman paper, 260
World Health Organization (WHO), 29, 30

X

Xanhtomonas, 7, 14
    campestris, 5
    oryzae, 9, 15, 16
Xenobiotic organophosphorus pesticides, 267
X-ray, 63, 65, 70, 79, 83, 173, 282, 309, 315, 324, 325, 330
    crystal data, 315
    crystallographic studies, 315
    crystallography, 70
Xylan, 258, 356–361, 363, 364, 367, 368
Xylanase activity, 220, 356, 360, 368
Xylanolytic activity, 365
Xylose, 237, 251, 356–359, 363, 364, 367, 368

Y

Yeast cells, 112, 113, 125, 126, 302
    cultures, 110, 113
    extract malt extract agar, 221, 226
    grown, 110
    metallothioneins, 126
    strain, 300

Z

Zwitterionic phospholipids, 13, 14